ADVENTURES AMONG ANTS

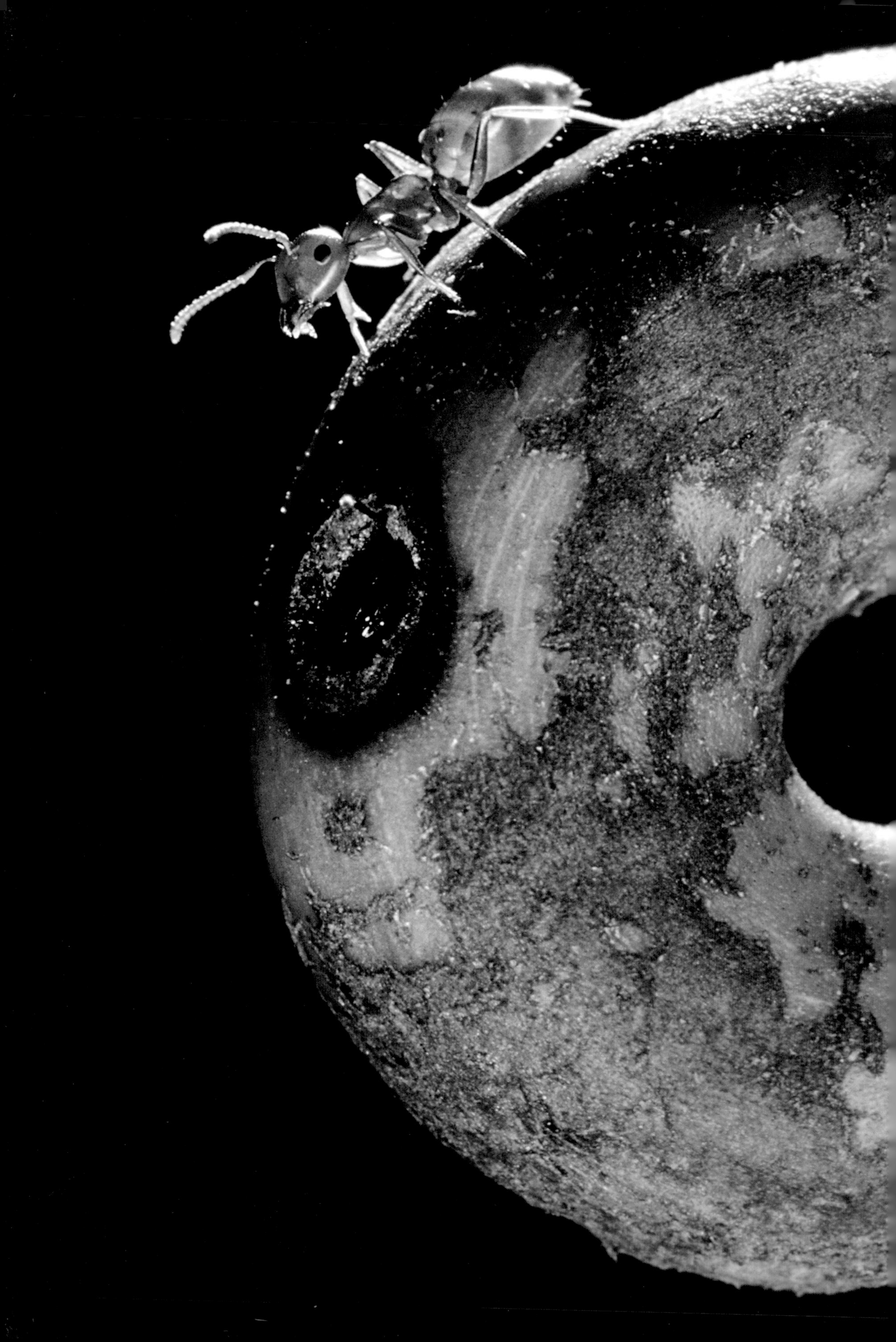

MARK W. MOFFETT

ADVENTURES AMONG ANTS

a global safari with a cast of trillions

University of California Press *Berkeley Los Angeles London*

The publisher gratefully acknowledges the generous support of the General Endowment Fund of the University of California Press Foundation.

University of California Press, one of the most distinguished university presses in the United States, enriches lives around the world by advancing scholarship in the humanities, social sciences, and natural sciences. Its activities are supported by the UC Press Foundation and by philanthropic contributions from individuals and institutions. For more information, visit www.ucpress.edu.

University of California Press
Berkeley and Los Angeles, California

University of California Press, Ltd.
London, England

Title page: A Bornean carpenter ant, *Camponotus schmitzi*, traveling along the spiral base of a pitcher plant. The ant fishes prey out of the liquid-filled pitcher of this carnivorous plant (see photograph on page 142).

Design and composition: Jody Hanson
Text: 9.5/14 Scala
Display: Grotesque Condensed
Indexing: Victoria Baker
Printed through: Asia Pacific Offset, Inc.

Library of Congress Cataloging-in-Publication Data

Moffett, Mark W.
Adventures among ants : a global safari with a cast of trillions / Mark W. Moffett.
p. cm.
Includes bibliographical references and index.
ISBN 978-0-520-26199-0 (cloth : alk. paper)
1. Ants—Behavior. 2. Ant communities. 3. Ants—Ecology. I. Title.
QL568.F7M64 2010
595.79'615—dc22

2009040610

Manufactured in China

19 18 17 16 15 14 13 12 11 10
10 9 8 7 6 5 4 3 2 1

The paper used in this publication meets the minimum requirements of ANSI/NISO Z39.48-1992 (R 1997)

This book celebrates a triumvirate of extraordinary human beings:
Edward O. Wilson, and his elemental joy in the naturalist's life;
Mary G. Smith, and her success at giving the field sciences their grandeur;
and Melissa W. Wells, and our partnership in this life of adventures

contents

Amazon Ant, the Slavemaker

Leafcutter Ant, the Constant Gardener

Argentine Ant, the Global Invader

introduction travels with my ants

A pale morning in June 4 AM
the country roads still greyish and moist
tunnelling endlessly through pines
a car had passed by on the dusty road
where an ant was out with her pine needle working
she was wandering around in the huge F of Firestone
that had been pressed into the sandy earth
for a hundred and twenty kilometers.
Fir needles are heavy.
Time after time she slipped back with her badly balanced
load
and worked it up again
and skidded back again
travelling over the great and luminous Sahara lit by clouds.

ADAPTED FROM ROLF JACOBSEN, "COUNTRY ROADS,"
TRANSLATED BY ROBERT BLY

My first memory is of ants.

I was down in the dirt in my backyard, watching a miniature metropolis. A hundred ants were enraptured with the bread crumbs I had given them, and they enraptured me as they ebbed and flowed, a blur of interactions. I marveled at how they sped into action when an entrance cone collapsed, or when one found a crumb or wrestled and killed an enemy worker. I could see that ants addressed problems through a social interplay, just as people did.

Years later, I met a group of Inuit children who had been brought by a special program to Washington, D.C., from a remote village in Alaska. Expecting the kids to be awed by the wonders of modern civilization, the welcoming committee was taken aback when the children fell to their knees to gape at a gathering of pavement ants, *Tetramorium caespitum,* pouring from a crack in the sidewalk. Alaska teems with charismatic megafauna like bears, whales, wolves, and caribou, but these children had never seen an ant. The awestruck boys and girls shrieked with delight as the ants circled and swarmed at their feet.

Ants are Earth's most ubiquitous creatures. They throng in the millions of billions, outnumbering humans by a factor of a million. Globally, ants weigh as much as all human beings. A single hectare in the Amazon basin contains more ants than the entire human population of New York City, and that's just counting the ants on the ground—twice as many live in the treetops.[1]

It's a part of our psyche, the need to care passionately about something to give one's life meaning: team sports, a just cause, wealth, religion, our children. Ants and I were destined for each other. As a junior high student back in 1973 I was enticed to join a science book club by the offer of three books for a dollar. One of my choices was *The Insect Societies,* and it riveted me from the moment I cracked its cover. Even today, its musty, yellowed pages bring a rush of memories of steamy summer days in the small Wisconsin town where I spent my childhood climbing maple trees and snaring crawfish and frogs. The book used a thicket of technical terms like *polydomy, dulosis,* and *pleometrosis* to describe ants, bees, wasps, and termites and featured exotica on every page. To me, the activities of these insects were every bit as mysterious as those of the long-lost peoples depicted in ancient petroglyphs. It would be twenty years before I experienced an approximation of that early, tingling thrill, when, in Egypt's Valley of the Kings, I scrambled over shattered rocks in the newly unsealed tomb of Ramses I, carrying a torch so I might find and photograph scarab-beetle hieroglyphs.

The dust jacket of *The Insect Societies* showed the author, Edward O. Wilson, in a natty dark suit standing in a laboratory at Harvard University, where he was a professor of zoology. "Mr. Wilson," the jacket said, "has published more than 100 articles on evolution, classification, physiology, and behavior—especially of social insects and particularly of ants."

I was a practicing biologist long before I acquired that book, however. My parents remember me in diapers watching ants and insist that I called each one by an individual name. When a little older, I cultured protozoa from water samples from Turtle Creek. I bred Jackson's chameleons—Kenyan lizards with three horns, like a triceratops—and wrote about the experience for the newsletter of the Wisconsin Herpetological Society. One school night during the dinner hour I received a call from a zookeeper in South Africa. Having read my work, he wanted my advice on chameleon husbandry. Mom's casserole got cold as my family stared at me, a socially insecure fourteen-year-old, explaining over the intercontinental telephone line how to maintain a safe feeding area for newborn lizards.

When I was in my second year as an undergraduate at Beloit College in Wisconsin, Max Allen Nickerson—a scientist at the Milwaukee Public Museum whom I knew from the Wisconsin Herpetological Society—invited me to join him on a monthlong expedition to Costa Rica. I was in heaven, about to live the dream of a boy who grew up on stories of early tropical naturalists. Finally the gear I had gathered over the years could be put to use in the pursuit of science: magnifiers, nets, bug containers, plastic bags for frogs, cloth sacks for snakes and lizards, boots thick enough to stop a snake bite. Over the next two months I helped to catch everything from a Central American caiman to a deadly coral snake.

One day as I wandered alone in the rainforest, lizards squirming in the sack hooked over my belt, I heard a barely audible sound that was subtly different from that made by any creature I had met so far. For me, that sound would prove as portentous as the rumble of a herd of elephants: it was the noise of thousands of tiny feet on the move across the tropical litter. Looking around, I spied a flow across the ground in front of me—a thick column of quickly moving orange-red ants carrying pieces of scorpions and centipedes, flanked by pale-headed soldiers equipped with recurved black mandibles that were almost impossible to remove after a bite. These were workers of the New World's most famous army ant, *Eciton burchellii.* Later that same day, I would be awestruck by an even more massive highway of ants, several inches wide, formed by the New World's most proficient vegetarians—leafcutter ants hauling foliage home like a long parade of flag-bearers.

In the two years that followed I went on treks to study butterflies in Costa Rica and beetles across a wide swath of the Andes, where I spent six months marching over plateaus of treeless *páramo* habitat and scaling rocky cliffs at 15,500 feet. I began to get a taste for the life of the seasoned explorer.

But I wanted more. I wanted to study the ant.

On returning from the Andes, I steeled myself to write a letter to Edward O. Wilson, whose *Insect Societies* was still my bible. I got back a warm, handwritten note encouraging me to drop by to see him on my way to the Woods Hole Oceanographic Institution on Cape Cod, where I was about to take a course in animal behavior.

Beloit is a small college. Its atmosphere is progressive and informal, and the students know their professors by their first names. So when Professor Wilson opened his office door, I greeted him with "Hi, Ed!" and gave him a hearty, two-fisted handshake. If my presumptuously casual attitude offended him, he didn't show it. Within minutes, this world-famous authority and recipient of dozens of top science prizes (he had already won the first of his two Pulitzers) was spreading pictures of ants across his desk and floor and exchanging stories with me as if we were boys. We talked for an hour, and I left with my head full of ideas for fresh adventures.

When I was a child, my heart was with the early explorer-naturalists. I studied the adventures of the insightful Henry Walter Bates and Richard Spruce, the brilliant Alexander von Humboldt, the groundbreaking Alfred Russel Wallace and Charles Darwin, the wildly eccentric Charles Waterton, and the incomparable Mary Kingsley. I admired these brave field scientists for their appetite for adventure, and I envied them their era. In the nineteenth century, entire regions were still uncharted. Most of Borneo, New Guinea, the Congo, and the Amazon were still labeled UNKNOWN. By the time I started exploring, in contrast, most of the Earth had been mapped and claimed, although since then I have managed to set foot in a few places where no outsider—and in the case of Venezuelan *tepui* mountaintops, no person—had ever walked before.

But I also read the books of Jane Goodall, Dian Fossey, George Schaller, and other living field scientists. I had lunch at Beloit with Margaret Mead, who banged her cane for emphasis as she recounted her experiences with exotic tribes. I recognized in these scientists a sense of adventure grounded, like that of the early naturalists, in a desire to know the unknown—but not by conquering it, as some early naturalists had, but rather by understanding it. Their fervor was infectious. John Steinbeck captured the attitude perfectly in *The Log from the Sea of Cortez,* a chronicle of his adventures in the Gulf of California with his longtime friend the biologist Ed Ricketts: "We sat on a crate of oranges and thought what good men most biologists are, the tenors of the scientific world—temperamental, moody, lecherous, loud-laughing, and healthy."

That's what I wanted to be.

When I arrived at Harvard in 1981 to begin graduate school under Professor Wilson, my first priority was to find a species worth studying for a Ph.D. in organismic and evolutionary biology. I knew where to search for ideas. Harvard is famous in scientific circles for its collection of preserved ants, the largest in the world. Located on the fourth floor of the Museum of Comparative

A scanning electron micrograph of the marauder ant *Pheidologeton diversus* depicting the normal behavior of a minor worker riding on the head of a major. There's a 500-fold difference in body weight between these two workers.

Zoology, where the profusion of mothball crystals was rumored to keep the entomology professors alive to a ripe old age, it had been founded in the early twentieth century by the legendary myrmecologist, or ant expert, William Morton Wheeler, and later expanded by the equally legendary William L. Brown Jr. and Edward O. Wilson. (After finishing my degree, I was privileged to spend two years as curator of that collection.)

One day I spent hours rummaging through hundreds of the naphthalene-scented cabinets searching for the least-understood specimens. From childhood, I have had an eye for all that is quirky in the natural world. In those cabinets, accordingly, I was drawn to the ants with oddball heads and mandibles, curious body shapes and hairs. I wondered what their bodies said about their lives and habits.

Continuing my search the next day, I came upon three drawers labeled *Pheidologeton,* a name I had never heard before. The glass tops of the drawers were dusty, and their contents were in disarray. The dried specimens, glued to small wedges of white cardboard that in turn were affixed with insect pins to foam trays, had obviously not been looked at for many years.

I was struck at once by the ants' polymorphism—that is, how different they were from one another in size and physical appearance. As in most ant species, the queen was distinctive, a heavy-bodied individual up to an inch long. But it was the workers that gave me an adrenaline rush. While the workers of many species are uniform in appearance, in *Pheidologeton* the smallest workers, or minors, were slender with smooth, rounded heads and wide eyes. The intermediate-sized workers, or medias, had larger, mostly smooth heads, and the large workers, known as majors, were robust, with relatively small eyes and cheeks covered with thin parallel ridges. The wide, boxy heads of the majors were massive in relation to their bodies, housing enormous adductor muscles that powered formidable mandibles.

I had never seen anything like this. The minor, media, and major workers didn't look like they belonged to the same species. The heads of the largest workers were ten times wider than those of the smallest. The biggest majors, which I came to call giants, weighed as much as five hundred minors. The energy and expense required to produce these giants—and to keep them fed and housed—must, I thought, be immense, which meant they must be of extraordinary value to their colonies. I left the collection that day certain I had found something special: few ants display anything close to the extreme polymorphism of *Pheidologeton*.

As a student I knew that the best-studied polymorphic ants were ones I'd seen on my first trip to Costa Rica—certain *Atta,* or leafcutter ants, and New World army ants such as *Eciton burchellii*. These ants have some of the most complex societies known for any animal, giving them an exceptional influence over their environment. Their social complexity is due in part to the division of labor made possible by their varied workers, which, with their differing physical characteristics and behavior, can serve different roles in their societies. Called castes, these classes of labor specialists focus variously on foraging, food processing or storage, child rearing, or defense, such as when large individuals serve as soldiers. Given its minor, media, and major castes, I suspected that *Pheidologeton* would be a treasure trove of social complexity.

From reading the books of Jane Goodall and other modern naturalists, I had developed the view that the best path to a career in biology was to find a little-known group of organisms and claim it, at least temporarily, as my own. I could then, like an old-fashioned explorer studying a map in preparation for a voyage, pinpoint those regions most likely to yield rich scientific rewards. Buoyed by this belief, I decided *Pheidologeton* would be my version of Jane Goodall's chimpanzee.

I soon found that my point of view was outdated. All around me, starry-eyed students who had come to biology because they loved nature were becoming lab hermits, indentured to high technology. Watching my fellow students, I realized that too much of modern biology represents a triumph of mathematical precision over insight. Sure, laboratory techniques allow for unprecedented measurements, but what good are those streams of numbers if it is unclear how they apply to nature? One thing I'd already absorbed from Ed Wilson was that much could still be done with a simple hand lens and paper and pencil. I was determined to spend my life in the field.

In the fall of 1980, I proposed to Professor Wilson that I would journey across Asia to investigate *Pheidologeton*—which I confidently proclaimed would be among the world's premier social species. My enthusiasm, if not my charts and graphs describing the species' polymorphism, won him over. I received his blessing and, within days of passing my oral exams, boarded a plane bound for India. Over nearly two and a half years I would visit a dozen countries without a break, vagabonding through Sri Lanka, Nepal, New Guinea, Hong Kong, and more.

Since then, ants have led me to all the places I dreamed of as a child. That's far more than can be described in a book, and so I focus my narrative on a few remarkable ants. I start with the marauder ant, taking my time both because this species was my own introduction to ants and because it exemplifies behaviors that come up repeatedly, such as foraging and division of labor. Thereafter, subjects are organized, in a crude way at least, by the ant approximations of human societies throughout history—from the earliest hunter-gatherer bands and nomadic meat eaters (army ants), to pastoralists (weaver ants), slave societies (Amazon ants), and farmers (leafcutter ants)—ending up, at last, with the world-conquering Argentine ant, with its hordes of trillions now sweeping across California.[2]

In this book, I will consider what it means to be an individual, an organism, and part of a society. Ants and humans share features of social organization because their societies and ours need to solve similar problems. There are parallels as well between an ant colony and an organism, such as a human body. How do ant colonies—sometimes described as "superorganisms" because of this resemblance—reconcile their complexities to function as integrated wholes? Whose job is it to provide food, dispose of waste, and raise the next generation—and what can ants teach us about performing these tasks?

To find out, let's begin our adventures among the ants.

a brief primer on ants

Anatomically, ants are like other insects in having three primary body sections: head, thorax, and abdomen—though the addition of a narrow waist gives ant abdomens extra mobility, enabling a worker to, for instance, aim a stinger or repellant spray from her rear end.[1] Almost every ant has pores near the rear of the thorax through which two metapleural glands discharge phenyl acetic acid and other fungicides and bactericides, required for a healthy life in the soil.

Ant antennae are elbowed at the midpoint so they can be manipulated like arms, though unlike the individual's jaws, often called mandibles, they can't grip. Ants keep their antennae moving for the same reason that we scan with our eyes: to monitor the environment. Beyond their elbows, antennae are flexible and endowed with sensors for touch and smell, senses more valuable for most ants than sight. An ant's compound eyes use many adjacent facets to produce images that are put together by the brain into a mosaic view. The eyes of most ants have little resolving power, though there are certain exceptions: inch-long Australian bulldog ants are so visual that I've watched them station themselves near flowers and seize bees out of the air.[2]

Mandibles are the prototypical tools used by ants to manipulate objects, and they are toothed in different ways to serve the needs of different species. Many ants can also grip eggs with spurs on the foreleg above the foot, in much the way that squirrels hold acorns with their paws.[3] Each foot, called a tarsus, is flexible and multisegmented and clings to surfaces not with toes but with two terminal claws and cushiony adhesive pads.

Ants are highly social. They are classified in the order Hymenoptera, as are wasps and bees, and some of these insects, such as the honeybee and the yellow jacket, are highly social as well, as are all the members of another insect group, the termites.[4]

The smallest known ant colonies, of at most four individuals, are those of the minuscule tropical American ant *Thaumatomyrmex*.[5] Colonies in the tens of millions are typical of some army ants of the African Congo. Supercolonies, like those of the Argentine ant currently battling for exclusive control of southern California, have populations in the billions.

Ant sociality, like that of the social wasps, bees, and termites, is expressed through a division of labor in which offspring that do not reproduce, called workers, assist their mother, called the queen, in caring for her brood, their future siblings. Despite the characterizations of Disney and Pixar, any ant recognized as an ant is female; males do exist, but they are socially useless and resemble wasps rather than ants.[6] When I call an ant "she," therefore, I'm simply reflecting reality. During their brief lives, males perform a single duty: they fly out of the nest, mate with a virgin queen (often several mate with one queen), then die. The queen will live much longer, starting her own nest and producing offspring for years from the sperm collected in this one mating flight. Because ant colonies are meant to be permanent, she and her workers, who live anywhere from a few weeks to a couple of years, will stay together. The only exception to this rule occurs when the workers rear the next generation of queens and males that depart their mother's nest to produce the next generation of colonies. When a colony's queen dies, with some exceptions we shall see later, the colony dies with her: her workers become lethargic and gradually expire.

In large part because colonies contain relatives, ants are altruistic, working without focusing on their own prospects for reproduction, which in any case are usually near zero. Edward O. Wilson and Bert Hölldobler further argue that colonies can be unified beyond familial bonds, as happens with humans. This allows for the success of colonies in which workers have multiple parents, including more than one queen.[7] That's not to say there can't be discord in a colony. Among some ants, for example, workers, though unmated, can lay eggs that develop into males. Such workers form a pecking order in which those at the top forage less, receive more food, and are more likely to lay eggs.

Instead of maturing gradually, like a human, an ant hatches into a larva, the stage during which growth occurs; after a quiescent pupal stage, the adult ant emerges. A female ant's size and accompanying functional role (or caste, such as queen, minor worker, or soldier) are largely determined by how much food she is fed as a larva, though temperature has an influence at times, and genetics can also nudge a growing individual toward a specific function.[8] Queens and workers (and different workers in polymorphic species) are distinctive in appearance because body parts develop to different extents depending on the individual's size. Adult ants do not grow, but workers tend to perform different tasks as they age. Young adults, identifiable by their paler color, remain in the nest and take on the lion's share of the nursing responsibilities (in

Opposite: A *Thaumatomyrmex* worker at Tiputini, Ecuador, using her long-toothed mandibles to hold her bristly millipede prey while she strips off its hairs before eating. These tiny, solitary foragers are notoriously hard to find.

most polymorphic species these are handled by the minor workers), cleaning and feeding the larvae and, in species in which the larvae spin silk encasements before transforming into pupae, helping the adult ants emerge from their cocoons.[9]

In addition to being highly social, ants are global, native to every continent except Antarctica and residing in virtually every climate. They have achieved universality by conquering Earth's most abundant habitat: the interstices of things, including the most secluded portions of the leaf litter as well as pores in soil, cracks in rock, and gaps and hollows in trees, right up to their crowns. As ants sweep through and conquer, they force other small animal species to the fringes of this prime real estate.[10] Ants sprang to prominence at the end of the Mesozoic Era, as the dinosaurs neared the end of their reign and when flowering plants first exploded in number, providing generous and distinctive crannies suitable for ant foraging and habitation, not to mention tasty seeds, fruit, and other edible plant parts and the insect prey that feed upon them. Housed and fed for success, ants have reigned over the landscape ever since.[11]

1

strength in numbers

We tracked marauder ant trails on steep forested slopes, accompanied by the "wish-wash" sounds of hornbills in flight and mournful calls from a green imperial pigeon. As nightfall approached, we made our way back to the village of Toro, in a valley of brilliant green paddy fields at the edge of the forest. My guide, Pak Alisi, invited me into his home for tea. "You know," he said, "here we call the ant you study 'onti koko.' That means you always find many together."

Yes, I agreed. With the marauder ant, the group is everything.

FIELD NOTES, SULAWESI, INDONESIA, 1984

"We have three kinds of ants here," declared Mr. Beeramoidin, the forestry officer at the village of Sullia in India. "A black one, a big red one, and a small red one that bites."

I was twenty-four, a graduate student on a quest for the ant I had reason to believe had one of the most complexly organized societies in existence. A column of dust-speckled sunlight emblazoned a rectangle on the floor too bright to look at directly—a reminder of the intense dry heat outside. It was late November, and I was worried my choice of season wasn't giving me the best weather for ant hunting.

As Mr. Beeramoidin spoke, his round, bespectacled head rocked from side to side. I had learned that this meant his attention was friendly and focused on me, and though I had only been in India a month, I had already adopted the same habit. I also found myself chewing betel nut, wearing a Gandhi-style *lungi* around my waist and flip-flops known locally as *chapels* on my feet, and using words like *lakh,* meaning a hundred thousand, to describe the number of workers in an ant colony.

Rocking my head in turn, I told Mr. Beeramoidin it was likely that scores of distinctive ants lived within a stone's throw of his office, though even an experienced person would need a strong magnifier to tell many of them apart. I sought just one of them, *Pheidologeton diversus,* a species to which I later gave the name "marauder ant."

In 1903, Charles Thomas Bingham, an Irish military officer stationed in Burma, provided detailed and theatrical descriptions of this ant. In one memorable passage, he wrote that "one large nest . . . was formed under my house in Moulmein. From this our rooms were periodically

Previous page: Marauder ant major workers serve as heavy-duty road equipment. This one in Singapore is gnawing at a twig, which she later dragged off the trunk trail.

invaded by swarms, and every scrap of food they could find, and every living or dead insect of other kinds, was cleared out." The locals found the swarms overpowering. "When these ants take up their abode in any numbers near a village in the jungles, they become a terrible nuisance. . . . I knew of a Karen village that had absolutely to shift because of the ants. No one could enter any of the houses day or night, or even pass through the village, without being attacked by them."[1] In spite of the vividness of Captain Bingham's report, the group remained a biological mystery.

I had arrived in India in the fall of 1981, primed to explore the social lives of the minor, media, and major workers of *Pheidologeton diversus*. My first stop had been Bangalore, more specifically its prestigious university, the Indian Institute of Science. My host was Raghavendra Gadagkar, a professor whose subject was the social behavior of wasps. He believed in learning from experience and smiled at my naïveté and youthful enthusiasm. Rather than teaching me how to eat rice without utensils, in the local fashion, for instance (the nuances of handling hot food bare-handed are many), he dropped me at the door of a local restaurant, recommended I order the "plate meal," and came back for me an hour later. During that first lunch I spilled more than I ate.

Bangalore was going through a dry spell, and I had trouble finding any *Pheidologeton*. Raghavendra recommended I try the Western Ghats, a chain of low mountains famous for its forests and wildlife, just inland of the western coastline of India. On the road from Bangalore to the coast was a village named Sullia. I was told it had a forestry office where I would find both accommodations and advice.

The next day, I learned a basic fact about Indian bus drivers: they were trained to accelerate around blind curves as if suicide were a career expectation. After a stomach-churning ride, I was dropped at the drowsy center of Sullia. I hoofed it to the forestry office, where I was delivered into the presence of Mr. Beeramoidin, who listened attentively to my explanation of ant diversity and then told me the guesthouse was full. Afterward, out under the roasting sun, my nerves jangling at the thought of the harrowing six-hour ride back to Bangalore, I kicked a tree in frustration—and got my first taste of *Pheidologeton diversus*. Hundreds of the tiny minor workers stormed from the earth, the major worker among them looking like an elephant among pygmies. Even Mr. Beeramoidin gave an impressed whistle, conceding with an enthusiastic rocking of his head that Sullia may be more of an ant haven than he thought.

Struck by my preternatural ant-locating skills, Mr. Beeramoidin promised to find me a place to stay. An old man with a limp appeared. The two men conducted a rapid-fire conversation in the local Kanaka language, then the old man guided me down the road to a tiny room next to a mosque. Except for a thin sleeping mat, it was bare: no toilet, water, electricity. That night, I lay for hours watching geckos in the moonlight. Awakened at dawn by the call to prayer, I hobbled to my feet, rubbing my fingers across the areas where the mat's reed latticework had impressed a design like a city map into my flesh.

Finding ants in the dry forests around Sullia proved as arduous as it had been in Bangalore. That first morning, the ants in front of the forestry office had vanished, as had Mr. Beeramoidin, whom I never saw again. I decided to comb the forests, but they were desiccated. It wasn't until the fourth day of looking that a diversionary hike at the edge of town through a watered

Minor workers of the marauder ant riding on an especially large major (a "giant").

plantation of stately oil palms brought me luck—a batch of *Pheidologeton diversus* crossing my path. I fell to my knees, thrilled to finally find some of Captain Bingham's fabled swarming ants, and began inspecting the *diversus* column.

First, a marvelous sight: a major worker was careening along carrying a dozen minors, much like the elephant whose *mahout,* or trainer, had given me a wave from the back of his pachyderm soon after my arrival in Sullia. Except the ant passengers didn't appear to be giving instructions to their beast of burden. Why were they there? I could see no evidence that the minors were cleaning or protecting their mount. I decided they were probably hitching a ride for a simple and practical reason: it takes less energy to ride than it does to walk. The smaller the individual, the more energy walking takes. Being bused by large ants saves the colony energy.[2]

While I was in the entomologist's "compromising position," my nose practically brushing the frenzied ant workers that scurried beneath me, a young man of about my age walked up. Oblivious to my rapture over the ants, he started a conversation by saying his name was Rajaram Dengodi, which he explained meant "King God of All Mankind," and inviting me for lunch. It turned out he was the son of the plantation owners and lived with his parents at the edge of the palm grove. When I arrived at their low whitewashed house, he proclaimed that I'd be sharing his room for the month.

Despite the grandeur of his name, Raja was a low-key fellow with no apparent ambition other than to strum his guitar. But he proved an admirable companion and was eager to learn about ants. During that first week, I mapped the plantation and decided where to concentrate my search. Then Raja and I set about following the activities of the local *Pheidologeton diversus*. It quickly became evident that the colonies were huge. We saw several migrations with dense legions of ants moving their larvae and pupae to new nest sites, which suggested the workers numbered in the hundreds of thousands.

We also witnessed the hunting and harvesting of meals on a massive scale. The workers carrying food moved along well-demarcated roads that remained active day after day. In time, I would learn that these tracks had as many functions as human road systems. Ant specialists call such persistent routes trunk trails. The marauder ant's trunk trails are substantial structures, with a smooth surface an inch wide. Along them, the ants craft soil walls or even a complete roof of soil. The trails frequently lead belowground, especially where they cross dry or exposed stretches of earth.

Hundreds of ants, and sometimes more, crossed back and forth on those trails every minute. In one extreme case I recorded eight thousand workers per minute climbing a cacao tree to flow into and out of a rotten pod over the course of a full day. Marauder ants excel at plundering large foods, such as fruit or carcasses, that take them a while to devour. But these expeditions represent only a small portion of their efforts. At any time, day or night, I could see them traveling from the trunk trails in ever-changing, reticulating networks, or, as Captain Bingham described them in Burma, in swarms. These extended into vegetation and leaf litter, where the ants' activities were hard to document.

I confirmed the observations of early naturalists that marauder ants can harvest seeds in bulk. More impressive, the ants returning to the nest labored by the dozen to cart centipedes, worms, and other creatures that, if viewed through ant eyes, would appear bigger than dinosaurs

Marauder ants subduing a frog in southern India.

to us. A few dozen minor workers, each about 3 millimeters long, easily hefted the head of one of the doves the Dengodis had tried to induce me to eat after they found out Americans eat meat. Later, Raja and I saw a seething mass of workers rip up a live, 2-centimeter-long frog, pulling its twitching body taut to the ground and then flaying the meat. Raja and I studied the action with both horror and a newfound respect. That was the day I named them marauder ants.

Though Sullia was in no danger from the ant swarms, it was easy to believe Bingham's report from Burma that droves of this species could overwhelm a village. Raja enthusiastically told me how the ants would sometimes pour into the family pantry and make off with supplies of rice and dried condiments.

At dinner we reported to Raja's parents about the marauders' feats of predation, which I described as astonishing, particularly because the workers have no stinger, the weapon with which many predatory ants—especially those species in which the workers carry on alone or in small groups—disable victims. Mr. and Mrs. Dengodi, who took everything I said with great seriousness, no matter how eccentric the subject, listened as I explained that the marauders' success with gargantuan prey seemed to rely on a coordinated group attack in which workers, individually inept, pile on high and deep, biting and pulling in such numbers that the victim doesn't have a chance.

I could attest personally to the effectiveness of that approach. While watching the frog, I'd made the mistake of standing in a throng of marauders. The sheer volume of the minor workers' bites was enough to drive me away, with one major lacerating a fold of skin between my fingers.

This scale of operations brought to my mind the most infamous raiders of all: the army ants.[3] As a teenager in America's heartland, far from any jungles, I had devoured popular descriptions of army ant swarms killing everything in their path. The stories often relied on florid writing, most famously in an unforgettable story by Carl Stephenson, first published in a 1938 issue of *Esquire,* "Leiningen versus the Ants": "Then all at once he saw, starkly clear and huge, and, right before his eyes, furred with ants, towering and swaying in its death agony, the pampas stag. In six minutes—gnawed to the bones. God, he couldn't die like that!" Although this is hyperbole, army ants do have an appetite for flesh and a coordinated battle plan that depends on sheer force of numbers.

Like many army ants, marauders have no stingers. Rather than incapacitating prey with stings, they mob it. This gang-style predatory attack is just one element of both ants' complex routine. How much deeper did the resemblance go? I knew that currently there are as many species of ant as there are of bird—perhaps 10,000 to 12,000—and that the marauder and the army ant are no more closely related than the hawk and the dove.

Convergence is the process by which living things independently evolve to become alike, as a result of like responses to similar conditions or challenges. The wings of bats, birds, and bugs are convergent because they are limbs that have been independently modified to function in flight; the jaws of humans and the mandibles of insects are convergent because both can be used to hold objects and chew food. If the marauder ant and army ants proved to be alike in how they hunt and capture prey, it would be a similarly marvelous example of evolutionary convergence. That day in Sullia as I watched the ants dispatch that unfortunate frog, I made a decision that would affect the first years of my budding professional life: I would study the kill strategy of the marauder ant. I would make that my quest.

FEEDING THE SUPERORGANISM

Standing in a Sullia field on a tepid afternoon, with Raja's guitar providing an incongruous musical accompaniment to the massacre at my feet, I felt like a general observing his troops from a hilltop and trying to make sense of the skirmishes below. My brain was whirling: one moment, trying to picture what it's like inside one of those tiny, chitinous heads; the next, envisioning all the ants at once, forming a kind of arm flung over the ground with fingers that were rummaging through the soil and low plants.

The nineteenth-century philosopher Herbert Spencer was the first to treat in detail the simultaneous existence of these two levels, individual and society, and in 1911 the ant expert William Morton Wheeler came up with the term *superorganism* to describe ant societies specifically. Both men saw an ant colony not merely as an individual entity, as one might think of a bank or a school, but more specifically as the exact equivalent of an organism.[4] They could readily make this point because others had already described the human body as a society of cells.[5] The superorganism concept took on real meaning for me as I watched marauder ants. Before coming to India I had read an essay by the physician and ant enthusiast Lewis Thomas, who took Wheeler's writings to heart:

> A solitary ant, afield, cannot be considered to have much of anything on his mind; indeed, with only a few neurons strung together by fibers, he can't be imagined to have a mind at all, much less a thought. He is more like a ganglion on legs. Four ants together, or ten, encircling a dead moth on a path, begin to look more like an idea. They fumble and shove, gradually moving the food toward the Hill, but as though by blind chance. It is only when you watch the dense mass of thousands of ants, crowded together around the Hill, blackening the ground, that you begin to see the whole beast, and now you observe it thinking, planning, calculating.[6]

Like a more traditional organism, a superorganism is most successful when its activities are carried out with maximum productivity at the group level. Consider the cells of a human body, an assembly of trillions. Although these cells may be doing rather little as individuals, collectively they can yield results as intricate and choreographed as a dancer's in a corps de ballet. I developed a feeling for a marauder colony as an organism. I watched as the ants worked together like the organs in a body to keep the ensemble healthy and stable, with their trails serving as a nervous system used by the whole to gather knowledge and calculate its choices. With mindless brilliance, this colony-being established itself, procured meals and grew fat on the excess, engineered its environment to suit its needs, and fought—and on occasion reproduced—with its neighbors. I imagined that, given enough time, I could watch each superorganism mature, spin off successors that bred true through the generations, and die.

How do the members of an ant superorganism supply food for the whole? Unlike the body of an ordinary organism, a colony can send off pieces of itself—the workers—to find a meal. Regardless of species, once an ant detects food, her searching behavior stops and is replaced by a series of very different harvesting activities: tracking, killing, dissecting, carrying, and defending. In the majority of species, an ant can mobilize others to assist her. This communication practice is known as recruitment and usually involves chemical signals called pheromones. Often, a wayfaring ant releases a scent from one of a battery of glands on her body, a mixture that serves to stimulate or guide her nestmates. The mobbing of marauders at prey reflects the speed and effectiveness of their recruitment.

I'd known about recruitment, without having a name for it, since I was a child. At family picnics, I would drop a crumb in front of a lone worker. Within minutes, a hundred ants would be pouring along a column to the bread. Had I been able to inspect the successful hunter who first found the crumb, I would have seen her glide the tip of her abdomen on the ground on her return to the nest, depositing a pheromone that diffused in the air—a common, though not universal, ant practice. When ants form a line or travel in a column, they are tracking such a plume with their sensitive antennae, which they sweep left and right before them, in many cases while running faster for their size than any baying foxhound.

Each ant adds pheromone to a trail offering a good payoff, so the scent builds over time. Then, when the food supply runs low and the ants begin returning unrewarded, the pheromone is no longer replenished and the scent dissipates, attracting fewer ants. (Pharaoh ants have an even more efficient way to flag a route that has ceased to be profitable, signaling "don't bother" by depositing a different pheromone at the start of the trail.)[7] The chemicals required to convey a message can be minuscule. With one species of leafcutter ant, a thousandth of a gram of recruitment pheromone—a minute fraction of one droplet—would be enough to lead a column of workers around the world sixty times.[8]

Since traffic depends on pheromone strength, it is modulated by the ants' overall assessment of a trail's offerings—what we call mass communication. This technique can lead to what appear to be deliberate choices by the colony, despite the ignorance of the individual ants of such matters as the size of the food item they are visiting and the number of workers needed to harvest it. For instance, a colony will more quickly exploit a nearby food source than one farther away, simply because it takes less time for the ants to walk the shorter distance. This results in the quicker accumulation of the trail pheromone, which in turn attracts more ants to the meal.[9]

Among the Sullia oil palms, however, such subtleties of individual reaction and mass response were hidden to me. Instead I recorded seemingly spontaneous eruptions of ant multitudes followed by sudden mass retreats, like an arm that was extended, pulled back, and then extended somewhere else. What was going on?

I thought back to a similar eruption involving army ants that I had seen as an undergraduate studying butterflies in Costa Rica. I was awakened one morning to a rustling sound in the room of the hacienda where I was a guest. *Eciton burchellii* army ants were everywhere, moving in waves over the floor, flowing through cracks in the wall, falling from furniture while clinging to the backs of beetles and silverfish. I heard a plopping sound as an inch-long body landed on the carpet next to my bed: a scorpion cloaked in ants had dropped from the ceiling. The only reason no ants had swarmed my body was that each leg of the bed had been set in a dish of oil by the owner's wife. Thank heavens—I doubt Señora Perez would have appreciated my dashing naked, draped only in ants, into her parlor. I put on my robe and ran to the ant-free hallway, then waited out the ant raid over toast and scrambled eggs.

What had those ants been doing? In a word, they were foraging.[10] For all ant colonies, this search for food is carried out by multiple workers at once. But while the foragers of most ant species operate independently of each other, army ants forage together, much like a pack of wolves looking for elk.[11] Unlike a wolf pack, however, army ant hunting groups do not have a circumscribed membership. Thousands may be present in a raid, but different workers come and go en route to the nest, a search strategy called group foraging or (my preference, because there is no set "group") mass foraging.[12]

Many fierce predators dispatch difficult prey without searching for it in a group. In certain ant species, workers acting alone can both find and kill small vertebrates. Workers of one Brazilian ant dispatch tadpoles larger than themselves.[13] But most predatory ants cannot overpower such prey without help. Most commonly, a successful forager—called a scout when a few scattered individuals are doing the reconnaissance—recruits a raiding party, often guiding it for many meters to the specific site where she discovered the prey.[14]

Elsewhere in the Western Ghats of India, I saw this system used by *Leptogenys*. The tight pack of slim, glossy ants was moving through the dry litter at the reckless speed of an Indian bus driver. I followed and watched as they entered the mud galleries of a termite colony. The ants soon emerged, each with a stack of termites in her jaws. This regimented form of group predation was a joy to observe, as long as I stayed back far enough to avoid the needlelike stingers that *Leptogenys* use to immobilize their prey in one-on-one combat. Later I determined that this species employs scouts. These individuals then return by themselves to the nest and recruit a few dozen nestmates who together do the potentially dangerous work of mining and transporting the unwieldy termites.

Army ants employ a completely different foraging technique. Rather than proceeding with guidance to food already found, the workers sweep ahead blindly in a mass, the absence of a single target turning the whole raiding business into a gamble.

Some army ants regularly invade homes, and in the underdeveloped world their arrival is welcome (even though they force everyone out for an hour or two), for they clear out vermin such as roaches and mice. Marauder ants perform a similar service—though they also make a nuisance of themselves by absconding with grain and other human foods, as Captain Bingham recorded.

Indeed, it was impressive to watch marauder ant mobs take on centipedes and frogs in India. From those clashes I saw that, like army ants, marauders recruit members explosively as each prey item is found, then kill and cart off the bodies together. But to understand marauder ant foraging behavior, I needed to learn how they located their prey in the first place.

After a time, Raja tired of the ant bites and stayed home to practice his guitar. By then I had been in Sullia three weeks, surviving on sticky rice splashed with a red curry so spicy that it often left me panting. This diet kept me ravenous, and to sustain my energy I purchased caramels at a roadside stand. (The shop had more ambition than inventory, with the former evidenced by its name, Friendly Mega Supermarket Store, which was crudely painted on a board.) I surreptitiously devoured the candy at night, fearful of hurting my host family's feelings, and disposed of the paper wrappers down rodent burrows on the plantation.

One cool evening as I watched marauders rushing in the tree litter, as greedy for high-calorie food as I was, a vision of the ants as a superorganismic being crystallized in my mind. I began to think about the army ant stratagem of foraging in a "group." Within the superorganism, what does membership in such a group entail for an ant? Is it proximity? Among humans, techniques as old as jungle drums and as new as Twitter allow people to form groups without physical closeness. Conversely, being close to others does not automatically confer membership in a group in a meaningful way. Often enough I have joined a crush of people on a city street—quite a crowd, but not much of a group.

I had seen many ant species in which nearby workers show no semblance of joint action. Is proximity even less meaningful to ants than to people? In many ways, yes. The workers of most ant species cannot detect another ant's presence until they are virtually on top of each other. Army ants, legally blind by human standards, sense a nestmate only during fleeting moments of contact. In such times, the ants distinguish friend from foe, but what they learn is unlikely to play a role in the organization of their armies. Rather than responding directly to others, ants tend to react to information left by nestmates who may be long gone—to the webwork of social signals, such as pheromones, spread throughout the environment in an ant version of the Internet.

Think of household ants following an odor trail to a cookie left on a kitchen counter. What happens if I pluck out all but one ant? Her actions won't change an iota as long as she can track the scent. She continues to participate in a group effort to harvest food whether the trail is thick with ants or not. Could we define an ant as being part of a group when her actions are constrained or guided by the varied signals and cues arising from the actions of her nestmates, and as solitary when she acts on her own?[15]

As it turns out, army ants conform to this view of a group. The workers have negligible freedom to wander far from nestmates and any fresh chemical communiqués those nestmates have left behind; the superorganism never sends out lone pieces of itself, but droves of workers operate as an almost tangible appendage that stays attached, through a continuous flow of ants, to the main body. Some scientists point to other aspects of army ant life, such as their ability to catch or retrieve prey in groups, but it is this aspect of their behavior—how they forage, and not what they do after they find food—that sets army ants apart from other ants.

My goal became to determine whether the marauder ant uses the army ant group approach to hunting. In India, I documented the movements of teeming battalions, with the workers numbering in the tens of thousands. But such details as whether the raids relied on scouts were difficult to assess during the bone-dry weather I experienced there, which forced the ants to be cryptic and subterranean. I would continue my marauder ant studies in Southeast Asia, where the species was common.

Rajaram Dengodi, smiling dreamily as he strummed his guitar, saw me off on the bus to Bangalore. As I climbed the steps, the proprietor of the booth where I had been buying caramels ran over and gave my hand an enthusiastic shake. He had gone upscale, with fresh paint and a fancy poster of the Indian deity Ganesha. I wondered how much of my patronage had gone into subsidizing his new, neatly lettered, laminated sign: FRIENDLY MEGA SUPRMRKET STORE.

HOW TO HUNT LIKE AN ARMY ANT

A year later, in Irian Jaya, Indonesia, the dryness of southern India was long forgotten. The rain was so thick and the air so muggy in the Cyclops Mountains that I felt like I was walking in a bowl of hot soup. My kinky-haired guide, Asab, had to scream his customary question over the roar of water battering leaves: "Sudah cukup?" ("Had enough?"). In the heavy rain I could barely see the ancient Russian machine gun slung over his shoulder—protection, he had told me, from guerillas.

For two days the downpour was nonstop. I slept in wet clothes. My camera, though sealed in a plastic bag, somehow got waterlogged. I was often up to my waist in mud, making it difficult, at best, to locate ants. The few specimens I did manage to collect were washed away in the middle of the night, along with the majority of my toiletries. Fortunately, my other experiences in most of Southeast Asia were far more pleasant and productive.

I had embarked for Irian Jaya from Singapore, where I would be based for two years. In India, on my diet of rice and caramels, the weight of my six-foot frame had dropped to 138 pounds; since then I had gained back twenty pounds, largely from my time in Singapore. It was hard to resist a country so immaculate and orderly that bubble gum is illegal and so attuned to style that when the *Straits Times* announced that Paris fashions had shifted from red and white to black, all the girls were wearing black within the week. For anyone on a student budget, moreover, Singapore was a dream come true: *roti parata,* fried *kway tiao,* Hainanese chicken rice, Hokkien noodles, and ice *kachang* are just some of the foods from the hawker stalls near Orchard Road that I frequented. Of more academic consequence was the University of Singapore; I often found myself nursing Tiger beer with ruddy expat professors, feeling like a character in an Anthony Burgess novel.

I rented a tiny room in a high-rise from a Chinese family whose composition kept changing. Each evening, I would return from ant-watching to find their apartment in darkness and would tiptoe past a dozen or more people sleeping in rolled blankets on the floor. At sunrise, I would be awakened by soft Cantonese voices and an aroma of tea. We had no idea what to make of each other, they with their elegant apartment managed like an ant heap, and me, the muddiest human in Singapore, leaving a trail of ants wherever I walked.

Sir Thomas Stamford Raffles, the early-nineteenth-century British colonial agent who founded Singapore, was a keen naturalist. His love of nature is manifest today in a Singaporean fondness for parks and gardens. This meant that there were plenty of places to observe marauders, since they do well in deteriorated natural habitats and on human-altered terrain. Lawns and gardens and the weeds that colonize human clearings almost always contain abundant supplies of high-energy food. Plants in open spaces allocate more resources to rapid growth and dispersal and less to defenses against herbivores or competitors. That means they can support more plant feeders, and thus more of the predators that eat them, including insatiable omnivores like the marauder ants.

The Singapore Botanic Gardens, founded by Raffles in 1822 to display some of his own exquisite plants, offer plenty of marauders in a manicured setting where they are easily watched. I was introduced to the gardens as an ant haven by D. H. "Paddy" Murphy, a senior lecturer at the University of Singapore. A native of Ireland, Paddy is an autodidact, an entomological genius of a kind that normally falls through the academic cracks. Because he lacked a Ph.D., his prestige-minded colleagues didn't know what to make of the fact that when any entomologist visited Singapore, he or she called on Paddy. No matter what the researcher's area of expertise—whether some obscure group of crickets, plant lice, or marauding ants—Paddy would pull out the specimens in his collection and begin gently instructing about the local species. The visitor would leave enlightened, while Paddy seemed to soak up everything his interlocutor knew.

In addition to showing me the Botanic Gardens, Paddy took me in his battered white Nissan on expeditions to Singapore's watershed, the Bukit Timah Nature Reserve. After several hours rooting in the mud and stuffing specimens into vials, we would finish our day with a stop for a drink. Oblivious to our jungle-rat appearance, he'd drive to one of Orchard Road's fancy hotels and, shuffling into its gleaming five-story foyer, demand two Tiger beers from the bar, all the while holding his insect net like a national flag. Libation consumed, we would then retreat to his flat, where his wife, a chemistry professor of Indian descent, kept a motley herd of little dogs. Sitting at the kitchen table, Paddy would scrutinize the day's catch, never raising his eyes from the magnifying glass. Meanwhile the dogs, announced by a rumble of paws on the tile floor, ran in formation like a migration of African wildebeests, circuiting the house every minute or two.

After another round of beers from his fridge, Paddy would drop me off at what came to be my favorite part of the Botanic Gardens, a seldom-visited back section where I began to understand how deep were the convergences between marauder and army ants. The foraging behavior of both displays a specific set of characteristics that, in scientific fashion, form a sequence in my head. In brief, (1) the workers are tightly constrained by one another's activities, such that while individuals constantly enter and leave the raid on a trail to the nest, (2) those in the raid nevertheless avoid spreading apart, so that the raid retains its existence as a cohesive whole; in fact, (3) adjacent ants stay close enough together that communication between them can be virtually instantaneous. (4) This unit moves along a path that (5) is not controlled by any steadfast leader or leaders within it, (6) nor by scouts arriving from outside. Indeed, (7) their movement does not target a specific source of food, (8) nor is progress dependent on finding food en route, because the ants are drawn forward not just to meals but also to the land ahead; further, (9) the advance can continue across "virgin ground," because advance doesn't require cues left by prior raids. Finally, (10) all foraging is collective. No ant sneaks out to grab lunch on her own.

These features basically define what we mean by the words *group* and *forage* in these mass-foraging ants. The first three describe a particular sort of group in which proximity turns out to be essential: no marauder ant searches alone for any significant distance.[16] That means food never has to be abandoned while help is enlisted. Overpowered quickly, prey is unlikely to be stolen, or escape, or have occasion to defend itself. The other attributes describe a certain kind of foraging in which the searching group has no predetermined destination and need not take any particular course.

All these details took me months to work out in that back section of the gardens. I was largely hidden there from the heavy tourist traffic, though I do recall one passing wedding party that was shocked and then fascinated to see me on my hands and knees, counting ants performing the superman task of carrying a lizard egg. The bridesmaids lifted the bride's veil as she too stooped to take a look.

2 the perfect swarm

At the end of my first week in Singapore I had my first clear view of a swarm. It was late afternoon in a remote corner of the Botanic Gardens. Paddy Murphy sat nearby, smoking a "fag" and examining a silverfish on a tree. I had spent the previous hour on my hands and knees following a trail of marauder ants that were obviously on a foraging expedition, because they were bringing back all kinds of prey. And there, suddenly, near the base of a Brazil nut tree, was a throng of ants—shimmering with the movements of thousands in the cropped grass. I'd caught glimpses of such mobs in my Indian plantation's understory brush, but here was one open to scrutiny, a band of ants 2 meters wide and over 7 centimeters from front to back. At the back of this band was a V-shaped network of columns 3 meters long that resembled the web of veins in a human hand and, judging by the slaughtered prey being carried along, served the same purpose of conveying nourishment. The web converged into a single column that was the colony's aorta to the nest. Workers laden with plunder marched along this route all the way home.

Paddy came over and gave a whistle of astonishment. My hand sped across a waterproof notebook as I penciled a sketch of the action. Afterward, examining my drawing, I realized how closely it resembled illustrations of army ant raids. In fact, the swarm compared point for point with descriptions of the most extreme form of army ant attack, the swarm raid.[1]

In the terminology of army ant researchers, the advancing margin, where the workers meandered ahead of their sisters, is the swarm front. The swarm is the band of ants behind the front, and the fan is the network of columns farther back still, which converge to form the single base column that extends to the nest. Among army ants, swarm raids are peculiar to some New World *Eciton* and *Labidus* ants and to a few African *Dorylus* species known as driver ants. Skirmishes within a raid appear chaotic when viewed in isolation; but when a raid is seen as a whole, a sense of order and even aesthetic beauty emerges.

The phalanx of ants stayed in tight formation. This made the raid's anatomy easy to pick out—a boon to humans, whose noses are too poor to register the pheromone scents that the ants prefer to use for communication and that bind the raid together. A century ago, Herbert Spencer saw a "closeness of parts" of this kind as strengthening a society's similarity to an organism. After all, we recognize a dove or rice grain by its boundaries: each has an inside and an outside. The workers that form a marauder ant or army ant raid may be separate creatures, but they do not drift apart, and therefore they form an entity that is not only cohesive but also distinct and well bounded.

The same was true outside the raids, throughout the colony. Over the next weeks I would learn that while trunk trails and their temporary offshoots could extend for a hundred yards, individual marauder ants stay on these roads and seldom travel more than a few centimeters from their sisters.[2] All foraging, I determined, is done in a group: my observations revealed no rogue hunters. (I did come upon strays, though. Some were stragglers, sick or lame, on paths all but abandoned. Then there was the occasional isolated worker that was just plain lost. I spent hours watching these individuals stumble around. But even after I gave one lost marauder a bit of my lunch, she had no idea where to go with it. Presumably these forlorn souls wander until they die.)

Certain things became clear to me as I sketched the raid that afternoon in the Botanic Gardens. Within the raiding horde, there's little appreciable movement of any ant at the swarm front beyond the ground covered by her nestmates—no exploration of fresh terrain except for a stint at the front of the raid, which is the one time in a marauder ant's life that can be unambiguously described as foraging. The trailblazers at the front (too temporary and plentiful to be considered scouts, they are appropriately called pioneers, as they are in army ants) cross onto new soil. Pioneers don't appear to be specialists at this task; whoever reaches the front does the job. Nor do they press ahead and fall back with the precision seen in movies depicting Roman soldiers massed against the Gauls. Sometimes they wander a bit. In any case, their actions are restricted to the vicinity of their neighbors, and the raids as a whole have no ultimate destination.

Marauder and army ant raids differ only in degree. One obvious difference is their speed: marauder raids move at a measly 2 meters an hour, maximum, while army ants can travel ten times that fast, the record being 25 meters in an hour. Scale the ants up to human sizes, and that would be over 800 meters an hour for the

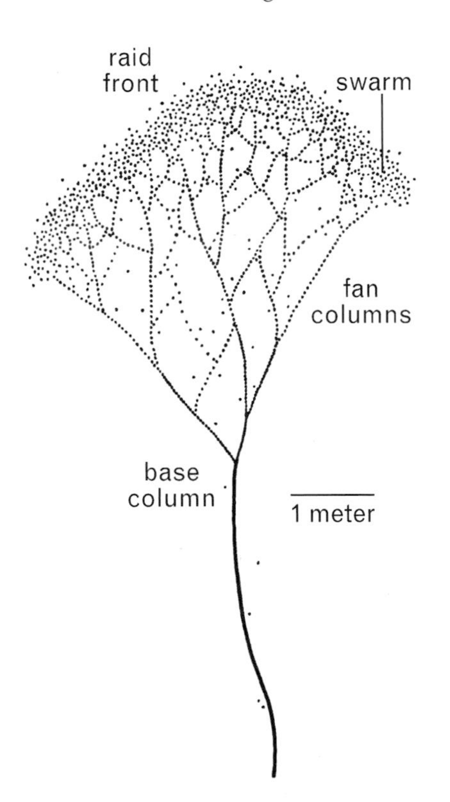

A marauder ant swarm raid, based on my original drawing, advancing toward the top of the page.

Workers of *Proatta butteli* seizing a wasp at the Singapore Botanic Gardens.

marauders, versus up to 8 kilometers an hour for an army ant raid. As a result of their slow speeds, marauder ant raids, which last a few hours, cover only 20 meters at most, while over the same time some army ant raids can traverse 100 meters or more.[3]

In Malayan rainforests to the north of Singapore I would later find a second swarm-raiding *Pheidologeton* species called *silenus,* closely related to *P. diversus* but with raids twice as fast, matching the raid speeds of a slow army ant.[4] Sluggish or not, the painstaking searches don't hinder the hunting prowess of these *Pheidologeton* species, particularly in the *diversus* marauder ant, which usually takes food in abundance.

I came to believe that there was simply no need for the mass of marauder ants to move along any faster. In fact, in the Botanic Gardens I came upon a different kind of ant nesting at the base of a withered tree that taught me that mass foraging might not require the group to move at all.[5] Lumpy beasts with unimpressive jaws, *Proatta* look incapable of doing anyone harm. Yet that day I saw three workers grab a wasp that must have outweighed them by a factor of fifty; trying to escape, it nearly lifted them all from the ground in an attempt to take flight. Nearby comrades, attracted by the commotion, seized the quarry by the hind legs. Then more nestmates, perhaps drawn to the site from a distance by pheromones, helped to pull the wasp into their nest.

Groups of the same *Proatta* ants also killed marauder ants that lagged behind on the base trail after a raid. What accounted for their success? While much of their effort is spent scavenging by themselves for all kinds of tidbits, the *Proatta* workers accumulate in such numbers within inches of their nest entrances that when an insect walks by, several ants are often close enough to pin it down. *Proatta* essentially stay in place and let prey come to them, using a group version of the ambush tactic employed by a human duck hunter hidden in a blind, or by solitary-living species such as the snapping turtle, which lies in wait for fish to pass by.[6]

As we have seen, proximity is not essential for ants to act as a group. But the *Proatta* behavior demonstrates that, as in the packed raids of the marauder and army ants, a high density of participants increases the likelihood that encountered prey will be caught. When their workers are close together, some ant species are even able to avoid active foraging almost entirely. Ants squeeze together inside their nests, and a Mexican *Leptogenys* species takes advantage of this density by giving their living quarters a scent that attracts the pill bugs on which they feed. The ants jointly kill and feast on these little crustaceans without leaving home.[7]

But by keeping on the move in a crowded mass, the marauder and army ants accelerate the odds of encountering dinner, compared to these sit-and-wait strategists. It is the difference between dragging a net through water and leaving it fixed in place. Both methods work, but a motile net almost certainly catches more fish during a given period.

ANATOMY OF A RAID

One morning as I watched the foremost workers in a marauder raid nose their way forward through the grass, Paddy leapt up behind me with an insect net. Suddenly the 12-centimeter-long praying mantis he was chasing landed with a shudder within my swarm. Paddy backed away with a mild expletive as the ants overcame the mantis. Some grabbed the wings by their edges and spread them out to their full green glory, while others took its head between their jaws until it cracked open like a nut seized by pliers. Soon the marauders were slicing and dicing the mantis with the cold efficiency of slaughterhouse employees.

Army ants, and particularly swarm-raiding army ants, are exceptional for their ability to consistently trap difficult, even dangerous, prey. I now saw that this attribute applied to swarms of marauders as well. To uncover the secrets of the marauders' predatory success, I began to study the moment-by-moment organization of their raids. By watching where the ants first advanced at the front and then doing a slow scan back to the base column, I discovered I could treat what I saw along the way as a chronological sequence. In practice, this wasn't necessarily straightforward. While the workers might be fearless with prey, they are skittish when it comes to other interruptions. They will retreat from a simple breath of air. To interpret their behavior, therefore, I stood as far off as possible. At times I used binoculars, once confusing a group of birdwatchers with my concentration on what must have appeared to be barren earth.

The ants in the narrow swarm behind the raid front seem to move randomly, going backward, forward, and sideways with respect to the front. There obviously must be a net movement ahead to account for the raid's progress, but it's hard to detect ants following one another in that direction. Few trails are evident, and the ants appear to be moving through a diffuse cloud of orientation signals. The swarm advances to new ground every few minutes, and the land it formerly occupied is taken over by the forward part of the fan as more ants begin to form columns by running along specific tracks. The fan is differentiated from the swarm by the fact that it has these columns, and where that demarcation is made depends on the observer's ability to pick them out. Farther back in the raid, the columns become fewer and busier, with an increasing proportion of ants on identifiable routes. Ultimately, at the back of the raid, the ants funnel onto one path, the base column.

Roughly speaking, each part of a raid has a different function, turning the ants collectively into a food-processing plant. Prey is located by the foragers at the front, subdued within the swarm, then torn up in the fan. From there, it is transported along the base column to the trunk trail and delivered to the nest, where most of it is ingested by the ants. Things aren't always so clear-cut, of course; as when kids crisscross the same ground on an Easter egg hunt, it is possible for workers in a swarm to find something the lead ants have missed, or for those in the fan to contact prey that's on the run from the other ants.

Mass foraging permits workers to flush prey and act in concert to catch it, like sportsmen engaged in a fox hunt but with the scale of operations increased a thousandfold. The downside of the ant stratagem is that the colony has to pack its greatest resource—its labor force—into an entity compact enough to cross a small area, instead of spreading those workers far and wide on individual search missions, as would a solitary-foraging species. The result is that the same number of workers finds less, but catches more.[8] How? The deployment of these ants maximizes the capture of quarry too large for solitary species, yielding an intake of food that compensates for the slow encounter rate. All ant colonies stash a reserve of workers in the nest, to draw from as needed. Marauder ant swarms are made up of such assistants, transplanted from the nest to the site where food has been discovered.[9] Keeping a reliable labor supply close at hand means that a raid can quickly respond to changing conditions—an essential component of success. Prompt conscription to the battlefront through explosive recruitment minimizes the time between the moment when workers first find prey and the arrival of reinforcements to pounce on it. No matter how fierce or capable the quarry may be, with no opportunity to make a getaway it will generally be overpowered by the rapidly escalating force of its assailants.

In his book on military theory, *The Art of War,* Sun Tzu recommended this stratagem in the sixth century B.C.: "Rapidity is the essence of war; take advantage of the enemy's unreadiness, make your way by unexpected routes, and attack unguarded spots."[10] The marauder ant, in its raids, has mastered this strategy beautifully.

HOW RAIDS BEGIN

The only time marauder ants are motivated to leave their manicured avenues and raiding paths to strike out as independent individuals, rather than in a coordinated raid, is in the face of disaster. Whenever I trod on a trail and mangled a bunch of ants, both the workers I panicked and their dead and damaged comrades released pheromones causing widespread alarm. The agitated survivors, whom I call "patrollers," rushed about in a frenzy, dispersing up to a third of a meter from their trunk trail. Each appeared to take her own path away from the trail rather than tracking those around her. The patrollers seemed to be in a frantic search for the source of the problem and would give my leg a serious chew if I didn't notice them in time.

Unless I bothered them further, all the patrollers would make their own way back to the trail within fifteen minutes. However, after stepping on marauder ant trails hundreds of times, mostly by painful accident, I noticed that on occasion a weak column would emerge from the bedlam and remain active much longer, advancing away from the trunk trail and branching here and there. Supplied with more ants pouring off the trunk trail, a minority of these columns would expand gradually into a wide, fan-shaped swarm raid that often reached 2 to 3 meters across—the largest I measured was 5 meters—and contained troops that pressed forward in concert. At this point, the ants no longer seemed to be looking for me but were again expanding out in a regimented hunt for food. So what began as a response to a footstep or possibly to a tree branch that had crashed onto a trail had transformed into mass foraging in epic proportions. Indeed, the more food the workers came upon in their journey, the more epic the raiding response seemed to become.

This was different from army ant behavior. Marauders raid both in columns and in swarms, with an occasional column expanding into a swarm raid. Army ants typically raid either in columns or in swarms, but not both. Most army ant species are column raiders, whose raids stay in narrow columns from start to finish, whereas swarm-raiding army ants spill directly from their nests in a broad swath, and the raid continues as a swarm throughout.[11]

What triggers the marauder ants to launch a raid? Though there is no scout to shepherd a raiding party toward any particular meal, I noticed that when a patrolling worker fell by chance upon some morsel, nearby ants converged on the site immediately—suggesting that the worker who made the discovery had released a pheromone—and with their arrival a new pathway soon formed. This episode of conventional recruitment to something tasty could escalate into a raid when an excess of ants coming to the meal continued to advance in a column beyond it, a response known as recruitment overrun.

Was food necessary to the process? Determined to get closer to the truth about how raids develop, I set up camp one weekend, coming as close as anyone has to roughing it in the Singapore Botanic Gardens. A tent would have called attention to myself, so I didn't bring one. In any case, my intent was not to sleep, or even to move from one spot. All I needed was a camp chair and a stockpile of grub—Grainut cereal, fruit and cheese, and jerky. Stationing myself just far enough away from a 50-meter trunk trail so as not to disturb the action, I watched a 2.7-meter-long segment of the route for fifty hours straight. The midday sun left me roasted. During the second night, a storm waterlogged my notes. But it was worth it. Twice during that period I saw a raid start spontaneously, with a column of marauders streaming out from the trunk-trail throughway without food or provocation. If that had been typical for the entire trail, the colony would have been spawning a raid every forty-five minutes.

I still needed to get a picture of how the raids related to each other. For a week, Paddy joined me at the Botanic Gardens to help me find out what the marauder ants were up to in the long term, in their choice of raiding locations. We mapped raids by marking each path with bamboo

Opposite: Self-portrait, after stepping on a marauder ant trail near Malacca, Malaysia.

skewers emblazoned with neon-colored flags. Within days, the ground around the trunk trail resembled my back after my one session with a Singaporean acupuncturist.

Most often, the raids crisscrossed the belts of land flanking the trunk trail. That is where the pattern became clear. Marauder ant raids moved readily both over virgin soil and across or along the course of prior raids, even ones from a few hours earlier. When a raid passed over an abandoned path, the foragers at the front seldom showed a change in conduct, neither avoiding it nor turning to follow it. On occasion, a raid seemed to retrace an old path a short distance; presumably, there's a latticework of residual scents from an old raid that must dissipate with time. But in general, each raid went its own way.

This differs from the activities of most ants, such as the seed-harvesting ants of the American Southwest. At first glance, the masses of harvester workers might be mistaken for an army ant raid as they pour out of their nest each day. Actually, though, they are less an advancing army than commuters caught in a traffic jam, reestablishing a trail to areas where they will then scatter to unearth seeds by foraging in the desert sand. Marauder raids resemble those of army ants in not being based on set courses; these mass foragers aren't obliged to retrace their steps, and they easily cross unfamiliar terrain. The actions of the individual workers may be severely limited, but those of the raid as a whole are not. This is foraging in the pure sense, invoking the freedom to search unknown terrain, in this case moving as one.[12]

MAKING SENSE OF ANT SCENTS

A year and a half into my Asian sojourn, I made my way by train and bus from Singapore to the island of Penang, Malaysia, where I stayed for a month at a delightful research station on the beach. On several occasions, and with little warning, I was asked by the station manager to vacate my bungalow with its one small bed: a VIP from the American consulate required it for the weekend with his twenty-something "daughter."

Thus evicted, I would take the opportunity to travel to another rainforest site on the island that was thick with marauder ants. Curious about how the ants communicate to stay in tight formation, on these excursions I studied the marauder ant's ability to produce chemical trails. With a field microscope, I dissected workers to extract two organs associated with their rudimentary sting: the Dufour's gland and the poison gland, both known sources of pheromone signals in other ant species. I used a fine forceps to tease free the ant's infinitesimal glands: thin, translucent sacs small enough to be an amoeba's luncheon treat. I then smeared the contents on the ground near processions of marauders, to lead the ants where I wanted them to go.

As I hoped they would, the workers dutifully followed the artificial runways. But their reactions indicated that the functions of the two glands differed. They followed the Dufour's gland trails steadily and accurately and for a long time, which suggests that its secretion is critical in establishing trails, especially stable ones like a trunk trail. By contrast, their response to a crushed poison gland was to run like mad and sloppily follow the route for several seconds. Their brief excitement suggested the poison gland's contents were reserved for inciting ants to capture prey or destroy an enemy.

This wasn't enough evidence to produce a set of sound scientific conclusions, but workers engaged in raids were too sensitive to my presence to allow for experiments on them. From watching the workers follow the scent trails I had drawn, I hypothesized that a marauder ant raid is prompted by two trail signals, as has been proposed for army ants as well. The majority of the routes within a raid must originate when workers at the front deposit exploratory trails: pheromones, likely derived from the Dufour's gland, released as a forager moves on a new path.[13] In addition, the workers in the vicinity of food lay recruitment trails that yield a massive response if the quarry—perhaps struggling prey—is attractive to many ants. Networks of columns materialize within the raid fan even when there is no food, however, suggesting that workers reinforce a selection of the trails from the front lines.[14] This would lead eventually (I hypothesize) to the accretion of Dufour's gland secretions into the base trail, and eventually, if that trail continues to be used and reinforced over time, into a trunk trail.

Recruitment signals come and go as food is harvested. The ever-present exploratory trails are the glue that binds individuals into a foraging group, the closest parallel in ant societies to the adhesives that join the cells of our bodies. The front-line workers' pheromonal scents keep the foragers immediately behind them close together and moving ahead as a unit, all the while leading the raid forward.

For both the marauder ant and army ants, the varied attributes of the raids—the cohesive advance, the lack of a target other than the general land ahead, the absence of scouts—seem unrelated. But these features are manifested in a simple series of actions so circumspect and tentative that in humans they might be equated with separation anxiety. Unless they are diverted to kill prey en route, the ants are committed to a single goal: to follow fresh trails leading ultimately to unexplored terrain. Each ant stays near her sisters on routes that draw her inexorably out from the nest and onward, eventually bringing her to the raid front. There, she encounters the first land that is barren of signals. In response she runs ahead, drumming the unmarked ground with her antennae and depositing a smear of pheromone that guides those behind her. She then returns hastily to her "comfort zone" within the pheromone-saturated land behind. Such timidity is crucial to keeping the troops functioning as a unit, the equivalent of human boot-camp training. It vividly contrasts with the pluck the same worker shows when she joins the wanton melee around prey.

In the marauder ant, as in army ants, every worker is in effect shackled to a nexus of social signals generated largely by individuals who happen to be nearby. Thus it is not so much the proximity of individuals but their lack of autonomy that makes the army and marauder ant superorganisms nonpareil. No matter how much individuality may be prized, there may be times when, for a society—ant or human—to function productively, it pays to march in lockstep.

OTHER ANIMALS THAT HUNT IN GROUPS

There are other members of the animal kingdom that mass forage. Some spiders are sit-and-wait socialists who weave a communal web. The more spiders, the larger the catch, with dozens bearing down to secure, say, a large moth.[15] Harris's hawks of New Mexico hunt in families of up to five, leapfrogging between perches until they see a rabbit. Then they converge for a simultaneous kill or attack it in relay. If the quarry finds cover, one or two hawks flush it out while others wait in ambush.[16] Among mammals, lions, wild dogs, wolves, and killer whales also hunt in groups, staying in range of one another while seeking prey too large or agile for them to catch unassisted. Some bacteria move in similarly voracious swarms called wolf packs, with pioneers advancing and retreating in army ant style.[17] By secreting enzymes together, they can digest prey far larger than a lone bacterium would have any chance of killing.[18]

Species that bring down large prey are not the only ones that forage as a group. Many bird species can mix together in a flock that, according to Ed Wilson, "behaves like a giant mower, leaving a pattern of well-trimmed areas juxtaposed to relatively untouched areas."[19] While birds act separately to glean insects, in a flock they can take advantage of their companions' guidance to avoid enemies such as hawks and to track the best bug-hunting locations.

Mass foraging can also be a tool for mass transit, as with cellular slime molds. After they eat an area clean of bacteria, hundreds of thousands of amoeba-like cells join together to produce a sluglike creature that resembles a blob of petroleum jelly. This slug can journey far greater distances than a single amoeba and can pass over pockets of air between grains of soil that would stop the lone amoeba cold. As it goes, the slug sheds individual amoebas, which feed on the local bacteria.[20] The slug is searching not for food, however, but for areas of low moisture and high illumination, where it casts off spores.

Another group, the "true" slime molds, grow by the expansion of one amoeba into a fan-shaped body called a plasmodium, which hunts for decaying matter. In high school, I kept an orange species that resembled a swarm raid shrunk to a few centimeters across. If there was little food, my pet crept over its Petri dish slowly but steadily. A sizable bonanza could bring it to a halt as it set about gorging itself; if a patch of food was more modest, part of the slug gathered to eat while the rest continued searching, its fanlike front reduced. A slime mold isn't as dumb as its brainlessness suggests: one variety can find the shortest route through a maze.[21] I admit, though, that a person must be very patient to find it interesting as a pet.

Some of the most army ant–like strategies are deployed by vegetarians. Workers of a few termite species spread out in a loose network while foraging, each walking ahead a centimeter or two and laying an exploratory trail before she retreats and another takes her place. The advance resembles the progression of a marauder ant raid, though it's less methodical and more dispersive than cohesive.[22] A forager who detects wood at a distance, likely by scent, will abandon its search and move straight to the food. Usually she explores the wood alone, then lays a recruitment trail back to the nest. Being defenseless and easily dehydrated, termites expire fast when lost. Staying in the columnar networks helps them find their way back home and hastens the construction of the galleries the termites require to survive on exposed ground.

Another vegetarian engages in mass hunts that have a protective as well as a nutritive function. Whereas an unaccompanied eastern tent caterpillar can easily lose its grip on a tree, several together will lay a silk mat that engages their feet and keeps them from falling. These leaf eaters then find meals in a procession, with the pioneers pushing ahead short distances before retreating, to be replaced by the ones behind.[23] A group can follow an old silk trail or strike out over new terrain. A lone caterpillar finding satisfactory greenery will lay an especially attractive—perhaps chemically stronger—recruitment trail back to the silk tent housing the colony, in some cases drawing out the entire population.

This is where all other animals that search for food in groups differ from ants like the marauder: whether caterpillar or bird, bacterium or wolf, individuals are fully capable of moving away from the pack or flock and foraging without companions. And with rare exceptions, "alone" in these species really means alone, because few animals have the capacity to recruit assistants from a distance. A few birds and primates call one another to food: for example, in Africa chimpanzees draw others to bonanzas of fruit in trees by uttering loud hoots, and pied babblers lead their novice fledgling offspring to feeding spots with a "purr" sound.[24] But such social actions are virtually unknown in most species, where signals such as the yelp of the coyote or the singing of whales more often function in maintaining appropriate spacing between individuals, in combat, courtship, or group bonding, or to keep pack members together when they are on the hunt, than in calling in the troops.

One rare exception is the naked mole rat, an African rodent with antlike colonies that include queen, small worker, and soldier castes. The worker rodents lay odor trails to the root tubers their colonies feed upon.[25] Another remarkable exception, involving a symbiosis between animals who have little in common, is the raven, who will call out to guide wolves to prey; the wolves share the prey with the ravens after the kill.[26]

COMPARATIVE MARTIAL ARTS

Even though marauder and army ant campaigns are directed at predation rather than military conquest, the byzantine structure of their pillaging and the frequency with which they do battle with other ants make it tempting to conceive of their "armies" in martial terms. Predation and combat have been linked in human history as well, the tools for one often serving handily for the other, with battles occasionally ending in cannibalism.[27]

Swarm raids compare neatly to the deployment of Roman heavy infantry and other early battalions that swept forward in a broad front. One Roman innovation was to spread troops a bit more widely than did previous armies, which gave each man a few square meters in which to defend himself. Though their workers are never far apart, marauder and army ants similarly tend to remain a few body lengths away from each other, right up to the front lines, a spacing most likely maintained by the ants in order to avoid treading on one another.[28]

Naturally, there are differences between the Roman armies and ant armies. Roman troops fell into formation only in times of active conflict, when soldiers on the front lines served as a defensive shield against another army open to view, protecting the soldiers behind them and slowing the advance of the opposing army before them. Among marauder and army ants, in contrast, the foremost workers serve as a contiguous search party to flush out prey. Rarely are the ants' opponents arranged in a similar configuration; rather, they are discovered and overtaken in sporadic fights.

Despite their tactical responsiveness to prey, marauder raids can seem regimented when compared to the flexibility of Roman legions. Deployed in formations arrayed three deep, the Roman troops could be reconfigured in response to changes in an enemy's assault. The phalanx might be preceded by cavalry that harried the enemy in advance, for example, or by scouts sent ahead to report on the lay of the land so that the day's plan could be adjusted accordingly.

My painstaking observations of the marauder ant raids left me with several unanswered questions. Animals as diverse as wolves, birds, and bacteria are able to mass forage in organized groups and then to move off in isolation. Why aren't marauder ants and army ants similarly able to employ long-distance scouts to assist in their concentrated raids? The risks a marauder or army ant scout might face would seem to be no different from those encountered by any kind of ant that searches on her own, entering a hostile world without backup. Wouldn't the rewards, for the group, far outweigh the risk to the individual?

Perhaps risk has little to do with it. Watching the marauder ants cart off fruit, seeds, and animal prey, I suspected that the unpredictable quality of their plunder simply made such reconnaissance pointless. Or maybe any tendency for an individual ant to scope out her surroundings—and in so doing wander off on her own—somehow interferes with the mass-foraging process, in which a total fixation on tracking the pheromones of the group is key.

Humans are accustomed to supervision and chains of command that encompass every level from presidents to petty administrators. Roman soldiers wheeled and charged under the direction of officers moving through the ranks. For certain ants, too, transient leadership roles do exist, in some circumstances—as with the successful *Leptogenys* scout I observed in India, who always stayed with the assembled troops, guiding them to the termites she found. What, then, of the leadership role of individuals in a marauder ant raid?

Once, at the Botanic Gardens, I attempted the near impossible: to follow an individual marauder minor worker entering a swarm raid. I picked her out because she was missing the end of one antenna. It was too difficult to focus binoculars on her, so I tied fabric from an old T-shirt across my face to keep my breath from disturbing the ants and got in close. I followed "Stumpy" for a minute through the tributaries of workers in the raid fan. She dashed wildly for a moment near the commotion of ants on a beetle larva—agitated, I surmised, by alarm pheromones released from the poison glands of the struggling workers—then kept going. Approaching the raid front, she wandered and finally entered a stream of ants, where I lost her.

Nowhere along her route did I observe other individuals guiding her, or her influencing other workers. As with army ants, the marauder ant is a species with no established leaders. If I could communicate "take me to your leader" to one of them, it's unlikely I would be shown the queen, who, like all ant queens, lays eggs but coordinates nothing. Nor does any of her workers inspire, cajole, or force the whole army to take a line of action. Proverbs 6:6–8 makes this point: we must "go to the ant" and "consider her ways, and be wise" because she does the job without "guide, overseer, or ruler." King Solomon must have been a devoted ant observer to reach this conclusion. In all likelihood, he grew up watching *Messor barbarus,* the dominant seed harvester of the Mediterranean, which indeed "gathers her food for the harvest," as the Bible tells us.

The hardworking ant described by King Solomon was likely a solitary-foraging seed harvester ant such as this *Messor barbarus* from the Kerman region of Iran.

A century ago, Harvard's erudite ant scholar William Morton Wheeler called army ants "the Huns and Tartars of the insect world."[29] But no myrmecologist has ever identified a Genghis Khan or Attila among them. At best, an individual in the raid may be momentarily better informed than others, giving her a brief and local influence.[30] That could happen when a worker at the front sends out recruitment signals to prey—but even then she is likely to be acting in concert with nearby sisters. No ant, in fact, can conceive of the raid in its entirety, know where it is going, or anticipate how the masses will respond when food is found or enemies encountered. A raid arises through a series of simple actions by each worker and others like her, in an engagement that can truly be described as "self-organized."

Humans constantly have to work around issues of self-interest that would otherwise impede the emergence of social institutions and infrastructure. Our clannish devotion to networks of kin and friends has proved particularly problematic in the context of modern warfare. The solution has been to divide armies into squadrons small enough for the troops to bond and be willing to take risks for one another.[31] Ant workers, of course, don't recognize nestmates as personas in the way I picked out the stump-antennaed individual,[32] and they never throw themselves in harm's way so that particular compatriots might live. What we perceive in ants as acts of heroism and devotion are really more akin to acts of patriotism. Since it is only the superorganism that matters, ant workers instinctively toil and die for the benefit of the colony, without recognition or recompense other than the remote possibility of augmented reproduction by the queen, the one member of the group who is indispensable. Mortality seems to be the basis of the domestic economy for prodigious, combat-savvy ant societies.[33] It is difficult not to think of the Spartan mothers who sent their sons off to battle saying, "Come home either with your shield or on it."[34] Brute force, apparently, is the key to tactical success for mass-foraging marauder and army ants.

3 division of labor

In the short grass of the Singapore Botanic Gardens, I dropped to my knees, then lowered myself to my elbows and, at last, to my stomach, eye pressed to soil, camera extended in front of me. My perspective standing up had been abstract, like that of a general assessing the movements of troops from a hilltop, where they were more pawns in a game than people engaged in a life-and-death struggle. Now, seen close-up through my camera lens, a marauder minor worker stood tall and solid before me, antennae moving as if to sniff me out. Her forebody was raised, forelimbs almost lifted from the ground, mandibles open. She was ready to pounce. Suddenly I saw the silvery blur of some creature, through my lens the size and shape of a tank, and the worker was yanked from her spot. I recognized the beast as a roly-poly, or pill bug, a quarter-inch multilegged crustacean presumably flushed at the raid's front lines.

My worker had seized one of the pill bug's furiously moving legs. Though knocked about violently, she managed to hold on. Two other minors, and then three more, grabbed the pill bug by other legs or the edge of its carapace. One whose head somehow got smashed released her grip and fell away. The others were strong enough to bring the pill bug to a halt. It tried to roll into a ball—a ploy that gives the bug its common name—but the tightly anchored workers prevented it from protecting itself. From the left, a media worker lumbered into view. She used her antennae to survey the scrimmage. Then she opened her club-shaped mandibles wide and struck. The pill bug's pale underbody went limp. Watching this skirmish conclude, I couldn't help but think about how groups of early humans brought down woolly mammoths using nothing but guts and some simple stone tools.

When I left Boston for Asia in 1981, I had a premonition that I would discover amazing things about the marauder ant—so amazing that my thesis committee might suspect I had concocted stories while smoking an illegal substance with an Indian guru. Knowing I had to come home with indisputable documentation, before I left for Asia I bought a how-to book on photographing supermodels, *Cosmopolitan*-style. With $230 in equipment that included a used Canon SLR, a macro lens, and three $15 flash attachments that gave me electric shocks, I miniaturized the glamor studio the book described by affixing the flashes to the front of the lens with a pipe clamp. By adjusting the strength of my lights, I adopted the concepts of "fill" and "hair light" to accentuate the gleaming exoskeletons of my minuscule models, defining each limb and chiseling every fiber on film.

During my travels in Asia, I used my camera to observe ants, triggering it whenever something happened that I wanted to examine later. In India, trying my equipment for the first time outside, I was stunned to see that through my lens, ants towered. Soon I was stalking them through the viewfinder with all the thrill nineteenth-century hunters must have felt tracking lions. With both quarries, the trick is to go unnoticed, to catch everyday behavior without being bitten—admittedly a more high-stakes proposition with a lion. Still, when tracking an ant in this way, I would forget her size, and she gained all the grandeur of the king of the jungle.

A minor worker stands a couple of millimeters tall. Photographing such a tiny insect requires concentrated effort and lots of illumination. When I focused the camera on my leg, my cheap flashes gave such an intense pulse of heat and light that smoke rose from my jeans. Fortunately, reducing the setting to one-quarter power solved the problem while providing sufficient exposure, but even then, the part of the picture in focus was often only a fraction of a millimeter deep—the length of a paramecium. With the flashes toned down, most ants ignored my "light cannon," especially when struggling with prey. Like a lion, an ant is easiest to approach and photograph when it is preoccupied.

In my six months in India, my photography budget was tight, but I took an occasional picture of marauders swarming, collecting seeds, and being harassed by hairy *Meranoplus* workers. Before I flew to Singapore to continue my work in Southeast Asia, I wrote the Committee of Research and Exploration at the National Geographic Society, which had given me a grant, to ask if they could develop my film. The committee's chairman, Barry Bishop—a member of the first American team to climb Mount Everest—kindly agreed. I put six rolls of Kodachrome 64 film in an express package and sent it off to him. Two weeks later, I was surprised by a Telex announcing that a writer from National Geographic was flying to India to meet me—about what, it didn't say.

A few weeks later, I left Sullia and traveled to Bangalore, where I was to meet the writer, Rick Gore, for breakfast at his hotel, Bengaluru, the finest in the city. By then, I had been living in rural villages so long that the hotel gave me culture shock. The corn flakes and coffee, though everyday American foods, were pricy by Indian standards, costing more than I spent in a week in Sullia.

Rick told me my photographs had gone to Mary Smith at "the magazine," who wanted to support my efforts, maybe even have me write a story for the magazine. I didn't know it at the time, but Mary is legendary for her work with such iconic scientists as the paleontologists Louis and Mary Leakey, the undersea explorer Jacques Cousteau, and the ape experts Dian Fossey and Jane Goodall. Why did she want to work with me? "She likes what you are discovering," Rick told me. "She also has no idea how you are making the ants look so glamorous."

I had no idea either. Up to that time the only photos of mine I'd seen were test shots I had taken of dead specimens back in Massachusetts, and they weren't anything to crow about. So a month later, when I arrived in Singapore, where Mary had sent the developed slide images, I was stunned. The ants that had been half visible to me through my camera in dim light were clear and crisp on film. Here were marauders confronting furry *Meranoplus*, sleek *Leptogenys* hunting termites, eagle-eyed *Harpegnathos* seizing crickets in mid-jump. Two years later, after my return to the States, when I met Mary, she compared my images to the visuals in the film *The Terminator.* "For you ants are huge, so they become huge for the rest of us," she told me. The photographs became part of my first article for *National Geographic* magazine.[1]

THE PLAN OF ATTACK

In Singapore, I splurged on flash attachments that did not shock me. To take in the mass-foraging pattern, I stepped back each day to observe the raids as a whole. But like a physiologist who examines muscle fibers to find out how humans move their fingers, I also came in close with my camera "microscope" to record the individual ants in action and learn the details of how they made their kills and harvested the victims.

These observations came as a welcome relief after months at Harvard measuring ants in museum drawers and categorizing them as minor, media, or major based on their frequency and size.[2] What I discovered in the field was that the slender minor workers form 98 to 99 percent of the population. Tiny, with heads about 0.6 millimeter wide, they are distinct. There are no intermediates between them and the other ants, which range widely and continuously in size. Within this continuum, there is a distinct peak in the numbers of ants at just over 2 millimeters' head width, and so these I called "media workers," and another peak at just over 3 millimeters' head width, for the majors. A few of the majors are substantially larger, with heads 5 millimeters wide or more—the size category I informally called the giants. The queen, who ordinarily stayed in the nest, had a smaller head than a giant, but a much larger body: she could be about 2 centimeters long.

Among different kinds of ants, I learned, work is divided up in two ways. In some species the workers are similar in appearance but flexible in their job skills, temporarily taking on any tasks as they arise, but the colonies of other species can also develop workers of different sizes to do different jobs on a more permanent basis. The former method allows colonies to adjust more rapidly to changing conditions, but it has its limitations: since the workers are identical and interchangeable, duties that require a specialized skill set may be poorly executed. Polymorphism—variation in size and shape, along with physiology and brain development—is an indicator of a more permanent specialization, and is the primary determinant of division of labor in the marauder colony. Because the workers of differing size are suited to a narrower set of tasks, they expand their activities, if at all, only under stress; in some ant species, for example, soldiers who ordinarily do little except fight will help tend the brood if other workers are taken away by a meddling researcher.[3]

From this, it has been determined that an extremely polymorphic species like the marauder ant is likely to have predictable labor needs, because the number of members in each physical caste, or size group, changes slowly, if it can be changed at all, based on the colony's requirements.[4] In fact, the size frequency distribution reveals something about how many ants of each caste a society requires, somewhat equivalent to the distribution of people in different job descriptions in a city.[5]

To pursue again the earlier metaphor, a colony can be seen as a "superorganism" that functions like the body of an organism, with the number of castes and the frequency of each being analogous to the number of types of cells and tissues and the size of organs. Ant species with small colonies are like the cells in simple organisms in that they have few labor specialists, but marauder ants are intricately specialized. Add the arrangement of the workers in space and their interactions with each other to the numbers and frequencies of the various workers, and one has the "scaffolding" of the superorganism, much as a body is built upon the number, location, and interactions of cells. The parallels are all the more remarkable since both the ant workers in a colony and the cells in a body communicate largely by chemical cues (hormones being a prime example for cells), the biggest difference being that workers are mobile and accumulate dynamically when and where they are needed, while most cells are fixed in place within the body.

Essentially all the participants in the raid front are the little minors. With my photographs, I was able to disentangle the blur of action as these ants brought down a nightcrawler or grasshopper thousands of times their weight. A single minor worker has no more chance of catching such a behemoth on her own than would an equally small worker of a solitary-foraging ant species. But she shares the front with other minors that contact prey at about the same time, and they pile on like tacklers in a game of American football. With this strategy, the chances of capture improve markedly: as in Swift's tale of Gulliver toppled by the Lilliputians, strength in numbers can't trump size.

It makes sense for a colony to produce a lot of minor workers and concentrate them at the front. If the prey were confronted by a single media ant instead, even one weighing as much as all those smaller tacklers combined, the larger worker would be less effective at subduing the worm or grasshopper. Though individually weak, minors working together simultaneously grab their quarry at different places and angles, making it hard for a victim to move. The prey is also more likely to slip by a single big worker than by a barricade of spread-out small ones.

Countless times I've watched a nightcrawler inching over the ground or a grasshopper resting on its green blade, minding its own business, as a swarm moves toward it with a whisper like a snake in the grass. If it doesn't respond by reflex, death is certain. At the touch of the first worker, the worm flips back and forth; the grasshopper makes its leap. But out of view in the vegetation, more ants are swarming in. About half the directions the flipping worm or leaping grasshopper could choose will land it deeper among the ants, while the other half will allow it to evade the ants by getting ahead of the raid. Blundering deeper is like colliding with a dragnet with a mesh of the width and strength approximated by the closeness and size of the ants; the more the worm or grasshopper struggles, the more the masses converge on it, as other ants are alerted and drawn into the fray. Soon all the little ant jaws hold their prey taut.

Avoiding the ants by moving ahead of the raid provides a temporary respite. The best hope for any creature is to dash to freedom to the left or right of the raid, and so carry itself out of the ants' path; but the distribution of ants must be difficult for prey to determine down among all the litter and plants on the ground, so taking this course may be a matter of chance. If the prey fails to chose the right direction, the army will advance to its new location and strike again. And if it escapes once more, a swarm may try a third time, or more. Because of their width, swarm

Minor workers at the front of a marauder ant raid in Singapore being cut to pieces while subduing a termite soldier.

A major worker crushing the termite after the minors pinned it down.

raids are most likely to repeatedly contact the flipping worm or leaping grasshopper. (A narrow column raid is different; its net is too narrow and weak, and most victims break free. The ants in column raids therefore reap mostly seeds and frail prey, though the raid may burgeon into a swarm if they find bigger spoils.)

Even escapees may not survive. I once saw a cricket rocket from its hiding place beneath a leaf. In a series of zigzag moves it ended up far from the raid, but a few ants still clung to it stubbornly. Their gnawing slowed it down, until at last its body convulsed. However, the ants that subdued it were now so far from their colony that they would die before ever finding their nest again.

Participants in a marauder raid seem to be forever in battle mode. They fight with a dogged precision that is chilling, and in large raids there certainly seem to be troops to spare. The minors show by far the highest casualties. The bounding cricket managed to chew a couple of the minors on its leg to a pulp before succumbing to the rest. On my way back to the raid, I saw minor workers puncturing a plump caterpillar, and one drowned in the jelly that oozed from it. Later on in that raid I saw a termite soldier with a burnished red head that dwarfed the minor workers surrounding her like a grizzly bear cornered by dogs. The termite's black jaws were sharp as knives, and each minor that came near was sliced apart as cleanly as if by a guillotine, until a dozen ants stormed her hindquarters and brought her down.

Like a war correspondent inured to tragedy, I watched hundreds of minors being sundered and smashed in struggles with prey, the horror of the slaughter magnified through my camera lens. By never straying from the task to save themselves, they displayed breathtaking devotion to their duty. It made me wonder about the advantage of psychological numbness in combat even among sentient humans. As one author wrote of the Civil War, "Soldiers perhaps found it a relief to think of themselves not as men but as machines."[6]

Such thoughts reflect how caught up I was in the drama of the moment, pressing the button of my camera each time a surprising event happened. I saw that the minor workers were able to stretch the legs of the termite soldier until she was spread-eagled *(click)*. By this time, the raid front had advanced beyond the victim, who was now deep within the swarm. Here the media and major workers roamed in numbers *(click)*. The large ants were as plucky as the minors, and they had the size and mandibular power to be worthy of the designation "soldier"; but by dint of their location, most of them joined the fray at the termite only after the prey was felled *(click)*.

My images transferred onto a storyboard that showed that inside the raid, after the minor workers immobilized the body, the medias and majors were a strike force that moved in to inflict what carnage they could. Small media workers fit into tighter nooks and crannies than the majors can reach, perhaps yielding a kind of division of labor in destruction.

The allocation of effort between the minors, which restrain prey, and the medias and majors, which smash it, is related to their respective locations within the phalanx. It's unlikely that special communications are used to get ants to these positions; instead, the minors reach the front lines first because they walk more nimbly than their larger sisters, while the larger ants are waylaid by their duty to crush prey farther back in the raid. Regardless, the role of minors at the front lines is clear. Only they and a few small medias secrete trail pheromones, testament to their importance in moving the raid ahead and summoning others to prey.

To a military historian, the marauder ant strategy evinces a classic use of personnel. Placing large numbers of abundant and expendable weak individuals in jeopardy at the front lines not only increases the catch but also minimizes the loss to the society overall. The Romans used a similar strategy at their battlefronts: instead of drawing from highly trained city dwellers, they largely conscripted farmers, who were available in droves and could be replaced at little social cost—a practice that continued at least into medieval times, when poorly trained men were, literally, used as cannon fodder.[7]

The minors' bold actions assure few large warriors being sacrificed, a sensible outcome given the expense of raising majors that can weigh hundreds of times as much as one minor. In a sense, the medias and majors are equivalent to the human warrior elite—physically stronger, superior fighters, often positioned behind the relatively inefficient front-line rabble. The human elite are provided with better weapons and training and protected by the most expensive armor, as tough as a soldier ant's exoskeleton.

The large workers are attracted to a prey's flailing extremities and dutifully hack off every moving leg and antenna. With the prey rendered powerless, unless its shape is awkward (like that of a praying mantis, which the ants will tear apart), the minors heft its body back in one piece. I once saw the ants retrieving a limbless gecko, which clued me in that they had taken it alive.

Dismemberment immobilizes but doesn't necessarily kill. Moving animal prey to the safety of the nest before the coup de grâce may reduce the chance of its being stolen by competitors or washed away in a storm. By keeping prey alive, the ants may also be able to preserve their meat (something that ants with stingers do by paralyzing their victims).[8] I learned of this strategy one day at the Botanic Gardens when I snatched a limbless katydid from marauders on the way to their nest. I put it in a jar and forgot about it until, two days later, I noticed its leg stubs still writhing. That night I dreamed I was that katydid, being helplessly transported to the bowels of the nest, to be digested at the ants' convenience by the protein-hungry larvae.

SPRINGTAILS

Marauder ants conduct raids to catch tough prey, but mass foraging helps them obtain other kinds of meals as well. The poorly armored minors, though not intimidating, are agile and have good vision. I've watched hundreds of them retrieve speck-sized jumpers called springtails.

Springtails are the rabbits of the insect world—fast breeding, abundant, and prodigiously jumpy. As the name implies, they use their tails as a spring. If one senses a threat, its tail, or furcula, normally folded under the body, snaps downward, launching the insect through the air.

Before exploring the marauder ant's tactics for capturing these motile creatures, let's first look at a very different approach. A speck herself, a burnished red *Acanthognathus teledectus* ant moves stealthily through the forest litter in Costa Rica, her long, pitchfork-shaped mandibles held straight to each side. Coming on a springtail, she slows to a glacial creep until two long hairs extending from her mandibles touch the quarry, indicating that her distance is perfect. Her jaws snap forward; their prong tips puncture the springtail and hold it tight. Quickly now,

With blows from her mandibles, an *Acanthognathus* trapjaw worker in Costa Rica repels a pseudoscorpion from the tiny hollow twig occupied by her colony. Behind her, a larva feeds on a springtail.

the ant slings her hind end under her body and incapacitates the prey with an injection of toxins through her sting, after which she hefts it overhead and carries it home.[9]

Acanthognathus displays the special skills required for solitary-foraging species to snare these speed demons. Success among springtail-hunting virtuosos depends on stealth and the use of mandibles as an unusual tool. Devices such as trap jaws and stingers are especially common among species with small colonies with only one kind of worker, such as *Acanthognathus,* whose workers so often need to act alone. Unlike with the antlers of moose or the tusks of elephants, their function is not to impress but to kill and butcher.

"Trapjaw ants" like *Acanthognathus* have evolved repeatedly among lone-foraging species. Typically, their mandibles are long, with pitchfork-like teeth only at their far ends, and they can open 180 degrees or more. In many cases, the jaws come equipped with trigger hairs. While the ants can be slow, their "bear-trap" jaws are not: the fastest muscular-driven action for any animal is achieved by the jaws of one group of these ants, *Odontomachus.*[10] These speed-biters nab insects and also ply their mandibles as defensive tools, striking them against the ground when harassed; the resulting recoil sends them flying head over heels to safety. In Surinam, I've seen schoolchildren, betting over candy, make a game of encouraging the *Odontomachus* ants' bouncing behavior while trying to avoid their searing stings.

Long jaws are great for catching prey but impractical at mealtime. Asian *Myrmoteras*, another group of creeping trapjaw ants that nest in any dark corner of the leaf litter, chew their prey from afar using the spiked tips of absurdly thin mandible blades that they can open an extraordinary 280 degrees. After chewing, they walk forward to place their mouths on the victim and feed at the oozing wound, then circle back to chew some more—the most awkward and labor-intensive approach to dining I have witnessed in all my travels.[11]

Acanthognathus have a partial solution to this logistical problem. While they use their long jaws to seize skittish springtails, they avoid the arduous dining experience of *Myrmoteras* by having a face like a Swiss army knife, with an entire arsenal of utensils at their disposal. To eat, they open their jaws wide, revealing a pair of what look like normal mandibles but are actually curved teeth, sprouting near the base of the longer bear-trap blades. The workers masticate their springtail meals to a pulp with these minijaws. As the small jaws are of a piece with the rest of the mandible, chewing with them sets the bear-trap blades waving to such a degree that feeding ants often knock over their neighbors.

Marauder ants have no elaborate built-in tools with which to seize springtails. Instead they must rely on commonplace, workaday mandibles (which have several small teeth along their forward margins, as do those of most ants). Furthermore, the marauder's massive, frenetic societies are at the opposite extreme from those of the slow, stealthy *Myrmoteras* and *Acanthognathus*. The tempo of an ant species tends to relate to its colony size.[12] Workers in small societies tend to be slow and cautious—a sensible way to approach elusive prey like the springtail on a low-energy budget. (Is the per capita energy quota of a small colony indeed likely to be smaller than that of a large one? Picturing a colony as a superorganism, a physiologist might predict that this would be the case. Since larger creatures are relatively efficient, burning fewer calories per unit of weight—or when measured microscopically, per cell—this gives them energy to spare.[13] We can extrapolate that the same would be true for superorganisms, resulting, for example, in decreased labor demands for each individual in a large nest.[14] If so, life must be precarious for *Acanthognathus* colonies—which, in my experience, are very rare, with no more than eighteen workers nesting in the rotted-out core of a single twig on the rainforest floor.)

How does the marauder ant, with its numbers and seeming chaos, nab the wily springtail? Lots of the ants seem to be doing the same thing at once, with sloppy overlap in their activities. But the effectiveness of large societies often has to do with redundancy rather than precision: although an individual ant may not be reliable, the density and overlapping actions of multiple ants ensure success for the raid. As each point on the ground is probed exhaustively, every critter, no matter how small, is rooted out.[15] Once flushed, a springtail leaps about as one ant after another frightens it. Sooner or later, one of the minor workers will snare the springtail and make the kill. The raid, in its entirety, becomes the colony's bear trap.

The effectiveness of this form of predation lies in exhausting the victim. Lions and wild dogs accomplish much the same thing. Although a solitary cheetah may have the edge on them in terms of speed, working as a pack the group predators can kill a gazelle that easily outruns them, wearing it down by chasing it sequentially, like relay runners, or by driving the animal toward an individual lying in wait. Marauder attacks aren't as subtle or as calculated, but given the ants' massive numbers, they may not need to be.

THOSE VORACIOUS OMNIVORES

The marauder ants' predatory skills are only part of the picture. "The voraciousness of these ants is very great," wrote a Vietnamese phytopathologist named Pham-tu-Thien in 1924. "We are dealing with a species whose greediness has fully developed its capacity for work." Pham recorded marauder ants consuming insects, seeds, and fruit.[16] What they take varies widely according to availability—they nibble on such oddities as leaves, flowers, bird droppings, and fungi when few other resources are available. But even when foods are bounteous, marauder ants tend to be wide-ranging gourmands.

Swarm-raiding army ants, often said to have among the Earth's broadest diets, don't compete with marauders in this regard. In particular, army ants are poor vegetarians, while marauders collect equal amounts of plant and animal material. Vegetable matter contains cellulose that many carnivores find indigestible. The only army ant approaching the marauder's omnivorous diet is south Asia's *Dorylus orientalis,* which, like the marauder, is considered an occasional agricultural pest—though it eats tubers such as potatoes, rather than the rice and other grains fancied by the marauder.[17]

The marauder ant species—*Pheidologeton diversus*—shows a proficiency at seed harvesting equal to that of many of its seed-harvesting relations in the group to which *Pheidologeton* belongs, the Myrmicinae, and I imagine the ancestor of *Pheidologeton* was like many of these relatives in eating seeds while scrounging for dead insects and perchance killing the occasional live one.[18] On my Indian palm plantation, instead of taking their seeds straight to the nest as they did prey, the workers established caches along trails, carrying grain down holes or under leaves, where it was stored or milled to an edible flour by medias and majors. The ants also harvested an herb called goatweed by dropping its seeds to the ground, where workers of all sizes congregated to chop them up for immediate consumption.

Marauders are even more organized when they harvest grasses, one of their pastimes in the Singapore Botanic Gardens. When a raid passes a fruiting grass plant, only the minor workers and small medias can climb the slim stalk. The first minors gnaw the attached seeds ineffectually, but productivity skyrockets when a media arrives. The ants now set up a little assembly line, in which the media extracts one seed after another and then appears to hand it to a minor to haul away. What is really happening, however, is that the minor, who is too weak to pull a seed free from the stalk on her own, snatches the seed from the media before the larger ant can depart with it. The media dutifully plucks another seed, which another minor grabs. With minor workers so numerous, a media seldom has an opportunity to exit with her find.

Windfall fruit and vertebrate carcasses draw much larger crowds that defend and often consume them where they lie. Tens of thousands of workers will dismantle a mango or a dead bird. When I spilled a bag of canary food next to a trail, the ants arrived by the thousands to carry off

300 grams of seeds in eight hours, ten minutes. Under ordinary circumstances, the workers never seemed to become finicky or grow tired of a food, but this overfed colony refused over the next several days to touch any more of the seed.

The one food source that marauders forgo is another kind of bonanza, the populous nests of social insects. Tackling well-fortified bees, wasps, termites, and other ant species requires a convergence of forces to break through the foe's weak points—a military tactic that marauders lack, though army ants display it in abundance.[19] Indeed, almost all army ants gang-raid social insects routinely; many species especially relish the eggs, larvae, and pupae seized from colonies of their ant relations.

Marauder ants don't just steer clear of social insect nests; they actively avoid making meals of them. When marauders kill another ant species in a skirmish, they cover the bodies with soil and abandon them. Despite this odd and unexplained aversion to cannibalism, the marauders evolved mass foraging in part for the same reason army ants did, as an aid in battle. They might not eat other ants, but they do compete with them for meals. The swarming multitudes in the raids that the workers at the front lines draw upon to subdue prey can also be used to overpower any rival that gets in their way.

Among combative ant species, known as extirpators, trumping competitors is generally a matter of preemptive control of resources. Arriving at the contested area "first with the most," as General Nathan Bedford Forrest said of battle strategy in the U.S. Civil War, these species succeed by assembling quickly and in abundance. After driving off more timid species, the ant troops can block other belligerent ants from building up at the site in sufficient numbers to fight back.

Because the marauder ant doesn't employ wide-ranging scouts, this species is seldom first to show up at a feast. But this doesn't present a problem: the raiding deluge overruns any competitor and keeps rivals at bay—even other extirpators, army ants among them.[20] Their tactics bring to mind the "rapid dominance" military doctrine proposed in 1996 by American military theorists. For humans, being on the offensive puts the enemy in a vulnerable position, giving the invaders a sense of invincibility even when it isn't justified.

> The key objective of rapid dominance is to impose this overwhelming level of Shock and Awe against an adversary on an immediate or sufficiently timely basis to paralyze its will to carry on. In crude terms, Rapid Dominance would seize control of the environment and paralyze or so overload an adversary's perceptions and understanding of events that the enemy would be incapable of resistance at the tactical and strategic levels.[21]

Marauders similarly take the offensive from the moment they contact alien ants, whether the foreigners number in the thousands or are just two carrying a seed. Often the minor workers blast forward in such abundance that other species fall back with hardly a fight. Even when

Opposite: An assembly line of marauder ants on a grass stalk in Singapore. A media worker extracts grass seeds, which the minor workers carry away.

clashes occur, the marauders triumph by using their first-strike capability. By mowing down enemies a few at a time as the raid advances, the minor workers suffer far fewer casualties than they would if they faced the opposition all at once, a similar outcome to that of the divide-and-conquer strategy of large-scale human military actions.[22] With the other side routed and unable to recruit assistance, the marauder ants' control of the booty is likely to remain absolute and uninterrupted from the moment of first contact. In Singapore I watched marauders steer hostile weaver ants up the tree in which this canopy species was nesting, and then the marauders gathered by the hundreds for a meal: they tore off the tree's bark, rotating bits of it between their mouthparts and forelimbs while sucking out the sap. This food ordinarily draws the marauder ants only in times of scarcity, and indeed at the time there had been no rain for a week.

From springtails and seeds to frogs and large fruit, marauders harvest a cornucopia. They are reminiscent of humans, who apply the dictum "because it was there" not only to climbing mountains but also to adding tasty morsels to our diets. Marauders and people are exceptions to the general rule that in the tropics, where so many different organisms live together, most species, like the springtail-hunting trapjaw ants, become specialists in a narrow niche to survive the intense rivalry for resources.[23] Marauder ants, in contrast, by interfering with all contenders for each meal and taking prey where others fail, exceed expectations by being geniuses at the competition game.

TRACKING FOOD FROM A TRUNK TRAIL

In Singapore's Botanic Gardens one day, I placed a meter-wide plywood board in front of a raid. The ants crossed it in swarm formation, which confirmed my suspicion that their raids don't depend on workers finding food or retracing old routes. Even so, I knew the ants were no fools—their raids slowed in areas with little to offer, the number of workers in them declining as the ants drained back to the nest until, if the dearth continued, the whole army would retreat. I decided to find out how the plenitude or distribution of booty changed an army's strength and direction.

The marauder ant's vegetarian proclivities made the job easy: it's more difficult to manipulate caterpillars and crickets than to move fruit and seeds.[24] Loaded with supplies from the grocery store on Orchard Road, I headed back to the Botanic Gardens and spread canary seed in a line extending from a trunk trail. It didn't take long for the marauder workers to leave their highway and flow along this line. They tracked the seeds precisely, continuing outward in a column even after they had passed the last seeds. I had launched my own raid!

Did the distribution of food affect how the raid progressed? I poured a seed pyramid ahead of a swarm. The ants continued forward for several minutes after contacting this jackpot and then drained back to the food, where they rapidly built up in numbers. The swarm raid now over, the excess arriving ants radiated from the seed pile in a network of branching column raids spread over several square meters (a process called recruitment overrun, described in chapter 2). I had seen marauder ants generate similar trail networks under trees dropping fruit, which they thus track down quickly. While column raids are ineffective for catching fast prey, these bifurcating formations shine when it comes to fanning a foraging populace out over large areas. Each time one of the weak raids in a network encounters a bonanza, any number of workers can be summoned within minutes from the trunk trail to seize and consume it.[25]

What if the enticements are less concentrated? My next approach was to scatter a few seeds in a meter-wide swath off to one side of a swarm raid that was crossing a field with little in the way of food. The raid turned and followed my swath its entire 15-meter length, even though I laid few seeds—one every 20 square centimeters or so, which would put three of them in an area the size of my palm. Somehow, raids track subtle changes in food density, even though the workers coming upon each seed are ignorant of the food distribution as a whole.

How does that happen? While the ants follow exploratory trails at the raid front, they are more attracted to any recruitment trails they come across, which lead to food. When there are more seeds on one side of a raid, ants must be drawn to them by the buildup of recruitment pheromones left by the successful foragers from that direction. New arrivals tend to follow the strengthened routes leading to the food-rich region, causing the raid to turn and track the seeds without any of the ants comprehending what is happening—a fine example of what artificial intelligence experts call *collective* or *swarm intelligence,* in which the raid viewed as a whole deals effectively with problems by adapting to changes in the environment. A.I. experts would describe the raid as "robust." Indeed, from computers to the natural world, scientists have found that seemingly thoughtful processes often emerge spontaneously from the integrated actions of simple-minded agents, like ants, with no need for leaders or any kind of management or centralized control.[26]

I went back to Orchard Road, depleting the grocery shelves of bird seed to continue my experiments. What mattered to the marauders seemed to be the relative abundance of food: when a raid was bringing in lots of other victuals, I needed more seeds to alter its course. The raids turned out to be smartly responsive to food in a variety of ways, branching or shifting in direction, width, and strength on the fly. Even though the absence of scouts made the raid blind to meals at a distance, the aggregate response of the workers to food at hand apparently enabled the raid as a whole to follow the food distribution in bountiful regions.

It's a subject of endless fascination for scientists that each ant can only proceed locally on the limited information at hand, and yet their societies manage to act globally. Darwin was right when he wrote that for all ants do with their modest endowments, "the brain of an ant is one of the most marvellous atoms of matter in the world, perhaps more marvellous than the brain of man."[27] But the true power of the mind of an ant emerges at a superorganismic level, when those brains join to produce colony-level actions to accomplish a goal. Lewis Thomas, the author who first introduced me to the superorganism idea in my youth, described an ant society as "an intelligence, a kind of live computer, with crawling bits for its wits."[28]

HOMEWARD BOUND

One afternoon it occurred to me that I could use the marauder ants' ability to track seeds to unravel a mystery. Every trail has two directions. How do workers select the correct way home?[29]

In most situations, the ants have no problem choosing a direction. Because workers ordinarily find food at the raid front—the end of the trail—every returning ant has but one way to go. Along the route, though, are junctions with other trails. Some of these don't present a problem: trails split at sharp angles, so nest-bound ants will make the right choice if they take the route that lies closest to straight ahead.[30] Still, in the labyrinth of trails between raid and nest, I saw many situations in which the ants could have made directional mistakes but rarely did. Why?

I realized that by pouring seeds in an arc, connecting one point on a trunk trail to another point farther along the same trail, I could give the ants a choice of two equally good directions back to the nest. I watched in anticipation as the troops rushed from the trunk trail to track the line of seeds along each end of the arc. Every ant who picked up a seed from the advancing front of either column then turned around and carried it directly back to the trunk trail. When the advancing armies met, the ants now had the option of completing the full loop, and they often did so if they had't picked up a seed. From their point of view, they were simply continuing as they had been going, away from the nest. A worker that picked up a seed *after* passing the site where the troops met would not turn around but rather would continue onward—a choice that, in any "normal" situation (not a loop), would have led her away from the nest.

The result was that all the seeds flowed away from where the armies converged. I called the trail segment within a centimeter or two to either side of this point the transition area because ants acquiring seeds in that stretch weren't consistent in their choice of direction. The transition area was usually near the middle of the arc, but I could change its location by laying down the seeds earlier at one side of the loop, causing the ants who found that end of the loop to travel farther than they did on the other side before the armies merged.

At first, I guessed that the ants had marked the trail with some kind of "arrow," as invisible to our eyes as the pheromone trail itself, which told their colleagues, "Go this way!" But that hypothesis crumbled when I waited until the seeds were nearly gone and the ants still moved around the arc with nothing to carry. I poured a new heap of seeds along the arc away from the transition area. If the trail contained a directional cue, all the ants taking seeds from the new pile should have gone in the same direction the workers had taken earlier when they took seeds from that spot. Instead, the ants proceeded to haul the seeds in both directions. While the workers were still retrieving seeds from the new pile, I poured yet another pile elsewhere along the arc. All the ants taking seeds from the first pile passed the second one and continued in the same direction they'd been going. But when ants began to pick up seeds from the second, newest pile, all of them followed the lead of the ants going past them with booty from the first pile.

Other experiments confirmed this behavior: ants picking up seeds took the direction of any passersby with food (and if there were none, they could go either direction). Were they being physically forced to go the same way, bystanders compelled to join a mob? No—the seeds weren't bulky enough, and the carriers weren't numerous enough, to inhibit ants from going whichever way they wanted.[31] Instead, it appeared the food-bearing ants were taking notice of each other's choices and deciding accordingly.[32]

As it turns out, this "go with the flow" approach is essential to the marauders' response to bedlam. Crush a marauder ant underfoot, and some workers, detecting alarm pheromones released by the body, rush off the trail on patrols in which they attack whatever they find. While the patrollers are in defense mode, the food-bearing ants do an about-face, clearing the disturbed area by rushing outbound along the trail instead of continuing to the nest. As laden ants farther along the trail confront this backflow, they turn and join the exodus, in this case propelled away from the nest by the urgent multitudes.

If the laden backtrackers reach the trail's end, they mill about before starting back to the nest. Usually they don't get that far: as the fleeing ants spread out more and more along the trail, their frantic pace slows to a normal gait, and they gradually start to turn around again under the influence of all the workers carrying food in the "correct," nest-bound direction. In either case, by the time the ants return to the point on the trail where the fracas took place—anywhere from five to twenty minutes later—the problem is long gone and the patrolling has all but ceased. It's now safe to go home.

Except in such emergency situations, traffic on busy marauder ant trails is well organized so as to avoid congestion. The scheme isn't to stay to the right or left, as on human thoroughfares. Rather, nest-bound ants tend to use the trail center, while the outbound ants stay to the sides. The center is easiest to travel, being concave from use, with few obstructions and the most concentrated pheromone. The inbound ants with their unwieldy loads end up there because they have difficulty maneuvering. Carrying nothing, outbound ants can quickly move to the sides of a trail to avoid their encumbered sisters. Similarly efficient patterns emerge among people, too. Think of how pedestrians will be diverted to the gutter as they try to circle around someone hefting a big package on a crowded sidewalk.[33] And during rush hour, without anyone thinking it out, clusters of pedestrians will move in alternative directions through bottlenecks—a pattern I have seen in marauder ants as well, where their routes head through a bottleneck underground.[34]

4 infrastructure

Through my camera lens, I closed in on a gray *Diacamma* worker with an elegant silver sheen striding along with what appeared to be a sense of purpose. I tracked her ascent of an embankment of soft soil. She went over the top and landed squarely among marauder ants following a trunk trail on the other side. Six minor workers pinned her in place as workers laden with food retreated; then a major arrived and executed her with a crushing blow, discarding the corpse just off the trail, where several minors buried her in the dirt as their food delivery operations resumed.

Marauder colonies maintain a fast, steady, well-protected flow of food and labor on their trails. Whereas small ant colonies, like people in small societies, are able to access and distribute the supplies they need without roads, larger groups depend on an infrastructure so complex that in the marauder ant it rivals human highway systems. The idea of a superorganism applies here, of course: whereas a microscopic organism like a microbe can rely on simple diffusion to distribute nutrients through its body, a large one, such as a human being, needs a circulatory system.[1]

A marauder ant major worker hefting a *Diacamma* ant killed after intruding on the colony's trunk trail. The discarded corpse was buried by the minor workers.

ROADS

Marauder colonies avoid both gridlock and species confrontations, like that with the *Diacamma* worker, when trails are in good shape. Highway construction efforts are part of the society's logistics, providing supply lines for fresh combatants on the front lines as well as streamlined routes for bringing home the plunder.[2] Trunk trails are well looked after—that's how they can be distinguished from the fleeting paths created by raids.

Each worker size class participates in the creation of the roadways. All the castes eliminate surface irregularities along a trunk: while the medias and majors chisel out embedded roots and pebbles, minor workers extract grains of soil, establishing the road's slightly concave shape in cross section. The dross is discarded along the edges of the trail, where it accumulates in embankments like the one the *Diacamma* walked over. When the ground is moist, the minor and media workers build up the ramparts into a complete soil cover, or thin-roofed arcade, fabricated from soil extracted from the trail surface or from mining shafts—blind-ended tunnels near the trail used specifically as quarries.

Members of the construction crews expend their efforts foraging for building material rather than food. It is likely that no communiqués pass among them.[3] Rather, like compulsive bricklayers unable to go by an unfinished wall, passing ants respond to the ongoing building project, and the structures emerge without any active collaboration. The portions of the walls that are suitably positioned and shaped along a trail attract the most attention from passersby bearing soil bits. As a result, the arcades rise to completion where they are most needed, without a blueprint, and damage to them later is repaired without fuss.

Accomplishing large projects without communications is called stigmergy. The marauders' approach to building has been duplicated by robotics experts, who have discovered that it's cheaper and easier to achieve a goal such as piling up small objects with a group of simple robots responding to the work done thus far than with one large, more intelligent robot.[4] Stigmergy is at work in such websites as Wikipedia and Google as well, where many people add their insights to the statements and choices of others.[5]

Major workers of the marauder ant serve the role that humans reserve for heavy-duty construction equipment. I have called the largest of these individuals "giants" since the day I first saw one lumber from that nest in Sullia to the cheers of Mr. Beeramoidin and other forestry officials. Imagine a man and an elephant working together to build a road; the size difference between the giant and the minor worker is nearly ten times that great.

Relatively scarce, the giants tackle jobs that, though infrequent, require their prowess. While the smaller ants are so omnipresent that their jobs invariably get done, removing just a couple of giants from the work crew can cause a trail to degrade.[6] Fallen objects such as twigs and leaves snarl traffic and must be cleared for the roadway to remain open for use. When one of these giants arrives at such an obstacle, she pushes beneath it, then lifts her head high while standing on tiptoe. Ultimately, she shoves the object to one side, if not on the first attempt, then on the second or third, in a manner similar to that used by elephants to clear human paths.

When the soil roofs of the arcades sag, the large marauder ants respond to the pressure against their heads as they pass underneath with the shoving technique as well. Captain Charles Thomas Bingham, the Irish officer stationed in Burma, called the majors "the trowels and rammers of the Ant's Public Works Department." Their actions raise the drooping arcades and conceivably increase their structural integrity by binding the soil particles. The soil covers are finely granulated on the outside and are smoothed internally by the majors' battening.

In addition to enclosing their roads, marauder ants build thicker soil edifices over prize fruit or meat bonanzas, structures that facilitate the business of feeding. Workers guard the outer walls while others eat in a narrow gap between this exterior layer and an inner scaffold, which absorbs any moisture in which the diners might otherwise become mired.

Covered-over passages and encased food bonanzas are kept tidiest in areas of dense litter or vegetation that provide physical support so that less caretaking is required to maintain them. To what use is all this effort? Not, it seems, as protection from the elements. The earthworks fall apart in rain, and disintegrate when the earth is dry. Arcades are thin enough to puncture with a tap of a finger, which means a route is weatherproof only when it travels through an underground tunnel, perhaps dug and then abandoned by other animals. Alternatively, near the nest the ants may make a subterranean route of their own: over time, construction crews can scratch away so much soil from the trail surface that the highway sinks from view, at which point the ants seem to be able to construct a thicker, rainproof cover that becomes flush with the surrounding land.

A marauder ant trunk trail with soil sides and partial soil cover, extending through the leaf litter in Johor, Malaysia.

DEFENSE

The main function of this relentless building is defense. Because trunk trails extend for dozens of meters, they travel through territory controlled by other ant species. Marauder ants must therefore be organized to protect the trunk trails from aggressive neighbors or even from hapless passersby such as the *Diacamma*.

Strangely enough, when the soil ramparts are absent or breeched, the job of defending the trails goes to the most expendable ants in the colony—the maimed and the decrepit. At the spot where I saw the *Diacamma* killed, a row of minor and small medias stood along either side of the trail, ready to fight off any more of her comrades who might wander by. Marauders darken with age, changing from creamy brown to a dark cocoa color, and I could tell that many of these guards were old from their near-black integuments. Amputees and the infirm struggled to stay upright as they jabbed at additional chance intruders from the *Diacamma* nest nearby.

Among ants generally, the risks taken by workers tend to increase with age, demonstrating that their long-term value to a colony diminishes as they get older. Months-old fire ants engage in fighting in battles with neighboring colonies, for example, whereas weeks-old workers run away and days-old individuals feign death.[7] The old and wounded marauders often serve in the worst occupations, such as trail guards. They also throw garbage from the nest onto the community refuse heap, or midden, where they work until they fall over, their bodies joining the rest of the colony's waste.

For the marauder colony bothered by *Diacamma,* all the fuss over the contentious stretch of trail became moot within hours, after an arcade had been completed: the *Diacamma* workers could now walk over the trunk trail, blissfully ignorant of the industry below them. If a trail should sink underground, it is as protected as a passage in an army bunker, safe even from human footfalls.

Bulwarks constructed over trails and provisions prevent battles among competing marauder ant colonies as well. Where they are absent, combat can last a day and engage thousands of minor workers, which pour along the line of contact between the armies. Sometimes the tangle of ants stretches a meter wide. Compared to the free-for-alls that erupt during prey capture, the fights unfold with extreme care. At first the minor workers examine each other more like dancers than combatants. Brawls begin when pairs interlock mandibles, then grapple for several minutes before disengaging and maneuvering for better position. Fatalities escalate as additional workers pull on one of the locked ants. Fighters can tuck their hind ends beneath their trunks, making it difficult for others to grasp them at their fragile waists. Meanwhile they wave their abdomens in an action called stridulation, in which a ridged surface like a nail file on the undersurface of their abdomen rubs against a scraper located below their slender waist to produce a rasp like the sound made by strumming a comb; it is barely audible when a large worker is squeezed lightly and held up to the ear. This may be a call for help, though ants, being deaf, detect the rasps only as vibrations through the substrate. After some minutes of struggle, one of the ant's limbs will pop off like the arm of a medieval torture victim stretched on the rack. Slowly, surely, the workers pull each other apart.

Among ants in general, most lethal fights are variants on this hand-to-hand combat. Some species avoid prolonged tussles, instead taking a hit-and-run approach, inflicting damage fast and then dashing away. Many of these use a sort of chemical mace, spraying insecticides from their abdomens. Otherwise ants have not developed techniques to safely inflict damage from a distance—a development in human conflict that began with the invention of the spear. In one remarkable exception, workers from cone ant colonies stop their opponents from foraging by surrounding the enemy nest and dropping stones down the entrance and onto their heads as they attempt to leave, a nonlethal, but effective, technique.[8]

Which marauder ant colony wins? One especially sizzling afternoon in the Singapore Botanic Gardens I conducted an experiment with bottles of spray paint. By spritzing a different neon color lightly on the traffic moving along the trunk trail, I was able to mark a small portion of several colonies' worker populations. Three days later I came upon a battle between two of the nests. Scanning the thousands of grappling ants, I watched as the pink colony's larger battalion eventually swamped the greens, which retreated. With only a hundred casualties on both sides, there was no further commotion. In fights between honeypot ant nests in Arizona, special "reconnaissance workers" move through the battlefield to assess the size of the opposing armies, then draw out more troops or organize a hasty retreat depending on the situation.[9] I have no idea if that's how the greens "knew" to give up—that the odds were against them. But at some point, the green army clearly decided to leave the field of battle rather than fight on.

Because marauder ants lack scouts that could monitor intrusions around their nest, conflicts among them have little to do with territoriality—the control of land. Fights occur only by accident, when one colony's raid contacts the raid or exposed trail of another, and may be avoided, even near a foreign nest, when trails are sealed over. Because the size of a marauder ant army is likely to increase the closer the battle is to its nest or to the food it is defending, the colony with the most at stake usually swamps the other and wins.

THE NEST

Like most ant species, the marauder ant is a central-place forager, meaning its food supplies are funneled to a single central nest, which houses the queen and her brood. It is here that the society invests most heavily in defense, which makes excavating a marauder ant nest excruciating for scientist and insect alike.

But I had to do it. Studying marauder ants without looking in a nest would make as much sense as studying people without looking in a house. I also knew the ants well enough not to take my first attempt at snooping into their home lightly. Thinking the whole business out in advance, I selected my combat gear: long pants, a long-sleeved T-shirt, a pair of tightly woven socks, and tough boots. Arriving at the Singapore Botanic Gardens, I tucked the shirt into my pants and the pants into my socks and advanced toward the nest with a sharp-tipped shovel. Hovering over the nest, I breathed deeply for a moment before slamming the shovel into the earth. I had loosened a tiny chunk of nest soil; immediately a mass of enraged ants poured from the entrances. I threw the soil to the side and dug in again. And again. It took only a few tossed scoops before workers of all sizes had swarmed over my shoes and socks and up my pants and shirt to the first exposed skin they could reach: my neck and wrists.

When I could no longer tolerate the hundreds of bites, I ran to where I was out of range of the nest and scraped the ants off my skin and clothing. Then I grabbed the shovel anew and leapt back into the fray.

Repeating this cycle a few times, I found that the horde of minor workers pouring from the expanding gash in the soil had moved out many meters. Adding to the problem, the ants knocked off my body had spread out to the safe havens I'd used previously. Eventually, I had to sprint away from the nest to find a moment's respite from the desperate defenders.

The thrill of a dig is in locating the queen. A marauder queen is a good runner, and during the time it takes to excavate a nest she's likely to have been on the move, which makes it hard to know where she normally resides. On my first excavation, she was cloaked by an entourage of workers of all sizes and as a result—ouch!—a pain to catch.

It is yet another example of the value of media and major workers that battles between their colonies are fought only by the minor workers, while the larger ants—the heavy artillery—enter full combat mode only at the most desperate hour: when the nest is threatened. This distinction makes military sense. In 1914, the British engineer and military theorist Frederick Lanchester proved the advantage of outnumbering the enemy, even using troops of inferior quality, when battles are fought in large-scale formations. Hence, the minor workers, which per capita require little in the way of resources for a colony to rear and maintain, form a kind of "disposable caste" for both combat and predation. During colony conflicts, fights are one on one, making for a battle of attrition in which quantity trumps quality.[10]

After witnessing an excavation, Paddy Murphy claimed to be in awe of my tolerance for marauder bites. However, the payoff was a delight: I got an inside look at the marauder's home life. Their nests are often at the base of a tree, where the colony takes over available hollows such as abandoned rodent burrows, cavities left when a root decays, or even buried jars—any space in the earth will do. Suspended among the heaps of workers in these hollows are eggs, larvae, pupae, and victuals such as seeds and legless animal bodies. Also present are smaller, outlying chambers dug by the ants themselves, an activity that results in telltale piles of soil around the tree base. These are near, but typically separate from, the ants' midden piles of seed husks and discarded insect parts.

In the outlying chambers are the pale callows, adult ants so young their exoskeletons haven't fully hardened. Here, the young minor workers take on the role of nurses, tending the brood. Also crammed in these chambers are major and media repletes—a special caste in the marauders, distinct from other medias and majors. With their bloated abdomens, repletes serve as living pantries, storing and then regurgitating liquid food to other colony members. (The food is oily, suggesting that repletes take their fill of oil-rich seeds.) Excavating my first nest, I saw the repletes much like fat cells in a human body and wondered how many a colony needed to stay healthy. The question still needs an answer. The repletes' liquid stores are only a part of their hoarded reserves: workers also stockpile seeds and insect flesh from recent catches. Because repletes' reserves do not spoil, it is possible they are drawn from mostly in exceptionally lean times.

It's unclear how individuals are chosen to take on the indolent life of a replete, leaving others to toil outside, but their lifestyles couldn't be more dissimilar. Repairing trails and mangling prey, "orthodox" medias and majors strut high on their legs. In the darkness of the nest, their replete sisters crawl with their bodies pressed to the ground or bury themselves among the brood, interlinked with the outstretched legs of resting minor workers.

Determining the population of a marauder nest—typically between 80,000 and 250,000 workers—can be an adventure. For an excavation in Thailand, for example, I traveled north of Bangkok to Tam Dao National Park in the company of primatologist Warren Brockelman, who was studying the brachiating ape known as the gibbon. At a station inside the park we heard stories of a local tiger that had grown so brazen that he would leap through open windows and drag out the bodies of his victims. Sleeping that first night in Warren's open lean-to, I was awakened in the dark by the sound of rumbling breath. At sunrise Warren pointed to tiger prints in the dirt.

Later that morning I joined Warren to watch a pair of pileated gibbons sing a duet in a little valley. Noticing a marauder trunk trail nearby, almost invisible in the heavy leaf litter, I fell to my knees and for over an hour inched along the trail, grateful to forget all those vertebrates, whose simple behaviors, like leaping through windows, made the marauder ant's superorganized throngs seem all the more awe inspiring, small or not. Oblivious to the time, I continued until I found myself near the crest of a hill. At the crest was the columnar trunk of a dipterocarp tree, and at the base of the tree, marked by the discarded soil and ant trash spilled around its buttressed roots, was the nest, at last! For several heedless seconds, I scrambled around the tree on my hands and knees, in the classic "compromising position." Then something caused me to look up, and there, just two yards away from me, was a bull elephant. Wrinkled and gray, he stood absolutely still and silent, with his right forefoot lifted as if he'd been about to step forward. For that moment only his eyes moved, the eyelashes rising and falling in a blink. When he turned and crashed into the forest, all I could think of was how unfathomably larger than an elephant I must appear from an ant's perspective.

After recovering from the unnerving but thrilling encounter, I exhumed the ant nest with a foldable camp shovel, put several kilograms of ants and soil in a plastic bag, and stumbled back to Warren's jeep, a half-hour away. That evening at the park hostel's dining room, I convinced the cook to let me put my bag of treasure in the kitchen freezer. I needed to freeze the ants—thus incapacitating or killing them—in order to separate them from the soil so I could make my counts. I went to bed satisfied by a good day's work.

But the next day a different cook was on duty. No one had told him about my ants, and he'd removed the bag from the freezer and placed it on the floor. The ants had revived, cut their way through the plastic, and stormed the kitchen. I managed to round them up after an hour, enduring countless bites to my fingers. Grateful that my knowledge of Thai curses was meager, I also managed to mollify the new cook with two Singha beers and many compliments on his stir-fried *pad see ew.*

The painful bites and Thai curses were forgotten once I had tallied the data and gone to relax in the hostel's dirt-floored canteen, where I ate sticky rice under a ten-watt bulb while trying to impress two girls, on holiday from Australia, with how cool it was to be an entomologist (and when that didn't work, a *National Geographic* photographer). But I'd forgotten one of the most important lessons of marauder ant research: one worker always stays behind, after a skirmish, waiting for the proper moment to exact revenge. This time it happened midway through my meal. I started to howl and slap myself, and the girls disappeared.

BREAKING CAMP

Marauder ants are often on the move, and it is here that their roadways again play a role. I have come across dozens of migrations in which the whole society relocates, using the trunk trail for its exodus. Such operations are vaster than any raid. Colony members that normally wouldn't venture from the nest—every egg, larva, and pupa, every swollen and cowering replete, every delicate callow worker—join a caravan that proceeds as far as 80 meters to a new nest site. The enterprise involves a staggering protective force of workers exploring almost to the span of my hand from the trail flanks. Two to six nights are required, with the convoy taking a break during daylight hours.

Only once have I seen the queen in a migration, and that was in the Malayan species

Pheidologeton silenus, which is similar in many ways to *P. diversus,* the marauder ant. It was near midnight. I had been sitting for six hours in a particularly water-saturated corner of dense rainforest at Gombak Field Station in peninsular Malaysia, watching ants hauling their brood. Suddenly, there she was, part of the convoy, marching along with her stout body and strong legs as if she were designed for a life on the run. Escorting her was a tight retinue of several hundred minor workers. Some of them rode on her body; others ran in a mass a couple of inches ahead and behind her and on each side. The emigration column swelled as she passed, with the entourage flowing at exactly her pace. So quickly I had no time to pull out my camera, she disappeared where the trail led into the dripping brush.

Why move? Changing house can be a time-consuming chore. The honeypot ants of the southwestern United States, who laboriously carve nest chambers into tough desert clays, seem to never move: perhaps their expenditures on home construction are too high. For others, migrations occur only after a dire circumstance, such as the flooding of the nest or attacks by a vertebrate predator. But with the marauder, when I expected a migration to occur, it often didn't, and vice versa. I documented migrations of colonies that were eating well (in one case, dining on daily servings of bird seed supplied by me) but then inexplicably moved to a barren area. Conversely, colonies often stayed put even after I had dug up part of the nest for study.

The frequency of marauder colony migrations remains a mystery. My best guess is that colonies move a few times a year on average, but because I couldn't watch colonies around the clock, I could not be sure the colony at a site was the same or had changed since I had last been there. Several times after observing one colony migrate, for instance, I saw another move into its abandoned nest, which made me wonder if the ant colonies were like human families upgrading their homes. One colony moved 8 meters and then two weeks later relocated to its original location.[11]

Similar to marauder ants, though at the opposite extreme from homebody honeypot ants, are the nomadic army ants, which have been characterized as unique for the frequency, predictability, and organization of their migrations. Describing the transient domiciles of African army ants, the Reverend Thomas Savage reminded his readers in 1847 that "a man's dwelling indicates the nature of his employment."[12] While the large colonies of other ants require intricate nests, and like large human populations are hard to move, army ants avoid investing in substantial shelters so that they are as prepared to change locations as are Bedouins with their tents. Many army ants use abandoned chambers under objects or beneath the ground, as the marauder ant does.

New World *Eciton burchellii* army ants take this trait to an extreme: their nests are called bivouacs, because the only physical structures are the bodies of the ants themselves, as a half million or more gather under a low branch and form a hanging, bushel basket–sized mass of interlinked bodies. (Other army ants form similar chains within their underground enclaves.) As Reverend Savage might have predicted from the simplicity of *Eciton burchellii* encampments, the colonies of this species and a few others migrate with great regularity, every day for weeks at a time. It's thought that as their armies became more effective in the ancient past, army ants tended to exhaust the supplies of prey near their nest, forcing the evolution of such roaming behavior in response to a recurrent need for fresh hunting ground.[13] No surprise, then, that one army ant species has been shown to migrate more often if the colony is underfed.[14]

The plainly nomadic *Eciton burchellii* has been a research favorite and has dominated our perceptions of army ant behavior. The evidence suggests, however, that other army ants vary in their nest movements and that the species that relocate in a regular migratory cycle represent the minority. Some early naturalists, who had the luxury (rarely afforded modern researchers) of remaining for years at a site, recorded army ant colonies staying put for many months.[15]

If my assessment is correct, many army ants may be no more nomadic than the many other ant species that migrate periodically, and sometimes on a regular basis.[16] For species with colonies of a few individuals, relocations can commence, as they do with Bedouins, at any provocation, as appears to be true of *Myrmoteras* trapjaw ants whose nests in crannies in forest litter require little construction. At least some ants seem to be driven to pull up stakes by food requirements or outright hunger. A nomadic mushroom-eating ant in Malaysia, for example, changes its nest most often when its food runs low.[17] With the marauder ant, the connection between migration and foraging remains uncertain, but in all likelihood this species doesn't eat itself out of house and home as often as the more predatory army ant, permitting extended residences in one place. As with other wayfaring ants, when the need for a migration arises, the establishment of reliable thoroughfares to the new nest is the cornerstone of its success.

CREATING A NEW COLONY

Marauder ants and army ants share a common strategy for mass foraging and to some extent a proclivity for moving nests. But they have very different strategies for establishing new colonies. In most garden-variety ants, the young queen flies from her mother colony to mate. The foundress snaps off her wings upon arriving at her destination and then digs a nest chamber. In it, she rears her first crop of workers unassisted. (In some species the queen forages at this stage, but more commonly she doesn't leave her chamber and survives off her body fat.) As soon as these few ants mature, they take over all the labor and begin the first tentative foraging expeditions, leaving the queen to lay eggs, basically the only task she will accomplish for the rest of her life.

The queen of an army ant colony, in contrast, does not grow wings or fly away. Instead, through a system known as fission, one of the queen's daughters inherits half the colony and takes it as her own; the other half goes its own way with the original queen. Because even a start-up colony has thousands, if not hundreds of thousands, of workers, army ants never have to deal with problems of a meager labor supply. From its inception, a colony can always count on a huge contingent for its raids.[18]

How marauder colonies get their start, however, is a mystery. This much I know: their queens are tough, and they are excellent runners during migrations, but otherwise they don't resemble army ant queens in that they do grow wings and fly from the mother nest. After mating, they dig a nest chamber and attach their eggs to a hairless patch on the underside of their abdomens, a behavior unknown for any other ant. They carry the eggs around by tucking the abdomen partially under their bodies, which forces them to stand awkwardly high on their legs.

Unfortunately, despite my repeated, frustrating attempts to observe a colony's establishment, the queens I managed to follow didn't survive to rear workers. Nor did I find a marauder "starter" nest, or any nest with fewer than tens of thousands of workers—again, despite many long searches. So this part of my fieldwork remains tantalizingly incomplete. I'm eager to find out how a juvenile colony, lacking multitudes of ants, gets its food. While mass foraging becomes obviously advantageous when there's a labor force that vastly outnumbers its quarry, perhaps droves aren't necessary for success beyond that achieved by an ant foraging alone. If ants use a buddy system, even two workers, or at least a small group, might travel together to gang up on prey. Whether the developing marauder ant colony employs such a strategy is at present pure conjecture.

My failure to locate small nests suggests that extremely few new marauder ant colonies survive. Indeed, I often saw young queens, after their mating flight, being killed by workers from an established marauder nest. A queen's survival likely requires her family to grow swiftly until it has a safely large population—a rare event. Fortunately, we can gain clues as to how colonies mature quickly from the marauder's relative *silenus*. At Gombak Research Station, I excavated *silenus* nests, put the nest soil and all the stray workers in a bag, froze the bag, shook it up, counted all the ants in a tenth of the soil to estimate colony size, and then looked through the remainder of the bag for any queens. My colonies each contained one queen and from 64,000 to 127,000 workers. But I also found a nest of a few thousand workers and twenty-three wingless queens that clustered together amicably. The founding of a nest by a gathering of queens is called pleometrosis. With multiple queens laying eggs, the worker population no doubt increases rapidly, perhaps giving the colony a head start in foraging as a group.[19]

The *diversus* queens that I kept together likewise showed no aggression toward each other, though why each foundress attaches her eggs to her own body remains unclear. All the marauder colonies I dug up contained only one queen; if pleometrosis is common in this species, and in *silenus*, too, the number of queens must decline with time.

In contrast to marauder ants, army ant workers cull their queens before they mate. Typically, they raise several new queens, and when half their number depart with the chosen one to form a new colony, and the old queen goes her own way with the other half, the excess queens, blockaded by the workers, are left behind to die.

It seems the marauder ant workers likewise do the deed of disposing of excess queens, but in their case this occurs much later in the life of the colony. I learned this at the Botanic Gardens as I tallied workers who were repairing a damaged thoroughfare. At one point I noticed a group dragging a dusky object out of the nest and along their trail. Extracting it from them, I found in my hand a wingless queen with the worn mandibles and the near-black pigmentation of an aged animal. She was very much alive but had apparently outlived her usefulness to the colony and was being evicted. Twice more I saw the same event at nests sizable enough to suggest that marauder ant societies can retain more than one queen for a long time. Allowing these workers to finish their job, I watched them abandon the struggling queen at the side of their trail or in the garbage heap.

Calling the female reproductive ant a queen is a poor metaphor because ant royalty does not lead, and unlike human monarchs, they sire their minions. Nevertheless, the two varieties of queens share a characteristic: with royalty comes favor but also great peril.

5 group transport

It was late in 1983. For the final leg of my doctoral fieldwork, after traveling without a break for twenty-nine months, I had ended up in the Philippines. I had just arrived at the base of Mt. Apo, on the southern island of Mindanao, at almost 3,000 meters the highest mountain in the Philippines and cloaked with forests. On exiting the bus, I found José, a self-proclaimed guide who, an hour into our walk, broke his silence to speak of the need for revolution while patting what he claimed to be a gun in his waistband.

On my way to Mindanao, I had faced riots against the government outside the Manila airport. I now recalled a U.S. government advisory that Mindanao was the center of the Communist movement and unsafe for travelers. It occurred to me for the first time since I had left Massachusetts nearly two and a half years before that it might be pleasant to experience Christmas at home again—or even be reminded of what the holiday celebrated.

Mist and tree ferns gave each vista of the flanks of Mt. Apo a Jurassic Park flavor. While José talked, I located my old friend the marauder ant. Among their legions, fifteen minor workers—the chief food-delivery caste—carried a centimeter-long blue sphere through dense brush. When I picked it up, more ants poured from a crack in the sphere to expose the remains of a bird embryo.

Whether it's political rebels or ants fetching home a bird egg, a successful social operation requires the coordination of individuals. The orchestration of a marauder ant raid is an obvious example of how organization emerges from collective masses within a superorganism. By moving in closer to watch the individuals as if they were tissues within organs, I could document equally compelling examples of social integration, involving smaller ant groups within a swarm.

Marauder ants are successful at "marauding" in part because workers can work together to haul provisions along their superhighways—or expel a rejected queen from the nest. Group transport is the carrying, dragging, lifting, rolling, or burying of a burden by multiple individuals.[1] We've all seen group transport, if only when a few ants pilfer crumbs at our picnics. But the average interloper at a picnic represents only a crude example of the sophisticated group transport accomplished by the marauder.

GROUP TRANSPORT AMONG ANIMALS

From watching the marauder ants for months while in Asia, I had become fascinated with group transport and began to investigate which other creatures could accomplish this basic task. I discovered that group transport is as scarce in the animal kingdom as using tools or hunting in a group. Such task-oriented cooperation is particularly rare in nonhuman primates, in part due to the overwhelming drive of each individual to keep food for himself and in part because they seldom deal with large objects in nature, although captive chimpanzees will reach food by carrying a branch together to use as a ladder or by group-dragging a box.

Even for species that cooperate to kill prey, jointly moving a cumbersome meal requires a delay in gratification that few can tolerate. Litter mates of rodents such as rats will sometimes jointly convey food they had initially fought over. Lions and also hyenas, wolves, jackals, African wild dogs, and other dog-family members typically feed in a free-for-all or along dominance-hierarchy lines, but occasionally they may jointly move meat to a shady spot or protected den. Even then, they tend to act like competitors who just happen to be pulling their dinner in the same direction.[2]

Nonhuman mammal societies usually contain one or two dozen individuals, with a few dozen at most. It's often simpler for such a small group, for example a pack of wolves, to travel to the kill site than to take a large carcass to a more desirable setting. Similarly, in the few ants with small colonies that eat big prey—such as New Guinea's *Myopopone castanea,* which feeds on blubbery insect larvae—the entire society may up and move to the food after it's been killed.[3]

Outside of ants, the best examples of group transport of food are found among other arthropods. Certain spiders gang together to build a web and even capture prey and care for their young; in some species, the spiders jointly drag prey to the protected interior of their web, where all feed. Then there are mated couples that procure an item of food too massive for one of them to manage alone. Pairs of some dung beetles roll and bury a dung ball on which the female lays eggs. A male and a female carrion beetle join forces to bury the corpse of a small vertebrate such as a pigeon or squirrel, which becomes food for their larvae.

But ants reign supreme at this form of altruism. Why is this so? For one thing, there is little antagonism between nestmates over food, reflecting how well ants work together generally. Also, as central-place foragers, ants take food back to a busy nest where much of the fare is consumed by the growing larvae in their protected nurseries.

Beyond that, the availability of a large labor force and the use of chemical trails make it practical for large ant societies to assemble transport crews and thereafter coordinate the direction in which they move the food. At the same time, the scale of activities in these species puts a premium on workers handling heavy items efficiently. Compared to lions, who take large prey every few days, a marauder ant colony may bring down thousands of food items bigger than the workers in the course of a single day.

Aspects of ant anatomy also simplify group transport. Their forward-directed mandibles are more effective in lifting burdens cooperatively than are the jaws and limbs of most other social animals. (Humans are a significant exception: with our upright posture and opposable thumbs, we are experts at group transport.) An ant's center of gravity is also low relative to that of large mammals, providing easier balance in group retrieval.

Portability is the minimum requirement for group transport of food. Ideal objects for transport, such as seeds or prey, come in solid packets just a few times heavier than their carrier—neither so small that lone individuals could carry them nor too large, soft, crumbly, or mushy to lift. It's always possible to cut up an item that is too large, as long as the material can be carved into portions the right size for a group.[4] Even marauders are unable to pick up rotten or soft fruit, which they ingest on the spot. (As a tool-based alternative to conveying liquidy meals, a slender New England ant sops them up with dried bits of plant material that the ants slice from nearby debris, drop on the food, and then take to their nest. Feeding from these items must be like squeezing soup from a sponge.)[5]

There are occasional examples among animals of altruistic transport of objects other than meals—most often other members of one's own species. Dolphins and gray whales will hold a weakened or injured companion at the surface to breathe. Elephants will work together to lift a fallen comrade to its feet. These mammals will also crowd around to help a disabled individual walk or swim from place to place, though the stricken beast often moves in part under its own power. This is a behavior I have never seen among worker ants; even if she is one of the relatively valuable soldiers, a wounded worker is left to hobble on as best she can.

Although workers of the Arizona honeypot ant, *Myrmecocystus mimicus,* group-transport other workers, they do so with a less friendly intent: in this species certain individuals are repletes, which have abdomens swollen to the size of a small grape with honey that they regurgitate like living spigots to nestmates. After a battle, groups of ants from the victorious colony will drag the vanquished repletes to their nest, where, hanging from the ceilings of the chambers, they are condemned to a life of slavery.[6]

During nest emergencies such as floods, ant workers rescue the brood, which can't escape danger on their own; they group together to move the cumbersome ones. Workers also help the queen get around, especially if she's wounded, as is true for the queens of another group of social insects, the termites. In South Africa, attending a conference of entomologists who spent their days rummaging through elephant excrement for dung beetles, I shattered a half-meter-tall, rock-hard nest mound of *Macrotermes* termites to expose a chamber containing their grotesquely rotund queen, who was over 5 centimeters long. A rescue party of workers immediately surrounded her and pulled her vast bulk out of view.

SHARING THE LOAD

On Mt. Apo, José and I settled down on the trail for a moment to watch the marauder ants at work. Although, as Frank Sinatra laments in the song "High Hopes," "an ant can't move that rubber tree plant," all ant species are celebrated for the loads they can bear—even as singletons. A marauder ant minor worker is no exception, carrying up to five times her mass. It's not that she is particularly muscular. Rather, it's a question of proportion. Total strength is determined by muscle thickness, which is proportional to the animal's height squared, while weight is proportional to the cube of its height. This means that unwieldy vertebrates end up with too much body weight for too little muscle. Galileo worked this out in 1636, writing that "a small dog could probably carry on his back two or three dogs of his own size, but I believe that a horse could not carry even one of his own size."[7] This formula explains how ants have the power to carry striking weights.

Through group transport, marauders take this excess-weight capacity to unparalleled levels. Not only do they haul food together, but they also gang-transport brood during a migration or in emergencies, the corpses of enemy ants that they dump near their trails, queens endangered during pandemonium at the nest, hunks of refuse bound for the colony dump, and, on occasion, obstructions on a trail or chunks of soil for making their arcades. In each case, they work in groups with greater effectiveness than any other living thing.

The capacity of a minor worker to carry five times her weight on her own sounds impressive, but in Singapore I had figured out that a 10-centimeter-long earthworm, like the largest ones I saw being heaved whole balanced between a hundred ants on the ant trail on Mt. Apo, would require a thousand ants if it had been cut in pieces and carried off one ant at a time—yet the ants gang-transporting the burdens on Mt. Apo were slowed to only about half the speed at which they hauled items by themselves. Even when the worm was five thousand times the weight of a minor worker and ten thousand times as voluminous, a gap was usually visible between the cargo and the ground: marauder ants lift rather than drag burdens. In human terms, that would be equal to getting friends together to run at breakneck speed while lifting overhead 250 tons, which would likely amount to far more than the contents of all their houses combined—an utter impossibility for a human.

Before coming to the Philippines, I had conducted an experiment. Although my breakfast ritual in Singapore was to have *roti prata* at an Indian food nook, for days when I needed to get going before 7 A.M. I kept a supply of a cheap Australian cereal called Grainut (which, being virtually inedible, has since gone off the market). One morning I was sitting in the Botanic Gardens next to a marauder ant trail eating said cereal, when I decided to crush some chunks among the ants and document the outcome.

As it turned out, the bigger the chunk, the more efficient they were, and the more food each ant was able to move along the trail in a unit of time. Beyond a certain size chunk, however, efficiency declined. It was apparently a matter of geometry: with increasing size, the weight of

Marauder ants carrying an earthworm at the Singapore Botanic Gardens. Two minors ride on the prey as "guards."

the chunks increased faster than their circumference, until there wasn't enough space to accommodate the number of ants needed to lift the food. For the cereal, this occurred with chunks requiring more than fourteen ants.

Looking closely, I could see how the heavier items would cause the ants problems. The porters tend to space themselves evenly, but the bigger a burden is, the more tightly packed they become, until they are barely far enough apart to avoid treading on each other. An earthworm, which is lighter per unit volume than the Grainut chunks and, being long and slender, offers more space for porters, could weigh more overall than a cereal chunk and still be transportable, because more ants could gather around it.

Thanks to the marauder's skills, few foods need to be diced for carriage—this labor-intensive step can be delayed until the catch is in the protected confines of the nest. Once its flailing limbs are removed, dinner is sped away posthaste. Among marauder ants, more than 80 percent of the colony food supply is brought to the nest by groups of ants. The rest consists of small items carried by individuals.

AMATEURS AND EXPERTS

My surveys of the animal kingdom had shown that group food-carrying exists in only 40 of the 283 known genera of ants. Of the remaining species, some restrict themselves to small prey that do not require this skill (such as *Acanthognathus* trapjaw ants that prey on tiny springtails). Even among ants with a well-suited diet and adequate means for communicating the location of meals, there are species that fail at group transport because of poor coordination: they end up engaging in a tug-of-war, though clumsy retrieval can occur if perchance the workers pull in the same direction. Otherwise, they eat the food where it's found or divide it into single servings and cart those away. Even that requires some cooperation, since all but one worker has to let go of each piece before it can be moved. In species adept at group transport, the workers are able to postpone dissecting and consuming the food while they coordinate as a group to move it.

In 1960, John Sudd of the University of Hull studied a British big-headed ant that performed badly in a freight-hauling group, often working at cross-purposes and dragging prey rather than carrying it. But given time, Sudd observed, the workers modified their behavior in such a way that the force they exerted on the food generally increased until they got the job done.[8]

It turned out the adjustments they made, such as changing the angle at which they applied force or shifting from pushing to pulling, were identical to the changes they made when hauling food alone, and these changes led, as they did for the solitary ant, to the food being moved. In other words, the British workers succeeded in group transport by behaving as if none of the others were there.

Programmed to replicate this kind of coordination, a group of simple robots was able to move a large object. European scientists even used these so-called swarm-bots to stage a mock "rescue" of a child by dragging her across the floor. Another team used tiny swarm-bots scented of cockroach to influence the roaches' collective decision about where they would gather to hide from the light.[9]

In contrast to Sudd's ants and the simple swarm-bots, marauder ants are unambiguously

With modest colonies of a few thousand, *Daceton* ants in Venezuela have developed only rudimentary cooperation in the transport of food. Here two workers have pulled so persistently in conflicting directions that the moisture has been wrung out of this caterpillar. Flies sneak in to drink from the oozing meat.

cooperative. Moreover, the behavior they display when moving food as a group is seen only during gang retrievals. A worker carrying a burden on her own walks on all six legs, grasping it in her jaws. If she joins a group, however, she places her forelegs on the burden, then presses her head against its surface, jaws open, but she does not use her jaws to grip unless she can hold on to a projection such as a limb. She walks with her remaining legs as she and her nestmates transport their load.

What about this technique makes marauder ants excel at gang retrieval? Picture several people hefting a box by thrusting upward, not only with the palms of their hands but with their foreheads as well. By pushing a load up, forward, and against each other, the clustered ants balance the weight effectively among themselves. Army ants use a different technique, lining up to straddle a burden under their bodies rather than encircling it. Either way, the groups cancel the rotational forces that solitary porters contend with when they lift a burden in front of them. Anyone who's felt a heavy box twist out of his hands has experienced this force, a problem that disappears when another grasps the object on the other side.[10]

While participating in a lift-and-carry operation, each of the marauder ant "porters" performs a slightly different task. As when several people haul a piano, an ant's movements depend on where she is located relative to the direction of motion. Workers at the forward margin walk backward, pulling the burden. Those on the trailing edge walk forward, apparently pushing it.

Ants along the sides shuffle their legs sideways and slant their bodies in the direction of travel. The ants sort out their roles during a few minutes of turmoil, then whisk the item off with effortless grace. When a media worker joins in at the front or back ends of large booty, she appears to be adept at guiding the group around obstructions or through shifts in the trail course, performing another valuable role in the transport team.

BUT IS IT TEAMWORK?

Should we consider these groups teams? Dictionaries define the word *team* as a group organized to work together, which could apply to many social situations among ants. Although in many team sports there is a set roster for each game, with ants, under most circumstances, the participants change and are interchangeable.[11] We saw this for raids: marauder ants come and go while the quarry is being subdued, and similarly to and from the raid as a whole. By comparison, transport groups are more stable, though ants may leave or join a group when, for example, an object becomes snagged, at which point the participants must sort out their movements relative to each other afresh.

Often, members of human teams divide the labor, doing different things at once to get the job done. Although ant workers cannot recognize each other as individuals in the way human teammates do, many marauder ant activities—among them killing prey, attacking alien ants, and maintaining trails—probably conform to the American football model.[12] In some cases, different worker castes play specific "positions" and concentrate on distinct tasks, as when minors hold down prey while medias and majors shear its limbs. In other situations, all participants belong to the same worker caste and show flexibility in how they do their jobs, as when minor workers perform differently in the transport group depending on where they are located around the prey.

One species of wood ant shows the ultimate division of labor in a transport team, with a degree of leadership exceptional among ants. Among *Formica incerta,* common in New England fields, when a successful forager can't move an item of food herself, she attracts ants in the vicinity or recruits some from the nest. Unable to assess the size of her find, she may not gather a suitable number of individuals. If not enough helpers arrive and she needs to leave to find more, those already on scene—even if they have already started carrying off her find—will wander away as if the food weren't there. Only the original food finder can keep the team motivated, and only she can go for more help. She must be present to guide the transport team from start to finish. Outside her role with this particular meal, of course, there is nothing special about her. If she is later recruited by another scout, she goes to work as a regular worker, while the individual who located the food becomes the supervisor for its retrieval.[13]

Several years ago in El Salvador I watched workers of the army ant *Eciton burchellii* chop a scorpion to pieces. I could see that the workers fell into different positions as the transport groups came together, but they didn't adopt behaviors specific to teamwork. This occurred because the media workers had trouble lifting an unwieldy hunk of the tail. Then one of the less common but bigger and stronger "submajor" workers arrived and was able to straddle it in the

A "submajor" *Eciton burchellii* army ant hefting a chunk of centipede while a smaller media worker behind her lifts its dragging end. The minor worker lying below them in a pothole along the route serves as "living road fill."

classic army ant manner and start it moving. Immediately one of the smaller medias crowding around was able to fit into the cramped space under the abdomen of the bigger ant, where she grabbed the scorpion's stinger, which was trailing on the ground. Thereafter the two functioned, as they often do in this species, as a team, with the forward ant doing the power lifting and steering, while the little one kept the back of the prey from dragging. Meanwhile, the scorpion's body was being carried by four ants: the same pairing of a submajor and a media handled the main axis of the corpse, with two more medias off to the side, helping lift the scorpion's appendages.[14]

The workers of small ant societies seldom show such collaborations, even of the accidental kind typified by army ants. Being dependent on individual initiative to get things done, each worker is likely to do fine on her own, often aided by special tools, such as trap jaws. Marauder ants serve well as an example of a large society in that the workers are more likely to complete tasks by toiling together or by sharing information with other specialists by means of the language of complex ant societies—chemical communication.

Humans are in some ways similar. Anthropological studies have shown that small groups of hunter-gatherers tend to be labor generalists, with everyone having the ability to be self-sufficient or near to it and pulling his or her weight with a wide range of work (beyond some sexed-based differences). In larger human societies and with increasing urbanization, a complex division of labor in which individuals have limited employment skills becomes more prevalent—as it is for workers in many ant species with large colonies too. This pattern has been understood in humans since 1776, when a Scot, Adam Smith, founded modern economics with his book *Wealth of Nations*. Smith saw specialization as necessary to the growth and development of societies because of the productivity resulting from each laborer's skill at his job and the reduction of time lost in switching between jobs.[15] But Smith also saw in this specialization the tragic "mental mutilation" of laborers, a decline in intellect from the repetition of menial tasks that he claimed must be countered by management from the state.

This deficiency can be observed for large ant societies as well, in which specialized workers are incapable of accomplishing much without the cooperation of nestmates.[16] A lone marauder ant is as hopeless as the urban sophisticate who, as in the movies *City Slickers* and *Romancing the Stone*, is dropped into a remote environment where he's incapable of caring for himself. In contrast to the simple interactions between individuals in ant species with small colonies, however, marauder ants show synergy in spades—not only at the emergent level of entire raids, but also more intimately, in the coordination of smaller, local teams. Group transport of food may be the most vivid example.

Synergy and faithfulness to the whole, not independence, are integral to the functioning of the most well-integrated organisms, just as they are with their social counterpart, the superorganism. A sponge, for example, though clearly an organism, is so simple that its cells often survive for a day or two when forcibly separated from the whole and can reunite to form a new sponge, whereas the cells in spilled human blood or a severed finger will perish, and usually in fast order.

Other animals have learned to work around the marauder's group transport finesse, as I saw for myself in the Philippines. Turning my guide's animated political conversation to the wonders of animal behavior, I pointed to a blowfly on a leaf to one side of the exposed trail. Gray and black striped, big and stocky, the *Bengalia* fly twisted on its perch to follow with its big brown eyes the teams of marauder ants moving food on the trail a few inches below. The fly flew down to touch one of these objects, a seed, then flew back to the perch, leaving that vegetable matter alone. Its next choice was a beetle; this time the meat-eating fly wrestled the prey from the porters and soared away, lunch taken care of.[17]

The marauder ants, however, have a fly defense system. Minor workers will ride on any large item in transit, with jaws open. They don't seem to be interested in eating the food. But when a fly descends, the riders, lifted with the booty into the air, rush in and bite hard. This forces the fly to let go of its meal and buzz loudly while hovering in midair, an action that, with luck, will knock off its tormentors. The fly has an additional tactic to safeguard it from this uncertain fate, though. After grabbing a food item, it immediately drops the piece several inches away from the trail before the riders can strike. Then it alights on a nearby plant and waits as the riding ants race off the food and back onto the ground, primed for action. The fly can now swoop down and depart with the unprotected morsel.

As José led me down the mountain late in the afternoon I decided that watching the interplay between ant and fly had been a fine note on which to end my travels in search of the marauder ant. The beauty of this performance, as elegant in its choreography as the peacock's dance, was in the fact that both the ants and the fly had brains small enough to fit on the sharp end of a pin.

african army ant

raiders on the swarm

Dorylus, Africa

6

big game hunters

It was January 2005, and I was in Africa again. I had already had many adventures in the sub-Saharan region. Years before, pursuing my love of frogs, I had hunted the goliath frog in Cameroon, hoping to set a new world's record with a 3.3-kilogram specimen but settling for one that weighed a little less and was a meter long with its legs extended. In Gabon, I'd surveyed ants in the rainforest canopy, working on a canopy raft—a network of pontoons placed 40 meters high on the tree crowns. (My fellow researchers and I were so bothered by the dominant *Crematogaster,* or "acrobat" ants, that we nicknamed them "Crematobastards.") In Ivory Coast and Senegal, I'd collected praying mantises, and in South Africa and Namibia I'd searched for spiders, flies, termites, and other curiosities. But this trip to Nigeria would be my first opportunity to investigate the similarities between the army ant and the marauder ant, my *Pheidologeton diversus*.

On the previous trips I'd stumbled upon Africa's infamous *Dorylus* army ants and especially noted the swarm-raiding species commonly known as driver ants for their habit—shared with swarm-raiding army ants in the New World and the marauder ant in Asia—of herding their prey before them.[1] I came to regret the stumbling. The *Dorylus* I encountered bit me fiercely when, distracted by the sight of a vervet monkey in the trees, I stepped on a packed file of the dark, shining workers. Their violent response reminded me of Ogden Nash's character study in his poem "The Ant":[2]

Previous page: The knife-blade mandibles of these trail guards of the driver ant *Dorylus nigricans* in Ghana can slice through vertebrate flesh.

The ant has made himself illustrious
Through constant industry industrious.
So what?
Would you be calm and placid
If you were full of formic acid?

The pain of a driver ant bite has more to do with mechanical damage than poison, though. Each chomp forced me to stop and extract a worker from my skin or crush with a thumbnail those gnawing at my clothes. (Their heads stayed embedded in the fabric through several washings.) Bites on a fingertip were so agonizing that pulling the ant off wasn't an option: when I gripped the offender between two fingers of the opposite hand, she would clamp down even more savagely on the delicate finger pad. In time I found a solution: I inserted the finger in my mouth and crushed the ant's head between my teeth, which immediately disengaged her jaws. The ant was about the size of a Tic Tac breath mint and just as crisp.

Munching on the insect, I detected a hint of nuttiness and a trace of the pungent sourness of formic acid. Driver ants are not as oily as the plump marauder food-storage repletes. Nor are they as tangy as weaver ants, which have a mentholated lime flavor and are served in India as a condiment with curry dishes. Certainly they aren't as desirable as honeypot ants, which Australian aborigines and southwestern American Indians find delectable.

Though they do not offer the candied delights of honeypots, driver ants are toothsome enough to chimpanzees to be one of their dietary mainstays, and that's what had brought me to Nigeria. Caspar Schöning, who at that time was studying driver ants at Copenhagen University, had invited me to join him while he organized a research project on ant consumption with University College of London primatologist Volker Sommer. Joining us would be Darren Ellis, a student of Volker's.

Just as different human cultures have developed different techniques and tools to kill prey, from spears to snares, the chimps across Africa have evolved distinctive traditions and methods to hunt ants.[3] (Until recently, this kind of cultural diversity was thought to be unique to humans.) The chimps at Gashaka-Gumti National Park, where Caspar and I were going, use branches as harvesting tools, stripping off their leaves and inserting them into driver ant nests or, possibly, poking them at ant trails; the technique is known as ant dipping. Putting the stick in their mouths like a lollipop, they peel away the furious ants, which cling to each other in strings. Eating ants sounds painful, but driver ants don't sting, and as I discovered, they don't bite your tongue if you chew fast. The chimps' use of sticks may have more to do with reaching deep into the nest in order to stir up as many ants as possible than with avoiding bites. In other places, chimps bravely reach into driver ant nests with their hands, which enables them to grab the delicious larvae and pupae.

But when driver ants are on the offensive, they're fearsome hunters. Caspar, a perceptive, gentle German with the physique of a welterweight fighter and blond hair cut military-style, was an enthusiast of the ant's prowess. The driver ant expert at Utica College, Bill Gotwald, had been told by a village chief in Ghana about a human baby that had been killed by the ants.[4]

I had heard that driver ants can even bring down a cow. Many African tribes believe that before a python feeds, it checks its surroundings for driver ants—which would flay the snake alive with their knife-blade mandibles if they found it too distended by a meal to escape.

As Caspar and I set out from Abuja for Gashaka-Gumti, on the border with Cameroon, we found ourselves in the midst of a desert storm. So much dust was being carried on Harmattan winds from the Sahara that our view on the two-day ride to the park was abysmal. The sun shone no more brightly than a full moon.

Jammed into a crowded jeep for the final step of our many-hour journey, we crossed a wide river just before reaching the field station inside the park. The water rushed up past the tires, but the driver plowed forward, knocking against rocks as the water continued to rise. A fellow passenger pointed to the spot where a student had gone for a swim the year before and been caught by a crocodile. His lifeless body was found under a rock, where the crocodile had stashed him for a later meal. Balanced precariously in the flatbed in the back, clinging to six other people, stacks of luggage, and tied bundles of squawking chickens, I concentrated on the happy fact that the air was clearer at this slightly higher elevation.

Once at the field station, a few low concrete buildings, Caspar and I dropped our baggage and headed up a footpath to start surveying the local ants. The undulating landscape was a mix of olive-green lowland forest and woodland savanna dominated by pale grasses. The earth was dust dry. Leaf litter crackled underfoot. On the way we passed handsome, colorfully dressed men and women of the Jibu and Chamba tribes. They gracefully balanced baskets of meat and fruit on their heads—trade goods that they would carry on the three-day walk to Cameroon. The air was split by the *pea-yaow* call of the putty-nosed guenon monkey.

Seeking moist ground, we left the path within an hour to search the taller gallery forests along a stream. Soon we were joined by chimpanzee expert Andy Fowler, a soft-spoken Englishman with a dependable wry smile, and Darren Ellis, a thin, bespectacled American, who began a dialogue with Caspar that continued nonstop for three weeks as they hammered out the protocols Darren would use for his master's thesis. He was studying the stick tools made by chimps, the driver ant's response to the tools, and the importance of these army ants to the chimp diet. This last topic required that Darren count the ant parts in any chimp dung he could find—a task that turned out to be even less pleasant than it sounded.

Andy led us to a driver ant nest he'd spotted two days earlier. From the base of a toppled-over kapok tree spilled pyramids of soil that extended into a streambed. We saw some old, weak, and wounded individuals withdrawing to the garbage heap to expire. Some were being captured by workers of a black-and-coffee-colored acrobat ant, who were waving their heart-shaped abdomens in excitement at the easy repast. Bagging vulnerable workers in the trash piles or the wounded left behind after an army ant raid is an industry for certain ant species.

This driver ant was *Dorylus rubellus*. As I watched their dead and dying, I thought of how my mentor, Edward O. Wilson, had figured out how ants recognize their deceased kin. In 1958, working at Harvard, Ed and two colleagues proffered ants squares of paper soaked with a series of foul chemicals associated with decay. One compound, oleic acid, yielded a full-blown necrophoric response, inciting workers to haul the paper to the trash pile. When the researchers daubed oleic acid on a live ant, her determined nestmates dragged her off as well. Until she licked herself clean, the unfortunate individual was repeatedly thrown back onto the midden with every attempt to enter the nest.[5] Smell like the dead, and dead you must be.

Searching along the streambed, I found the driver ants' trail, which was exposed for 3 meters before it climbed an embankment and disappeared into the forest. A few workers were carrying chunks of insects, probably collected in a raid deep in the forest, slung under their bodies in classic army ant style. What I saw next caused me to drop to my knees. The drivers looked and behaved so familiarly that it was easy for me to believe I was in Asia again watching a stream of marauder ants. Part of the reason was the posture of the guards, who stood or strolled near the trail in marauder ant fashion, high on their legs, with raised heads and open mandibles.

I soon spotted differences from the marauder, though. An inordinate number of medium and large driver ants, rather than the small workers, had taken guard roles. The traffic on the trail also seemed chaotic, seldom sorting itself into lanes. Contrary to some descriptions I had read, the ants ran side by side with their feet on the ground, not atop one another. Workers stepped on those next to them, however, and big ones strolled over small ones—common practice in polymorphic species, preventing traffic snarls behind workers that come to a sudden halt.

On portions of the route, the guards interlinked in a way I had never seen before: hooking their clawed feet together like some horror-film version of armored cheerleaders forming a pyramid, they welded a lattice over the traffic below them in a defensive shield bristling with jaws. Unlike marauder ants, driver ants have no eyes, and all their joint activities are truly examples of the blind leading the blind. Watching the guards link one leg to the next, I wondered how they sensed the hordes beneath them in order to correctly position their bodies. This shield tore whenever it was snagged by the food in transit below, which meant the basketwork of ants constantly had to be rebuilt.

Food also got snagged whenever the trail passed under leaves. Lacking the major-caste road crews of the marauders, the drivers could get past these obstructions only after relentless buffeting from the brute force passage of workers below, except at one moist site, where the ants had eschewed their bristling covers of live guards in favor of marauder-style earthen barricades.

I was lost in thoughts of road construction when two giant forest hogs, hirsute and high as my chest, ran into the streambed ahead of me. They gave me a look of evident horror, then dashed off noisily. An hour later, an impressively virile olive-colored baboon mock-charged, fangs bared, while his females walked behind him bleating nervously.

THE DRIVER ANT RAID

Back at the field station that night I saw my first driver ant raid. It was advancing at full force outside the dining room, where the savanna grasses had been chopped to create a lawn. While the primatologists remained on the porch drinking Star Beer, Caspar, Darren, and I got on our bellies to ant-watch. The raid extended from the nest of a *Dorylus rubellus* colony at the base of a tree 15 meters distant. That meant the ants were just getting started: driver ants surge ahead for 80 to 120 meters before retreating. Behind a front about 7 meters wide was a prodigious swarm 15 to 30 centimeters deep. Small workers rushed through to reach the front, where they slowed down to search for prey before retreating. Larger ants were mostly located in the whirlpool of activity farther back in the raid. Great numbers of them stood in guard posture all around the reticulating columns in the raid fan, while others were busy killing, dismembering, and carrying the kills.[6]

A swarm raid requires substantial ant power. The regiment has to be packed tightly because there's no telling where the next kill will show up within the "net" of ants. Within this raid there were tens of thousands of ants, with two to five workers occupying each square centimeter of ground. They were so numerous that the sound of them rummaging in the litter or dropping from twigs was like the patter of rainfall. (Indeed, they did create a shower: driver ants scrounging for prey will climb up plants, but they save time by dropping to the ground instead of climbing back down.) In two hours, however, all this ruckus bagged the colony only two thimblefuls of invertebrate meat—ant-sized plant hoppers, centipedes, worms, and spiders.

Perhaps the raid was passing through an unfruitful stretch of land. Because driver ants and other army ants have been found to travel much farther during their raids than do marauder ants, I'd expected that they wouldn't be as sensitive as marauders to local shifts in food abundance. Their unresponsiveness to food distribution would encourage army ants to continue through a barren part of the landscape until they finally found prey, or until their distance from their nest ensured diminishing returns. Such doggedness may be important when a search party is concentrated in a raid rather than spread widely, as it is in solitary-foraging ants, especially when food is scarce and scattered, as it is during Nigeria's droughts.

Yet this seemed to be only part of the explanation, because there was prey around that the driver ants missed. The tens of thousands of ants we watched in those two hours took in less than several nearby packs of two hundred or so *Pachycondyla* ants that were on their evening excursions, recruited by scouts to catch termites. Grasshoppers, crickets, and *Pachycondyla* broke away from the trawling *rubellus* raid and survived, even when we increased the ants' chances by tossing the escapees back into the swarm.

I thought about the slowness of marauder ant raids and the meticulous care the workers put into combing the raided area to extract food; generally, far more marauder ants than driver ants return laden with booty.[7] Could that difference reflect the greater significance to the driver ant diet of tracking down, not lone invertebrate prey, but food that comes in widely separated bonanzas? This ant's typical meal is found in prodigious stockpiles: brood plundered from ant nests. Indeed, mass foraging in army ants likely arose as a strategy to effectively surprise-attack other ant colonies. Army ants are thought to have begun to regularly kill large, nonsocial prey like spiders and centipedes only after their ant-plundering colonies had evolved to huge sizes and developed the capacity for wide swarm raids.[8]

The overkill population of tens of thousands within a raid, most doing nothing but walking around before leaving again, may represent a particularly huge reserve force, to be drawn on when the occasional megameal is encountered. If big ant nests are the mainstay of the driver ant diet, there's a good chance that raids may go bust on some days.[9] Yet swarm-raider army ants can prowl over 10,000 square meters in the course of a week. That's equivalent to combing the length and the breadth of three football fields, an area that should contain plenty of ant colonies, large and small.

That night I sat under the brilliant Milky Way and made notes in my journal in front of the campfire. Finding windfall meals, I scribbled, requires that ants maintain the size of their raids even during periods when their take almost always ends up being small. What surprised me most about the driver ants was not the strength of the day's raid but something else I noticed: throughout the raid, workers were constantly going home early and empty-handed, while equal numbers streamed out to replace them.

Theodore Schneirla, the dean of army ant research, concluded that army ants are inefficient. When a raid is at full steam, each worker's transit between nest and raid can take an hour. Based on my records from that first raid and others I saw later on, I calculated that an aggregate of thirty hours' work time was lost to the ant society every single second from all this walking to and fro by thousands of ant workers.

Why doesn't each worker stay out in the raid until she has something to show for her efforts? Coming and going from the front lines, might workers spread among themselves the risks of the hunt? Or do workers tire out and plod home for food and rest as fresh troops stream from the nest to replace them? These notions were illogical. Rather than commute to the nest, workers could save an hour by doing their R&R within the raid itself. That's where the food is, and indeed many workers within the raid do stand around. I have said they serve as guards, but they sometimes look more like office workers stealing a power nap at their desks in the middle of a grueling day.

Pausing in my writing to watch a burst of falling stars, I thought of another explanation: maybe the ants roaming the trails make contributions outside the raid, such as building arcades or guarding the route. However, there were times when *all* the ants ran between raid and nest without pause. I was confident that most ants returning from a raid hiked all the way home, accomplishing nothing along the way but cardio exercise.

Feeling the weight of dinner in my belly, I wondered if the ants heading back to the nest were transporting hidden booty.[10] In the marauder ant, nest-bound workers often have abdomens bloated from drinking the syrup of overripe fruit.[11] But this kind of "tanking up" would not account for the huge numbers of driver ants going home holding no visible reward in their jaws. Where would the food needed to fill so many bellies come from? Not fruit, ordinarily: though driver ants do eat certain native forest fruits, in general vegetable matter is a minor part of their diet. With so few ants hauling prey corpses, the only way the homebound ants could have full bellies would be if most of the prey were being consumed on the spot. Though raiding workers do lap tasty juices off worm and larva prey, I saw no evidence of such activity on a large scale. The ants haul most flesh back to the nest to eat, ingesting during the raid only those foods they cannot cut into pieces.

Then again, remembering how the basket-laden people walking from Nigeria to Cameroon paused to chat with their returning friends—perhaps to compare notes on the value of the goods they had sold—I wondered whether the ants were sharing information. Through signals I could not recognize, returning ants might inform the nest of a raid's success in acquiring food or finding a new nest site. Fresh troops would then depart for the raid with updated knowledge of the colony's current requirements. This feedback might draw more ants into a raid, or lead to its retreat or to the start of a migration. If information does flow between raid and nest, that might help explain evidence for day-to-day differences in the duration and distance of raids.[12]

For societies as large as an army ant's, this hypothesis seemed reasonable. As I saw for the springtail-catching trapjaw ants, the care with which workers in small colonies conduct their business can reflect their limited operating budgets. A large ant colony almost invariably shows more frenzied activity and a faster tempo than a small one. Both ant and human societies can be more productive per capita as their size increases, despite all the mad rushing about: in large cities, people interact with numerous others, exchanging and creating ideas at a high rate.[13] There's a payoff for all their "type A" behavior. Workers in large ant colonies likewise glean information from the crowds around them. In the seed-harvester ant of the southwestern United States, for example, unemployed workers perceive how many of their compatriots are devoted to different tasks by the scent each passerby has picked up from her environment, which reveals the job she is performing—one of several known instances where ants show a capacity to accumulate evidence before making a decision. The workers then adjust their efforts accordingly, shifting, say, from nest maintenance to foraging when foragers are in short supply.[14] Some individuals' rough assessment of the labor situation may be mistaken, but a large colony can afford errors, and the "foraging for work" method enables ant societies to redistribute labor effectively without the need for a supervisor.

Among ants, who are acting without a leader, each individual responds based on the small amount of information available to her. But by gathering all those bits together, the superorganism as a whole behaves sensibly. The raid's features emerge from the collective decisions of the incompletely informed masses, each ant contributing so infinitesimally as to be essentially irrelevant to the outcome. Indeed, the organization of a swarm raid has been accurately re-created by a computer just by modeling the ants in terms of a single, simple set of behavior rules.[15]

Under many circumstances this wisdom-of-the-crowd is characteristic of humans, too—a valuable feature of human democracies. The average of a large number of decisions, even by individuals who are poorly informed, often turns out to be surprisingly smart and accurate. The U.S. military, for example, located a lost submarine with scant information by averaging the guesses of a variety of experts as to its fate, even though no individual guess was close to correct. This has been put forward as a reason to avoid, in the way ants do, an overdependence on a few leaders or "experts," whose judgment can be less reliable than that of a crowd.[16]

One result of the greater information flow in large societies is that larger colonies are more homeostatic than small ones, which is to say that they are more stable in their internal interactions and their relation with the physical environment, in much the way the health of a human body is maintained by a flow of information through our tissues generated by our endocrine glands and nervous system.[17] Large ant societies tend to have a more dependable influx of food, for example, and their nests have internal temperatures better regulated to suit the varied conditions required by developing brood—as in the sun-exposed nest mounds of temperate ants.[18] As I found out for the marauder ant, raids also appear tightly regulated and appropriately responsive to their environment.

While there may be something to this hypothesis for many aspects of colonial life, keeping such a vast labor pool in constant motion makes no sense as a way to run a business—or an ant raid. Perhaps the structure and momentum of a raid are somehow sustained by the manner in which the ants cycle between the nest and the raid front as an unbroken part of the superorganism. It may be that the incessant long hauls between raid and nest are a by-product of this dynamic, with which, in much the way humans have retained their (now useless) wisdom teeth, the ants have been unable to dispense.

RAIDING A NEST

Day after day, Caspar and I explored the terrain for driver ants, but at night the convenience of the dining room colony was irresistible. I'd eat yam tubers and ground cassava on the porch with my companions while gazing at the tree with the driver ant nest. With the monkeys asleep, the primatologists' fieldwork was over for the day. Ours, however, had just begun. Each night after feeding and watering ourselves, Team Ant would scoot over to check out *rubellus* in action.

On my second night at the field station, the colony raided away from the dining room, into an expanse of savanna extending along the station's border. This raid contained far more big workers, about the size of a marauder ant major but without their boxy heads, than the one the night before. I had no idea why. Could the ants adjust their work crews according to some labor need I couldn't perceive?

Stepping into the 2-meter-tall grass, I worried about not seeing the bloodthirsty ants before they saw me. (Army ant studies invariably suffer from a variation of the Heisenberg Uncertainty Principle: an observer may influence a raid simply by agitating the subjects with the slightest disturbance. If only myrmecologists could concoct an antigravity device that allowed them to hover over a colony without touching anything, not to mention an oxygen tank to keep them from breathing on the ants. I'd order the deluxe unit, with a personal air conditioner, because the ants don't appreciate sweat dripping onto them, either.) Fortunately, I soon found that I could watch these ants without their noticing because there were workers swiftly ascending grass stalks directly in my line of sight. With this early-warning system, I was able to stop walking and survey the ground before the more slowly progressing ants in the dense thicket there began to ambush my feet.

Then I noticed that the *rubellus* were driving carpenter ants, each one clutching a larva or a pupa, up to the tips of the stalks, where they froze. Because of the paltry space on a blade, as few as two or three driver ants were spread out behind the carpenter ants on a meter's length of stem. It looked as if each driver ant was pursuing the carpenter ants on her own, beyond the raid front, perhaps (given the ants' blindness) by using vibrations or chemical signals. As I scribbled in my notebook, a driver ant tackled the carpenter ant at the tip of one stalk. Both fell in a tangle into the maelstrom below, where the carpenter ant's pupa was snatched away and her head torn off before she was buried in driver ants.

Tracking an individual driver ant in the welter of a swarm is well-nigh impossible, which made the action on the blades of grass a great opportunity for me. I watched intently as one carpenter

ant after another met the same fate. (As one friend pointed out, it's cruelly ironic that the most popular product for killing ants is called Raid. Talk about living by the sword, dying by the sword.)

Rooting in the earth, I located the carpenters' nest entrance. Driver ants were emerging with carpenter brood, as survivors scurried to temporary safe havens with their charges. The larvae looted from the nest numbered in the hundreds. It wouldn't matter how many ants escaped death, though. They were doomed unless the queen survived as well. The colony could be resurrected only with her continued supply of eggs.

Based on the number of driver ants pouring out of the nest, however, I was sure that she had been killed and her colony vanquished. Then I recalled my dad, who had worked on an ambulance in his youth, complaining that people made up to be dead in the movies were often shown with wounds that, though horrific, wouldn't be fatal. The same was likely true of attacked ant colonies. In tropical American forests with high densities of army ants, every spot on the ground is raided on average at least once a day, so all the colonies have to be able to survive repeated attack.[19] Despite the devastation, the fact is that army ants largely cull rather than eliminate their prey. This culling may make the army ant's role in the ant community similar to that of a grazing mammal that crops grass just enough so that it grows back.

Do army ants practice sustainable harvesting, then? Not likely. I've never seen an army ant worker hold back on making a kill. After all, army ants don't control territories, so it's likely another army ant colony would reap the benefits of an earlier group's restraint. I imagine their assaults simply cease when they reach a point of diminishing returns.

Whether the carpenter ant colony was defunct or wounded, the local population of ants would have plummeted, giving other ant species a chance to colonize the area during the following weeks. Over this same time, the driver ants' other prey—the crickets, spiders, scorpions, nightcrawlers, snails, and varied kin whose numbers had been diminished through death or exodus—would crawl, hop, or slither back, repopulating the ransacked ground.

TRACKING PATCHES AND PREY

Before my departure to Nigeria, I had spent days rummaging through the Biosciences Library at the University of California, Berkeley, reading all I could about army ants, particularly African driver ants. I found that because of the time it takes a prey population to recover, it has been widely assumed that army ants avoid overhunting a site, perhaps warned off by the chemical trace of past raids. This idea was corroborated by the behavior of the most-studied army ant, the New World *Eciton burchellii*. This species has a highly mobile life organized around a huge, synchronized brood produced in a predictable pattern. Colonies migrate almost every day for a couple of weeks while feeding their hungry larvae, then settle in one spot for another two or

Opposite: In a desperate gambit, a carpenter ant retreats up a grass stalk with a pupa as driver ants raid the ground below in Gashaka, Nigeria.

three weeks until the clutch of tens of thousands of adult ants emerges. During this stationary period, the ants minimize the overlap between their forays to some extent by raiding in a starburst pattern around the nest, with each day's raid separated from the tracks of previous days like the spokes of a Ferris wheel.[20]

Most army ants don't lead such regimented lives. Their queens produce brood continuously, or at least in a less regular way, and their migrations and raids are less predictable and probably, in the case of the migrations, less frequent. Nevertheless, I expected that most army ants would avoid recently raided areas. I was therefore surprised, on my third night at the field station, to see a raid going on in the area near our dining room where I had documented my first raid by this colony only two nights before. As they had that last time, they dined on watermelon we had discarded, bulldozing their way into the fruit, an unusual treat for them. *Eciton burchellii* has been observed to raid night after night under electric lights that attract clouds of insects, demonstrating that this species can be motivated to deviate from its starburst raid pattern. Perhaps under certain circumstances it doesn't pay to give up on a dependable food source.

Patch is the scientific term used to connote a local food supply that takes a long time for an animal to consume, and *patchiness* describes an uneven distribution of resources.[21] Some patches are bonanzas, like a watermelon rind in one lump.[22] Once the recruits arrive at such a large object, they need search but little, if at all, to harvest it completely. Other patches consist of a cluster of scattered pieces; once that type of patch is located, the ants must seek out the constituent items, such as the brood scattered within a raided ant nest. The two kinds of patches aren't mutually exclusive. Our watermelon castoffs each represented a bonanza, but all of the pieces, scattered across the lawn, represented a patch, too. Some patches are restocked, as with the excess fruits that fall continually from a tree or are tossed daily from a dining room—these are patches that keep on giving. Ants finding a long-lasting food supply can save themselves a lot of foraging effort by returning to the site as long as visits pay off.

Shuttling individuals to resources in appropriate ways is a matter of having an appropriate transit system. For marauder ants, the grand pattern of food acquisition depends on the colony's durable trunk trails, which are rare in swarm-raiding army ants. My time in Asia had shown me that when most of the land within range of a marauder ant colony is barren with the exception of a verdant field rich in seeds and bugs, the colony's trunk trail will point to that field like an arrow.[23] In addition to accessing patches of widely scattered food like this over weeks, branches of a trunk trail can lead to concentrated bonanzas on which the ants may feed uninterrupted for days. Because the food present in another colony's jurisdiction is available only at the cost of war, the most stable routes also lead away from the trunk trails of other colonies. This configuration allows competing colonies to avoid each other even when their nests are close together.[24]

That third, moonless night in Gashaka, I set my alarm at two-hour intervals and groggily checked the dining room colony repeatedly to confirm that the ants spent until 5 A.M., well after the raid itself had ended, eating the watermelon down to its hard green casing. As a spotted eagle owl hooted in a nearby tree, I watched ants ply the same trails without pause until they devoured the feast, keeping the former raid route open, at least temporarily.

A driver ant emigration is being staged beneath this protective envelope of aggressive workers in Gabon.

Apparently, driver ants can gorge themselves at one place in much the way marauder ants do, albeit for a shorter interval. A colony Caspar studied in Kenya spent three days carving away at a colobus monkey corpse until it was reduced to fur and bone.

Driver ants also reuse abandoned trails that extend into patches rich with scattered food. I saw an example of this at the start of that third evening, when the first ants returned to the watermelon-strewn area near our kitchen. The initial sign of the impending raid was a trickle of workers crossing the lawn in a narrow file. I was puzzled, because this species should raid only in swarms, not columns. Then I realized that it was not a raid at all—instead the workers were retracing the path they had used for their raid two days earlier. The feeble vanguard turned out to be a harbinger of the swarm that arrived as a separate wave an hour later.[25] Just as the chemical traces of an old route could (in theory, at least) repel future raids if the original swarm had left little to eat behind, another old route may attract raids to a locale that remains desirable over an extended period—it remains as a persistent track, or trunk trail.

While trunk trails are invaluable for accessing food patches, they have their uses no matter how food is distributed. It's basic geometry. Marauder ant raids, for example, extend 20 meters at most from their starting point. If every raid departed from the nest, the colony would be able to hunt in a circle with only a 40-meter diameter, a region that could quickly be exhausted of food. But the trunk trails of most colonies are 30 to 60 meters long, vastly expanding the reach of their raids. In a sense, these thoroughfares are part of the marauder ant's nest: sturdy and safe and, even when no foraging is going on, holding a reservoir of ants that comes in handy when a raid is launched.

Army ants also employ trunk trails to access distant regions, and they can do so at breakneck speed. An *Eciton burchellii* colony in its stationary phase may develop several routes radiating in different directions. Rather than departing the nest in their usual swarm, the ants run in a line along one of their abandoned trails, retracing the trunk trail's path for 50 to 100 meters. Only then does foraging begin in earnest, as ants pour off the trail in raid formation to explore promising new terrain.

It isn't known yet how army ant workers mark their more productive paths to make them attractive for future reuse as trunk trails. In general, detecting an old route, and whether to follow or avoid it, should be easy for army ants, whose capacity to pick up scent is legendary. Some New World species will follow nearly anything, even a thin streak of water laid on the ground. This versatility suggests that their skills could even extend beyond recognizing the trails of their own species to tracking the trails of their prey. Workers at the front lines of a raid likely take whatever cues they find to lead them to a meal, scents left by other ants included. Caspar and I were focused on swarm raiders, but around the world most army ants mount column raids, which illustrate this ability well. Narrow columns don't have the ant power to take down big prey, which will escape them easily. Instead, most column-raiding army ants depend on finding ant nests, with their hoards of brood. Even a weak raid can stage an effective attack on an ant nest if the workers, recruited from the column network, are quickly able to accumulate in numbers.

There is a logistical problem, however: the nest entrance can be difficult for a raid column to find. Picture the column as an elongating line, and the nest entrance as a point on the ground; the odds of them intersecting aren't good. But if the column raiders avail themselves of their scent-tracking skills, their raid need only cross the trail of the other ant species. That's an easier proposition, because everywhere in a rainforest there is a tapestry of the pheromonal guidance signals deposited by all kinds of ants.

I have devised an experiment to determine whether army ants are tracking prey by following their scent. When I find a column raid in progress, I scrape away the ground surface ahead of their front line. If the ants are still in search mode, they will continue across the upturned ground without hesitation. But if they are in pursuit of a prey species of ant, my action will have removed the pheromone signals they are tracking and disrupt their advance. Twice on my visits to Barro Colorado Island, a research station operated by the Smithsonian in Panama, I stopped the column raids of an *Eciton hamatum* colony cold—until they picked up the scent again and continued on their way.[26]

Whether reused on occasion or not, army ant trunk trails are generally less ubiquitous, obvious, and persistent than those of the marauder ant. Army ants focus on discovering virgin hunting grounds through migrations and shifting raids, because their diets are dominated by ant colonies and large invertebrates that are slow to replenish. Because marauder ants regularly eat fruit, seeds, and small and large prey, they can generally gather food even in the frequently reraided areas near their nests. There may be limits to the utility of their trunk trails, though: typically after a few weeks, a colony abandons one highway and starts another. I haven't been able to prove it, but it may be they need a change of venue once they have depleted a region.

Conversely, the predilection among army ants for meals of large invertebrates and social insects may explain why their raids travel ten times faster and several times farther than those of marauder ants. With their great breadth of diet, marauders can afford to be slow and methodical in their searches, whereas a sizable army ant swarm must do considerable reconnoitering to take in enough of their widely separated social insect prey.

I took off for an hour to look for ants along the path to Cameroon, unable to stop thinking about how an aggregate of army ants, using the interweave of their columns and scents, explores the world much as an organism does. My mind turned to the driver ants' blitz of the carpenter ant colony. Two young men strolled into view, baskets filled with dried meat balanced on their heads. Judging from their stares and whispers, I must have been an odd sight indeed.

After years of living on foreign soil, I have developed a method to deal with this reaction, based on the assumption that, since I am the center of attention anyway, I might as well make things interesting. Whether faced by a crowd at a bus stop in Nepal, by kids leaving school in a mountain town in Bolivia, or by weary foot travelers in Nigeria, I walk back and forth like a professor in a class, carrying on out loud about whatever is on my mind.

"The swarm raid rolled over the carpenter ant nest as if it were nothing," I said to the two astonished villagers. "Just a tiny minority of the driver ants were involved in the kills. What does that tell us?" I turned dramatically to the young men, who were grinning at my animated speech. They gave my hand a shake and continued on their way, chattering.

My brief soliloquy led me to the obvious conclusion. The driver ant raiding enterprises were superorganism-level organs geared for much larger confrontations than those I had witnessed so far. Two days later, this suspicion was confirmed.

7 clash of the titans

Early on my first evening at Gashaka-Gumti, after the long day's search for driver ants, I collapsed on the hard earth outside my room at the field station and contemplated the parrots flying overhead. But then I became aware of movements in the grass, and I turned my head to witness a remarkable sight: a row of handsome, 2-centimeter-long cylindrical *Pachycondyla analis* workers, right next to my face.

Scientists studying *Pachycondyla* have determined that raids of species like the one at Gashaka don't proceed like those of driver ants and other army ants. Rather, they are led by an individual that has scouted a promising target population of termite prey and recruited a couple hundred workers to harvest it. I was witnessing that now. Traveling in a compact squad 2 meters long and two to four ants across, the workers in front of me marched at a steady pace of 1 meter per minute, following the leader to dinner. There were no stragglers. Compared to the swarm raids of the army ants, this raid seemed leisurely and orderly in the extreme—another example of how ants in smaller societies move slowly and with care.

Eventually the ants entered the brush, where I couldn't follow. Circling the field station, I saw more columns sallying forth, one from each of several nests that were apparently operating on a tight and synchronous schedule. Thirty minutes to an hour later, the ants reemerged from the brush and headed home in identical formations. Only now, each one held several termites, stunned by the toxins from the ant's stinger, bundled between her jaws.

Two nights later, a feeble column from the *Dorylus rubellus* driver ant colony near the kitchen, retracing the route taken by the raid of two days before, passed next to one of the *Pachycondyla* nests. Some of the *rubellus* ants stopped in the open-jaw guard position, preventing the much larger *Pachycondyla* from departing on their raid. Every minute, one or two of the besieged *Pachys* (as ant experts call them) stuck her head out of the entrance to the nest and jabbed at the tormentors. Occasionally a *Pachy* succeeded in grabbing a driver ant and pulling it below, where I could just make out the workers tearing it apart.

Every fifteen minutes or so, there was a surge of activity in the driver ant column, and *rubellus* workers poured down the *Pachy* hole—twenty, fifty, a hundred of them, a veritable blitzkrieg that must have prevented the *Pachys* from implementing a coherent defense. I envisioned the savagery below, the feckless *Pachys* massacred, their brood consumed. Eventually, though, the *Pachys* forced the driver ant horde to retreat. With a flashlight I made out, faintly, what I took to be the survivors in their bunkers, nursing their wounds.

One hour later, the column was succeeded by the full-bore swarm raid of driver ants, which swept through, swamping everything. The disoriented scramble that ensued reminded me that, compared to *Pachycondyla*'s methodical predation on termites, army ant raids seem based less on finesse than on brute force. I'm convinced driver ants have little success with prey smaller than themselves not because such prey isn't worth the effort—since so few ants take food most of the time, this argument hardly seems viable—but because the sightless rampagers are individually clumsy.[1] This again is what we expect with ants with a large colony size: faster movements and more inefficiency.

Even so, there were signs of order. As I watched the raiding ants struggle with a scarab beetle, others walked over or alongside them, delivering food to the nest. The frenzied workers on the beetle never seemed to confuse the movements of their prey with those of the dead insects being convoyed past them. Meanwhile, their food-hefting sisters managed to stick with their job and ignore the fighting, even though the ants embroiled with the beetle must have been discharging a powerful alarm pheromone. I couldn't imagine how they kept it all straight.

I turned my attention back to the entrance where the driver ants had poured into the *Pachy* nest. By now the swarm had departed, leaving the hole deserted, in ominous stillness. There was not one *Pachy* to be seen.

But in the morning, I saw the *Pachys* marching out once again, with no dead of their species in evidence. Remarkably, they had survived the *rubellus* attack. In fact, the *Pachys* must have been picking off the driver ants one by one and (I imagine) eating them underground the whole time the multitudes were passing overhead. Instead of facing their demise, the *Pachys* had beaten the odds with a classic maneuver: by taking advantage of the choke point at the nest entrance, they had greatly decreased their adversary's access and combat power. They had been in control of the situation the whole time, transforming the driver ant raid into the ultimate stay-at-home feast.

How could they take such a pounding? I found out by dropping a *Pachy* onto a driver ant trail. In one fell swoop, she was buried from view by a mass of driver ants. That was the sort of brutality I expected from an army ant! Convinced she was done for, I returned twenty minutes later to find driver ants still laying siege. I extracted the hapless *Pachy* with a pair of forceps and, shaking off all but two of her attackers, put her down for a look. She lay motionless yet looked intact. When I picked off the last two ants, she roused herself and ran off. She had been playing dead.

I surmised that the *Pachys* were too well armored to be killed—the driver ants' mandibles slip right off their exoskeletons. Because army ant raids pass by within ten or twenty minutes, a victim need only stay immobile and wait for her assailants to give up. On several subsequent occasions, I saw a *Pachy* escape after driver ants had restrained her.

Fighting back is rarely an effective way to survive an army ant raid. No matter how many of its workers die while catching prey, the raid never appears to retreat; the attackers just keep piling on—an advantage to having a humongous army. Some beetles and millipedes avoid death by exuding noxious chemicals; driver ants respond by burying them with soil and abandoning them with no harm done. Spiders and praying mantises avoid capture by New World army ants by freezing in position; unable to detect prey except by their escape response, the ants leave them unscathed.

Many ant species have evolved other defenses to give their colonies a chance to survive—it helps, for example, to be built like a tank, like the *Pachys*. Other species climb grass stalks in a gambit to carry themselves out of reach, or barricade themselves in their nest chambers. The workers of New World *Stenamma alas* mold a tiny sphere of clay that they then use as a door for their nest entrance. A worker closes this portal upon detecting predators—especially army ants. This is reminiscent of a defense used by the ancient Cappadocians, who lived in what is now Turkey and carved stone discs that they would roll across the entryways to their underground dwelling places when an attack was imminent. *Stenamma* take it a step further, constructing false entrances to blind-ended tunnels that lead their foes astray.[2]

TERRORIZING TERMITES

The next night the mood at Gashaka was depressed. Caspar and I had been at the station only four days, but several of the scientists had been counting the months. The malaise that can descend on people isolated for too long in the field had worsened that day, when one researcher had a recurrence of malaria and retreated to bed. Meanwhile, everyone had been on the lookout for chimp feces for Darren, and the samples were piling up; Darren had spent the miserably hot day trying to sieve ant parts from a single turd. He had a few driver ant heads to show for it, but at this rate, his thesis would require several unappetizing years.[3]

Anxious to escape the conversational doldrums at the dinner table, I checked on the nearby driver ant colony. I was surprised to find meter after meter of workers carrying hundreds of ghostly bodies along their route, which I reckoned were brood being transported in a migration. Then I noticed that they were heading in the wrong direction, from the savanna toward the nest. A look through my macro lens revealed that the cargo was *Macrotermes,* termites known for their castlelike nests of clay.

Compared to the small incursions that *Pachys* make on termite galleries, this reflected a battle royal going on somewhere in the dark. Here at last was a show of the voraciousness for which army ants are celebrated. My spirit soared with the primal recognition of "nature, red in tooth and claw," as Alfred, Lord Tennyson, described it. Was this the boom in the raiding "boom and bust" I'd been looking for?

Neither Caspar nor I had read about driver ants conducting an attack of this sort.[4] These swarm raiders were thought not to invade termite mounds, on the theory that they were unable to penetrate within. Termite capture was believed to be the sole expertise of other kinds of African army ants with more subterranean habits. (The Mofu people of mountainous northern Cameroon hold these belowground termite hunters in great respect, calling them the "prince of the insects." The villagers collect the workers in a calabash gourd and pour them out in their houses, then lay a trail of ochre on the ground; this is meant to guide the ants to the most termite-infested sites in their homes.)[5]

Termites, like ants, have a caste system that can include small workers and soldiers. Looking closely at the trail, I watched the ants hauling the termite workers, pale blobs about their own size. Once in a while, the corpse of a soldier termite would go by; it was also the size of a worker but had an elongated orange head and needle jaws. At rare intervals came the headless body or the rust-red, bodiless head of a second, larger kind of soldier. That night I roused myself every couple of hours for an update on the progress of the ants, checking out the action on the trail by flashlight beam. By morning, the booty of the ants included the rotund carcasses of the developing reproductives, the cockroach-shaped kings and queens (termites are in fact social roaches).[6] This evidence told me that the ants had breeched the colony's defenses and were now invading the nest proper.

Where was this rampaging taking place? Caspar followed the driver ant trail to where it disappeared underground near an eroded *Macrotermes* mound a couple of meters across. These termites have huge colonies and raise a type of fungus to help them break down their woody food. We dug a meter down into the nest, exposing dozens of chambers containing the termites' soft gray fungus gardens, each about the size of a softball. Crawling in and out of pits and holes in each garden were both adult termites and the younger, more delicate nymphs. As we dug, *Pachy* ants, homing in on a good thing, raided the exposed gardens and soon had stacks of termites in their jaws. There were no driver ants to be seen: either the troops were attacking from a different direction or we had the wrong termite colony.

All day and through the next night, the driver ants continued to drag the termites to their nest. But where were they coming from? Scanning the surrounding landscape, Caspar and I eventually found two more columns of the same driver ant colony in the savanna. Judging by

Driver ant workers investigating a termite fungus garden presented to them in an experiment in Gashaka, Nigeria. The same colony had attacked a large *Macrotermes* termite nest at this site.

the differing proportions of the small worker, soldier, and queen termite castes being moved along each of the three highways, it seemed they represented three separate attacks, on either different termite colonies or different battlefronts within the same termite nest. In the end we estimated the ants hauled away at least half a million termites, large and small—several flaccid kilograms in all.

An intriguing story, but how incomplete. In science, we learn by bits and pieces, leaving others to unravel further details. I could only guess at the scene that had unfolded somewhere beneath our feet. During the underground portion of their forays, the driver ants must have clashed with termite soldiers ganged along narrow access routes to their castle in much the way *Pachy* ant workers had shielded their nest entryway. But in this case the driver ants broke through and launched into wholesale looting of the corridors beyond, transforming a steady raid advance into a focused attack that continued for hours.

Since I hadn't witnessed the original killing spree, I decided to reconstruct it. I took a fungus garden from the termite nest that Caspar and I had dug up and deposited it next to a file of driver ants. There was no response. Apparently, the ants did not perceive the garden as a source of prey. Their assault began only after I had crumbled the fragile material to expose the termite workers inside. Then the driver ants infiltrated every crevice and pulled out dozens of the buttery-soft termites.

During their second night of gathering termite booty, I could hardly get close to the trail, which had become completely walled over by a bristling envelope of guards. By knocking the guards away, I succeeded in getting a view of the ants below, many of which were now carrying termite corpses away from the ant nest. They were transporting their own brood as well, slung under their bodies in the same way they carried food. The raid trail had become a migration route, and it was clear why. Caspar and I held our noses against the stench arising from the nest: under our feet, the termites had begun to rot. The driver ants were abandoning ship, taking with them any salvageable meat and leaving the garbage-thieving acrobat ants to scavenge the decomposing bodies left behind.

Army ants, including driver ants, often migrate along a prior raiding route, while at the same time conducting a raid along another. Near the migration's midway point, the queen makes her run, shielded by a retinue of workers. To see this happen requires round-the-clock diligence. I took advantage of this opportunity and lay down on my side at what I thought was a safe distance from the trail, with my headlight duct-taped to a nearby tree branch to cast a steady beam of light. Unfortunately, my days had been long ones. I recall noticing that among the migrating *Dorylus rubellus* ants were tiny workers I had never seen at any other time; presumably, they served as nurses. But then I fell into a dream about being a dwarf ant—only to be awakened soon after by the sound of my own scream, and the pain of *rubellus* ants embedding themselves in my arm.

Generally, the exodus of migrating driver ants continues round the clock for two to four days. The number of ants participating can reach into the millions. But this seemed to be a small colony, in the hundreds of thousands, and its migration was over by the second morning. Caspar and I located its new nest, 67 meters from the old site, and we pried up a rock for a view of the massive company beneath. The workers were piling termites in a larder 15 centimeters wide in a preexisting cavern.

This was the first time a driver ant colony had been found to stash its food. If their raids have been crafted by evolution to take advantage of the rare windfall, the ants should be masters at stockpiling an excess catch in this way.[7] However, whereas the *Pachy* ants sting their prey to keep it incapacitated but alive, the driver ant kills it, and corpses rot fast in tropical heat. Driver ants and other army ants also lack repletes and don't take the seeds that marauder ants can horde.[8] Napoleon observed that an army travels on its stomach—anticipating the idea of a superorganism by seeing the ensemble as an individual—and the same is true of an ant army: unable to keep a fresh larder, this colony was forced by its stomach to stay on the move.

PREDATION VERSUS DEFENSE

Two weeks later, my arms were blotchy from bites. I had stared at ants so long, I saw their flowing columns even when I closed my eyes. I felt like an obsessed FBI profiler investigating the habits of a serial killer. By now, the daily activities of *rubellus* ants had fallen into a predictable pattern. Raids began early in the evening and continued into the morning hours. After a raid ended, the flow of returning workers on the trail could proceed for hours, even on into a second night.

From my first day at Gashaka watching the open-jawed guards, I'd documented how protective the driver ants were of the commerce on their trails. The *rubellus* reacted even more hysterically to my presence than did the marauder ants. When I so much as breathed on them, the food-transporting ants retreated, while the other workers scurried off to patrol up to 30 centimeters from the trail. I used my mask, constructed earlier from a disintegrating T-shirt, to keep from creating a ruckus.

I had thought that standing guard over a procession and patrolling near it after a commotion must be part of the colony's defense, not part of its foraging behavior. And to some degree this must be true. Except for a few long-term trunk trails, all the trails of driver ants and other army ants are created during raids that have recently cleared the surroundings of food. For this reason, the numbers of guards or patrollers are out of proportion to the likelihood that those ants will find a meal; ipso facto, they will more likely serve to protect a column than locate prey.

This is unambiguously true for the majors of the New World *Eciton* army ants, which have fishhook-shaped mandibles suited only for suicidal defense against vertebrates—their jaws have to be pulled out of the skin with tweezers.[9] Except for these specialist saber-bearers, which never catch prey, there's no evidence that any army ants distinguish enemies from meals—a driver ant's actions don't differ whether she bites an entomologist or an aphid.

But no matter what the ant species, the line between defense and foraging can be blurry, because any concentration of ant workers has the capacity to serve as a snare for food. As an example, an insect might flee from the army ant raid front into the raid fan, where ants stationed as guards along the network of trails can participate in dispatching it. *Dorylus rubellus* on trails far from the raid reacted to grasshoppers, crickets, foreign ants, and striped mice in the same way they did to my clumsy presence: by patrolling and attempting to seize them. They caught two of the crickets and a carpenter ant, and in another case a grub that had caused no disturbance, and cut them up and carried them off as food.

The same thing happens during marauder ant patrols. In fact, for both driver and marauder ants, the workers on patrol appear to take on the movement patterns of those within a raid, absent an advancing front. The hypothesis I developed in Nigeria was that the only thing that stops a raid from developing after a disturbance is that the patrollers are soon drawn back to the overwhelming scent of the thoroughfare from which they came. But food can break the workers from this attraction: I scattered tiny bits of meat in front of the driver ants on patrol, and this was enough to set off a small raid from the side of the trail, as prey near trails often does with marauder ants.

The reaction of *rubellus* to disruptions along the trail is mild compared to their reaction to threats to the nest. Driver ants have a unique response to such disturbances, perhaps because, unlike in the barricaded constructions of big ant societies (weaver ants and leafcutter ants, for example), the populace often can be viewed from outside. One afternoon, Caspar and I tried the chimpanzee luncheon technique of jamming a stick in the midst of the ants visible within the wide nest hole of one colony. Workers poured out of the hole and began patrolling thickly within a meter of it. Others ran along the stick and cascaded in strings from the end. Within an hour the ants had closed the gap with a 25-centimeter-wide plug of their menacing bodies.

For this colony, our meddling resulted in an eviction. When I stopped by the next day, the ants were busily abandoning their nest, tracing a dense migratory route through the forest that

Workers of the driver ant *Dorylus nigricans* in Ghana transporting huge numbers of their colony's pupae during emigration to a new nest site.

shimmered with the gaping mandibles of the jet-black soldiers.[10] I sat down at a safe distance and took out my notebook. At one moment a driver ant colony can be rushing headlong into battle with a termite army a million strong; at the next it might be fleeing from a chimpanzee with a stick or the breath of a man on its trail. Advance or retreat, eat or be eaten—these are choices even army ants have to make.

The resemblance between patrolling near a trail and swarming in raids set me thinking about how easy it is to make assumptions about the function of behavior, which can lead to misinterpretation. This seems to have been the case for the South American ant *Allomerus decemarticulatus.* Its colonies occupy shrubs that have hollow pouches at the base of their leaves, making for multiple living quarters. The workers also build shelters along the plant stems, using fine hairs spliced off the plant and bound together with fungi and feces. These thatched roofs, it is claimed, serve as traps.[11] Reportedly, the workers reach through the gridwork of openings in the thatch to jointly ambush prey of a size and vigor normally caught only by army ants, pinning and dismantling grasshoppers against the platform as if it were a torture rack.

This notion of a "trap" implies that a grasshopper, for example, would avoid the ants if they were not hidden. This seems unlikely; I doubt grasshoppers could notice the minute ants of this species, particularly in mid-leap, let alone change course to avoid them.

During a research trip to Tiputini, Ecuador, I put the trap idea to the test. I hung a mosquito net over a plant with a thriving *Allomerus* colony, added a hundred grasshoppers and katydids, and sat inside for the next five mornings—an unusual case of using a mosquito net to keep

Colonies of *Allomerus decemarticulatus* build defensive covers over their trails. The workers are emerging from the circular entryways in these covers to catch an intruding *Pheidole* ("big-headed") ant.

insects in instead of out. Even after the grasshoppers settled down, they were indiscriminate in their movements, hopping from where ants hid under the structures to where ants strolled in full view to where there were no ants at all. When they landed among the ants, even on the structures, they got away unhurt. Certainly if the structures served as traps, they were inefficient ones.[12]

The *Allomerus* constructions run continuously from one nest pouch to the next on different branches of their shrub and contain a highway of workers commuting from nest to nest. Other plant-dwelling ants erect similar covers over their trails, even with similar holes through which they come and go to forage. Such arcades probably serve primarily to protect the enclosed traffic against enemies (out of sight being out of mind), as do the trail covers of soil built by marauder ants and many driver ants.

Indeed, the *Allomerus* workers at my study site didn't wait in ambush hour after hour at each "foxhole," as would be expected if the structures were sit-and-wait traps; when conditions were calm, most of the holes were usually empty. But that wasn't true at times of danger to the ants on the passageway within. After a day of pulling grasshoppers from my hair, I noticed interlopers of another ant, a species of *Pheidole,* or big-headed ant, climbing the plant to pin down a wounded grasshopper missed by the *Allomerus.* Upon the arrival of the *Pheidole* ants, the *Allomerus* workers began to guard each of the several dozen entrances to their arcade nearest the commotion caused by the intruders. These guards, aided by nestmates roaming the arcade surface, also caught and killed one *Pheidole* and carried it off.

Ants of many kinds will on occasion catch and kill enemies and prey along their trails, especially when workers are densely packed; it's a matter of overwhelming the quarry, as army ants do, through staggering numbers, a tactic that can succeed even for a timid species if their legions are great enough. In this way the organization of a superorganism can be more responsive than the tissues in a body: trail-bound workers can shift seamlessly in their behavior from transport to protection to predation. It's as if one's liver could change function when the heart is incapacitated, and pump blood.

8 notes from underground

After a four-hour drive in a sedan taxi crammed with five other people, including Caspar Schöning, I emerged barely able to stand. The driver had dropped us off in front of a low building, the headquarters of Nigeria's best-known national park, Cross River. There a young woman showed us to the office of the assistant director, who informed us that we would have to wait for the director before seeking our ants. After an hour in his waiting room, he ushered us into an expansive office, through a door labeled "S.O. Abdulsalam, Director, Esq."

Caspar and I explained to Mr. Abdulsalam, whose large frame was wedged behind an expansive desk, that we had just spent two weeks at Gashaka-Gumti, where we found just one species of swarm-raiding driver ant, *Dorylus rubellus*. After two days in long-distance taxis, we had arrived at Cross River with hopes of finding a greater diversity of army ants. Having listened to us intently, Mr. Abdulsalam declared that collecting ants at Cross River was a laudable and serious endeavor. Meanwhile, his underlings among the park staff arrived one by one and filled chairs around a long table. The director of tourism, the director of security, the director of the environment, the director of education, the director of research, the director of things-that-go-bump-in-the-night—in the end there was a baker's dozen of them, each putting on a dazzling smile for the boss.

Over the next hour the director of the directors proceeded to invite each man to demonstrate his rhetorical skills. Every one expressed his sincere belief in the Importance of Ant Research at Cross River, his surprise that no Ant Scientist had discovered the park before, and his general gratitude for our visit. The director interrupted occasionally to embellish a point, and when each had finished, he would sum up the speech we'd just heard, adding flourishes of his own that left no doubt as to who wore the oratorical crown. Somewhere along the line I managed to interject that, while we appreciated Mr. Abdulsalam's generosity, it was getting late and we had lamentably few hours in his fine park. No doubt, I explained, the primatologists from Gashaka (who planned to arrive tomorrow at noon, exactly when Caspar and I had to depart) could advise them after we left on unaddressed matters of every kind.

With this, Caspar and I were shown to the office of the director of security to pay the park entry fee of one dollar. There I learned one way to identify an official's position in the Nigerian hierarchy. While the director had summoned his secretary with an intercom, and the assistant director had fumbled with an old hand-cranked bell, the lowly director of security had to scream over a shoulder—even though in each case the woman had been sitting close enough simply to talk to.

Bureaucracy loves a vacuum. It takes root and flourishes in places where the cogs and mechanisms of governance are rickety or dormant. I have been confounded by procedural excesses in offices and shops and checkpoints throughout the world, but sometimes the curlicue of red tape takes my breath away. Perhaps that's why, as Aldous Huxley wrote, "However hard they try, men cannot create a social organism, they can only create an organization."[1] I left those offices thinking about the value we humans place in authorities and chains of command, despite their being so open to abuses of power and greed and so prone to failures of communication. Leaderless ant societies, by comparison, seem to be a universal success story, capable of mobilizing themselves as needed for any job.

By the time we were turned loose it was late afternoon, with a few low clouds rushing overhead. We hitched a ride from the headquarters into the park and, with no time to spare, hiked into the thickety forest. Immediately we found army ants pouring across the road. A little further on we encountered another species. Then another. Walking briskly, we discovered four species of army ants by sunset. Our most thrilling find was a *Dorylus mayri* driver ant raid, which resembled a swarm of *Dorylus rubellus* but scaled up a notch. There were millions of ants, and they ran blazingly fast, in a front that spread a full 20 meters wide and traveled high into the trees. Caspar assured me this species compared favorably in size and speed with *Dorylus wilverthi* of the Congo, which Albert Raignier and Jozef van Boven studied in the 1950s. Their

Belgian jaws must have dropped when they documented the largest army ant colony yet recorded: twenty-two million workers, by their calculations. To sustain this army, they estimated, the queen had to lay three to four million eggs each month—about a quarter billion eggs in her lifetime.[2] (Months later, on an expedition to Ghana with the army ant ecologist Bill Gotwald, I would come upon another of Africa's most awe-inspiring driver ant species, *Dorylus nigricans*, raiding with a swarm front 32 meters across. Crying out from what felt like hundreds of tiny vises embedded in my legs, I ran through a stream of flowing bodies visible for as far as I could see, an experience almost up to the standard of the 1954 film *The Naked Jungle*, in which Christopher Leiningen, played by Charlton Heston, finds himself "up against a monster 20 miles long and 2 miles wide. . . . 40 square miles of agonizing death!")

Caspar and I turned our attention to a batch of army ants pouring over a spot 3 meters in diameter. Their columns streamed under every object and reticulated through rotten logs. This "*Dorylus* species in the *congolensis-kohli* complex," as Caspar described it, was intermediate, meaning it carries out its activities just out of view within the leaf litter. Other driver ant species, such as *rubellus, mayri, wilverthi,* and *nigricans,* are surface active, raiding on the exposed ground and occasionally in trees, whereas the subterranean army ants live and die mostly hidden from view.

The portioning off of foraging activities by layer may help avoid strife between colonies. While army ants seem willing to assault almost anything in their path, they rarely fight one another; instead, their raids shift out of the way with little squabbling and minimal loss of life.[3] Given their cordial rules of engagement, the great myrmecologist Carl Rettenmeyer once proclaimed to me that army ants are the "civilized insect." I suspect army ants hold back because they are not in competition with their neighbors over territory (unnecessary given the wide and ever-changing expanses they roam). Even in local clashes between army ant colonies, a détente is sensible: whereas the marauder ant uses its raid muscle to rout rival ants, army ants literally devour the competition, including other ant species. Applying lethal force against other army ant species could result in mutually assured destruction. I'm amazed that species raiding many meters horizontally can, as one solution to this problem, separate their actions over so few centimeters in depth, from the surface to the leaf litter to the soil below.

SLICERS AND DICERS

That night, Caspar and I set up our tents near the park entrance. I shared with him my most memorable moment of the day: watching a mass of *Dorylus mayri* ants in the midst of a huge swarm raid remove the eyes and limbs of a pair of 3-centimeter-wide freshwater crabs—small, but rock tough. This kind of activity would be difficult for most tropical American army ants, whose plier-like mandibles tear flesh but are incapable of slicing it. Only a few species can manage to consume even a frog; the others have to abandon their vertebrate carnage uneaten.

The mandibles of African driver ants, however, are like sharp scissors, built for severing. I had guessed that might explain why driver ant raids are eerily quiet to those who have witnessed the swarm raids of the New World *Eciton burchellii,* which are accompanied by the chirp of "ant birds" that snatch prey that has been missed by the ants, the buzz of parasitic "ant flies" that lay eggs on the escapees, and the flutter of "ant butterflies" that feed on the droppings of the birds. In the driver ant raids I witnessed in Africa, these attendants seemed to be missing—perhaps, I hypothesized, because the driver ants would make mincemeat of them. However, I later learned that birds are actually present, but of a species that is quiet and circumspect.[4]

What, then, of the claims of popular stories, such as the one about Leiningen, that army ants regularly kill and eat vertebrates? Caspar told me that for all his months in the field, the only examples he'd seen of live vertebrates being killed by driver ants were a frog and a litter of helpless "pinkies," baby mice the size of the tip of a pinkie finger. In Cameroon several years before, I had helped tribal pygmies remove an antelope from a snare that had one flank partially carved away by driver ants. It must have been bleeding and restrained at the time of their arrival, making an easy target. Most records of vertebrate predation in equatorial Africa concern tethered specimens like that, Caspar said. It's no coincidence that people living within range of driver ants keep their babies on their bodies and let their livestock roam free.

So, exactly how good were driver ants at carving meat? Two days earlier I had conducted an experiment: I gave a *rubellus* ant colony a fresh lizard kill, dropping its 6-centimeter carcass right on the ants' trail. They showed no hesitation: in two minutes the body was packed with workers of all sizes trying to pry through its skin. Each stuck one mandible between the scales and squeezed hard; but again and again, the ant's opposite jaw slid uselessly over the surface. Two hours later the lizard remained unscathed. During the third hour, the ants began to carve off strips of scales, and three hours after that, most of the skin was peeled away. Altogether, it took the ants ten hours to reduce the lizard to a spinal column. An impressive perseverance, though no doubt a thin-skinned frog or pinkie would have been an easier meal.

To a much greater degree than marauder ants, driver ants process food where it's found—tearing up their prey to the extent that, to quote the coroner in *The Wizard of Oz,* "she is not only *merely* dead, she's really most *sincerely* dead." Their proficiency at butchering prey is essential to their livelihood, due to the fact that they heft burdens by slinging them under their bodies. Although this group transport technique is efficient in that it allows the transporting ants to easily pull together in a common direction, it means that all booty must be carved up until the pieces will fit between a worker's legs. The legs of *rubellus* are long, but not long enough to handle large prey whole unless it's skinny, such as earthworms and centipedes, which can be straddled by up to six ants—though in my experience, the workers beyond the first can be meddlesome assistants. As often as not, they climb partially or entirely atop the food, to gnaw at it rather than aid in its transport.

GENERALISTS AND SPECIALISTS

Our bedding prepared for the night, Caspar and I joined four rifle-bearing park guards dressed in green uniforms, berets, and canvas boots at a nearby concrete outpost for a dinner of spiced boiled yams, plantains, and chicken. We talked to the men about the diet of Americans and Europeans, and then about the African army ants and what they eat—something that probably differs by species. Most, such as the *congolensis-kohli* complex species, go after just about any prey: we had seen them retrieve caterpillars and grubs and even watched them storming the flank of a relatively vast grasshopper, which suffered them quietly for a minute before exuding a noxious snowy foam behind its head and leaping to safety. The swarm-raiding *rubellus,* which also feeds indiscriminately, is likewise a "generalist predator." There are other army ants, however, that have been found carrying only earthworms and have been described as "earthworm specialists."[5]

How can we tell if a species is a generalist or a specialist?[6] After all, an individual's diet depends on a cascade of contingencies. Once an ant finds prey, the choices she makes depend on her colony's needs, the time and energy required to pursue the prey, the risk involved (whether it is likely to hurt her or to waste her time by escaping), and the nutrients and energy it contains.[7] Many ants, for instance, give termite colonies a pass because termites are so well defended. Other species, such as Gashaka's *Pachycondyla,* display behavioral and anatomical adaptations that specifically aid in catching termites, such as body armor and strategies for recruiting a regiment and disabling termite soldiers—though a cafeteria experiment, in which scientists provide the ants with a buffet, might show that they eat other things as well. In fact, some prey are so defenseless that almost any ant may consume them: a helpless caterpillar, for example, must be hard to pass up. Finally, familiarity with local sources of food can also be a factor for some ant workers, just as it is for the person who grew up in Chengdu, say, and craves Sichuan cuisine for the rest of her life.[8]

But whether a species is a specialist or generalist is determined not simply by what it will harvest and eat but also by its foraging behavior. A specialist not only has the skills to catch a specific food but has a search strategy that targets its food source. Consider, for example, the specialist *Centromyrmex* ants, which live and die inside a termite nest. The workers have to be blind, tough, and strong-limbed to invade termite galleries, where they are unlikely to contact anything to eat but termites.[9] If, instead, *Centromyrmex* took their termites by foraging more widely, and if they readily transported such easy prey as caterpillars back to the nest, with their adaptations to capture termites they could be defined as "specialists" with a varied diet. We might liken them to early hominids, who made tools to hunt mastodon but, only occasionally having opportunities to use them, dined mostly on other things. In the end, whether a species is considered a food generalist or specialist depends on whether a researcher is interested in, among other things, its morphology, activities, habitat choice, ecological role, or everyday diet.

Driver ants concentrate their foraging on a specific stratum on or under the ground; yet within their preferred layer, most species are consummate generalists, scouring the raid front for any prey they can detect and catch, in whatever habitat they happen to be passing through, and with, as far as we know, little regard for costs and benefits. Their polymorphism, however, offers the possibility of specialization *within* a colony, because the workers can act almost like different species in their contributions to hunting and gathering. One way for this to occur is through size matching, in which, for example, small workers find or kill and harvest smaller prey than do larger workers. While there is no evidence that an army ant worker will pass up prey in favor of another more suited to her size, small workers are able to search tunnels too narrow for their larger sisters, which could open up unique opportunities for them.[10] Meanwhile, bigger ants can carry bigger items to the nest, though rather than selecting their booty by size, it's more likely each ant cuts a chunk appropriate for her to lift.

SUBTERRANEAN RAIDERS

Back at the Gashaka field station there was a species of ant that didn't put on anywhere near as obvious a show as *rubellus*. This ant, which belongs to one of the primarily subterranean groups of the genus *Dorylus,* was in an easily overlooked colony that I was lucky enough to locate and study. In the lingo of army ant biologists—of whom there are perhaps a dozen—these subterranean driver ants are simply "subs." Compared to intermediate army ants like the *congolensis-kohli* complex species, with their moderately long legs, and the even more lissome surface-active driver ants such as *rubellus,* which use their spindly legs to run fast on open ground while holding large objects beneath them, subterranean army ants have narrow heads and short extremities.[11] These are sensible adaptations. Long limbs get in the way in cramped quarters, whereas short, stocky legs are more suitable as digging tools, giving better leverage for moving soil. Compared to driver ants on the surface, which can be spaced several body lengths apart as they move along, the subs are often piled on top of each other within their narrow passages.

Subs live in terra incognita. Stefanie Berghoff, then a doctoral student at the University of Würzburg in Germany, has made the only attempt to date to study them in depth.[12] She went after an Asian species, *Dorylus laevigatus,* one I have from time to time seen crossing trails and moving under logs, typically massed as thick as porridge. By placing bait in buckets full of holes, Stefanie showed that the ants employ stable underground trunk trails to continuously access the same foraging areas for two months or even longer—a pattern previously unheard of among the army ants.

As with marauder ant trails, the *laevigatus* trunk trails are an expanded base from which to hunt en masse. Shifting networks of raiding columns extend from the trunk trails to catch invertebrates. At the front of each column, Stefanie found, pioneer ants led the advance in classic army ant style, with each ant replacing and surpassing others while presumably laying a short exploratory trail. Columns stayed beneath the surface, but on occasion they would emerge aboveground. Usually the column raids retreated after ten or twenty minutes, but when the ants in the column contacted a termite mound or palm oil in one of Stefanie's buckets, the route they were taking was reinforced by the workers exploiting the food bonanza until it transformed into a branch of the trunk trail, remaining active for twenty-seven days or longer. The workers harvested few termites at a time from the mounds, but they also attacked several ant species for their brood, along with worms and a wide spectrum of insect larvae.

Although *laevigatus* is essentially a column raider, Stefanie also recorded three full-bore swarm raids on the surface, all of them small for any of the army ants that swarm-raid full time, the widest being only 3.5 meters at the front. Each swarm advanced from multiple entrances, moving at the marauder ant's slow crawl of 2 to 3 meters an hour for up to eight hours, in each case set off by a colony's excited response to the baits of palm oil that Stefanie had placed nearby.

The workers of the subterranean army ant *Dorylus laevigatus* in peninsular Malaysia typically move in a dense mass.

The subs that I found myself observing at Gashaka, like the *rubellus* colony Caspar and I were following at the same time, lived on the grounds of the field station. This complex consisted of a couple of simple houses for the scientists and a concrete supply building with a veranda complete with two lounge chairs and a dining room table. Nearby was a dirt-floored, thatch-roofed kitchen. Most people would not have taken notice of a muddy basin a meter wide in front of this structure, where the cook discarded his dishwater, which was then slowly absorbed into the ground, leaving scraps of food. But for a biologist in a dry place, any source of nutrients and moisture is worth inspecting. And sure enough, here were the subs, smaller and a brighter orange-red than the *rubellus* workers. I spent hours watching them, pulling up a log for a comfortable seat. Each time I shifted a spaghetti strand to get a better view, the cook, keeping a safe distance, gave me an odd look—and who could blame him?

Watching the subs emerge from the earth made me wonder what it's like down below for them. It must be one thing to excavate dirt incrementally, as most ants do while building a nest, and quite another to prowl through the soil for food day in and day out. Human beings rarely travel *through* the earth: cracks and crevices suited to our size and locomotion are scarce. But for creatures the size of an ant, the soil offers a number of travel options. Often, half of a soil matrix consists of pores, which result from the imperfect packing of different-sized particles, large to small, from sand to silt to clay. Larger organic matter can cement these particles into dirt clods, or peds, with crevices, or voids, between adjacent peds. All of these provide an infinite number of passageways for the subs, and if one of the pores or voids proves too cramped to pass through, they can expand it by raising and lowering their bodies.

Particularly beneficial to subterranean raiders are the cracks that form as soils dry, along with the macro pores that arise from the biological and physical impacts of dissolving minerals, tunneling worms, and decaying roots. Generally such conduits are more continuous than pores or voids, though the raiders still need to remove debris to keep them clear for transit. During such excavations, the soil particles are pushed from worker to worker under their bodies in a kind of bucket brigade.[13] This was first observed in a South American army ant by the nineteenth-century naturalist Henry Walter Bates:

> In digging the numerous mines to get at their prey, the little *Ecitons* seemed to be divided into parties, one set excavating, and another set carrying away the grains of earth. When the shafts became rather deep, the mining parties had to climb up the sides each time they wished to cast out a pellet of earth; but their work was lightened for them by comrades, who stationed themselves at the mouth of the shaft, and relieved them of their burdens, carrying the particles, with an appearance of foresight which quite staggered me, a sufficient distance from the edge of the hole to prevent them from rolling in again. All the work seemed thus to be performed by intelligent co-operation amongst the host of eager little creatures; but still there was not a rigid division of labour, for some of them, whose proceedings I watched, acted at one time as carriers of pellets, and at another as miners, and all shortly afterwards assumed the office of conveyors of the spoil.[14]

Sitting next to the washbasin, roasted by the midday heat, I pondered Bates's observations. Stefanie's Asian *laevigatus* nested in interconnected chambers underground, with one queen and more than three hundred thousand workers. I envisioned her ants racing out from these chambers along trunk trails that resemble subway tubes crammed with pedestrians, and wondered how they conducted their raids underground, an operation impossible to observe. In my imagination, a small worker pushed into a crack in the tube wall, to be followed by others. From this humble beginning, columns of workers soon spread through the soil matrix. Looking closely at one of the columns, I could see the ants shove aside or remove objects from their path, creating a route through the porous matter. As they reached a crack, they roused an earthworm and drove it ahead of them. (Although the surface-swarming driver ants earn their common name from this herding behavior, there is no reason to suppose the same thing doesn't occur in species raiding underground.) The worm crawled into a labyrinth of abandoned root channels. The ants cornered it in a cul-de-sac, much as the *rubellus* forced the carpenter ants up grass stalks until they could go no farther. Their exertions focused thus far on advancing, the small workers had cleared a corridor just wide enough for themselves. While some of them restrained the earthworm, others went back over the raid path, enlarging it to make room for the larger workers who would be recruited to assist in the kill.

Barring cave-ins, such a passage will be available for later use. Because of the *laevigatus* ants' constant raiding, commonly foraged areas will eventually fill with their tunnels "until they look like Swiss cheese," Stefanie told me. By clearing their abandoned passageways during occasional raids, the army ant subs provide attractive living space—a limited resource belowground—for their future prey.

Everyday ants, not just army ants, build tunnels when constructing their nests, and these can yield "homegrown" food as well. As a child I had an ant farm, which the manufacturer provided with *Formica* "wood ant" workers and pupae. The earth I gave them for digging had prey in it. One day, I noticed a worm nose through the soil and enter one of the ant galleries. The ants normally came to the surface to eat bits of my dinner dropped on the area at the top of the farm, but they were able to grab the worm and consume it without leaving their chambers. I suspect this kind of thing happens all the time, even with species that search for food on the surface.

How does raiding underground compare to raiding on the surface? The capacity of subterranean army ants to swarm out in three dimensions vastly increases their prospects for locating prey, though the constant need to excavate raid pathways may slow the search. Yet because army ants often take prey larger than they are, the subs are unlikely to lose what they track. Their quarry, especially their larger quarry, is unlikely to find an escape route that the ants can't follow.

Driver ants such as *rubellus* are conspicuous for raiding in swarms on the surface, but that doesn't stop them from searching underground, too. Raiders investigate every cranny, occasionally dredging up earthworms, and as Caspar and I had discovered, they will also go underground to demolish immense termite nests embedded deep beneath their mounds. Even the trail leading from the raid to the nest may be replaced by alternative, subterranean routes. The first time I sat down to watch a *rubellus* trail, the ants had all disappeared by the time I looked up again from my notes. Then I noticed that they were avoiding me by going into and out of holes in the ground on either side of me. My meddling had set the ants to patrolling, and apparently they'd probed around for a preexisting tunnel that now served as a substitute route to bypass me. What a sneaky superorganism!

The difference in habitat between "sub" army ants and surface-foraging driver ants such as *rubellus* must therefore be one of degree. After all, driver ants nest underground. And since they migrate frequently to new sites, they must be skilled at exploring belowground to evaluate possible homesteads.

To determine if a cavity is suitable for a new nest, honeybees fly back and forth within it to ascertain its volume and condition.[15] Ants may choose living spaces in a similar manner. We don't know exactly how they do it, but one plausible theory is that a scout assesses the area of a potential nest by laying a segment of pheromone trail and wandering on. The larger the space, the less often she will bump into her own trail again. The method driver ants use to gather intelligence on nest suitability is especially mysterious, given how voluminous their cavities must be to accommodate millions of ants linked together like curtains of chain mail, a great way to pack a huge population into a compact, but still substantial, volume. Sometimes they make errors, as Stefanie wrote me. "I once watched an *E. burchellii* emigration in Trinidad. They moved into a cavity under a rock which was obviously too small. In the middle of the emigration the ants appeared to 'realize' this and frantically started to search for new bivouac sites. They would start new bivouacs (i.e., attaching themselves and forming small balls of workers) almost everywhere—including under my boot!"

From studies of other ant species, we know that the more attractive a site, the more quickly and easily a scout will be able to recruit nestmates. She gauges the site's desirability from her estimate of its size and from such other factors as the width of its entrance, its height versus its breadth, and the amount of light entering the space—a detailed evaluation that would seem to require some rudimentary cognition. With multiple workers gathering these kinds of data from different places, after a while more ants are recruited to superior locations, with each new arrival confirming the site's virtues for herself, or following another worker to a different location if she's unimpressed. The final choice comes about by quorum sensing, a kind of decision by voter turnout. When the ants recognize that enough nestmates have already accumulated at one particular site, the migration to the chosen real estate begins.[16] Surprisingly, this voting behavior is not known to take place within organisms, or at least not healthy ones. It is thought that cancer cells may use it to regulate their interactions.[17]

THE SOLITARY ARMY ANT

There by the kitchen washbasin in Nigeria, what struck me about the subs was that they seemed to dine night and day, not on live prey, but on debris and human castoffs. Driver ants will devour some refuse, but only as a supplement to their regular diet, and only for a short time. At the field station, *rubellus* driver ants had consumed watermelon chunks in meals that lasted several hours. On another occasion, Caspar and I came upon a forest encampment; probably erected by poachers hunting bush meat, it was little more than sticks lashed with vines to form a primitive shelter. The poachers were off somewhere, but *rubellus* were massed thickly on their dishware and feasted on cooking-oil residue for a couple of hours.

By contrast, the subs at the research station fed at the washbasin for more than two weeks. They must have maintained stable underground trunk trails to reach the basin, much as did Stefanie's Malay subs, but they emerged from a number of exit holes in frequently changing columns to access oily food scraps and drowned insects in the mud, and they went crazy over the neon-red fruit of palms I gave them from the nearby forest. Uncurbed by midnight chill or the direct midday sun, they continued to come and forage as long as the basin was moist from the cook's discards.

This army ant species was common: Caspar collected them at baits of margarine that he buried at several sites in the forest. The swarm-raiding *rubellus,* in contrast, seldom showed up at the margarine baits. Perhaps they are poor competitors underground or, with their longer legs, are more circumscribed in their movements down below.

By the third day of observation the subs began to strike me as odd.[18] While driver ants will feed opportunistically on fruit or corpses, they are overwhelmingly predatory. The popularity of the washbasin (which, with its kitchen dregs, made for an interesting cafeteria experiment) among the subs suggested that they were indefatigable scavengers. Moreover, whenever they came upon live prey, even frail ones, they were ungainly in making the kill. The half-dead flies I slapped from my neck managed to crawl to freedom, with only one or two workers clinging ineffectually to them for a time. The subs were more successful when their numbers were greater: one healthy fly made the mistake of landing on a busy column of workers and was captured—an example of a trail serving by chance as a "trap." The subs were also different from other army ants in that they seldom slung food, whether a rice grain or an insect, under their bodies. Instead, groups of up to six carried food between them by grabbing it at different angles in the manner of other species of ants. Perhaps the standard army ant protocol didn't work well in the confined underground spaces where they normally foraged.

The real surprise was that the species didn't seem to forage in a mass, at least not with the quick and cohesive advance of the trailblazers of other army ant groups, such as Stefanie's tightly packed *laevigatus*. This had to be the reason the subs were ineffective at catching prey. Rather, the workers would slowly—over a couple of hours—spread apart from a hole in the ground or one of their trails until they had scattered across an expanse 30 centimeters or more from where they had started. As they searched during the ensuing hours, some went as far as 10 centimeters from their neighbors—a long way, relative to their size and leisurely movements. Mind-numbing to document, this activity didn't involve a raid front or any other kind of regular progression.

The workers had no trouble, however, coordinating food harvesting. A strong column of them developed quickly after I dropped a dead cricket in front of a lone ant. But this behavior looked more like what I would expect from run-of-the-mill ants than from army ants, which recruit to food items during the advance of a tight raid. The pattern can be difficult to pick out from all the action, but to see recruitment in a swarming species like *rubellus,* it's best to find a worker that has accidentally become separated from her neighbors, perhaps among the stragglers from a raid or on patrol near a trail after a disturbance. Drop an insect in front of this

The army ant "sub" species at the field station in Gashaka, Nigeria, would forage for kitchen refuse, such as this pebble slick with cooking oil that they are carrying.

ant, and she runs in loops, releasing pheromones that set nearby ants to looping in the same manner until they converge on the prey.[19] In a rush, a column of ants extends to the quarry, and they carve it apart.[20] By strewing dead insects in the washbasin, singly or in numbers, I found not only that the sub recruited help far less explosively than her army ant relations, but that she wasn't doing so during the advance of a raid.

If the Nigerian subs hadn't belonged to one of the army ant subfamilies, I wouldn't have given them a second glance. After all, many ants have foraging patterns in which many independent workers "diffuse" outward from trails or nest entrances, laying recruitment trails when they find new food.

That thought brought to mind *Pheidologeton affinis*, a similarly unprepossessing and partially subterranean ant that I'd studied a few years earlier at the Gombak Field Station in peninsular Malaysia.[21] I had hoped to detect raids in *affinis*, since it is cousin to the marauder ant, my *Pheidologeton diversus*. At Gombak, in a low building in a clearing in a forest, I lived on instant noodles cooked in a room with no electricity. Bats swooped by my kerosene lamp each night as I wrote my notes. My clothes, washed in a bucket out back, never dried completely, which gave everything I owned a mildewed odor. One week, the kitchen tap yielded only a trickle of foul-tasting water. A brief investigation netted me the corpse of a snake that had plugged the pipe to the rooftop cistern. Feeling queasy, I sat down on my front stoop and watched the *affinis* ants, which were emerging from holes in the soil near the steps and crept along in several meandering columns extending into a patch of lawn.

I knew these columns typically lasted a day or two, with upward of five hundred ants passing by each minute, though most often their routes were much more ephemeral and lightly used. Sometimes a column would elongate by a meter or two over several hours. This happened in bursts after foragers came upon seeds from a crabapple tree, their most common food; the ants would then extend their route to the area with seeds. The surfeit of arriving ants would scout several centimeters in the vicinity, which often led to finding additional seeds.[22] If that failed to happen, their activity would fade away until the column retreated.

Dependent as the workers were on finding food, their progress never transformed into anything that could be mistaken for a raid. Moreover, they acted cripplingly shy. Retreating from confrontation, they were usually hopeless with prey, avoiding the centimeter-long lawn-dwelling caterpillars even when I dropped one right on their column. If a caterpillar didn't go far, however, ants would follow it for a few centimeters and give it a tug. Sometimes a media or major would arrive to crush it in her jaws, and minors and small medias would carry it off.

Day after day, my patience sorely tested, I spent hours putting little flags in front of roving individuals and groups, wishing I could figure out what *affinis* was up to so I could study the army ant–like *Pheidologeton silenus* in the forest nearby.[23]

Could what I saw with Nigeria's army ant sub be some kind of spread-out, sluggish raiding? Perhaps the behavior of the sub and the Malaysian *affinis* looks different when the ants are constrained to narrow channels in their favored habitat, the soil; certainly packed together under these circumstances they must catch roving prey that blunder into their columns, as the ants had done in Nigeria with the hapless fly aboveground. But looks can be deceiving, and I knew

Related to the marauder ant, *Pheidologeton affinis* collects only weak prey and tiny seeds (see the minor worker at bottom).

that finding workers at high density wouldn't be enough to establish that active mass hunting occurs. Army ant raids in particular don't involve scouts and are not sustained by constant food discovery. I specifically wanted to determine if and when both the Nigerian subs and the *affinis* ants lay trails, not just to food but to new ground—exploratory trails. I pined for a way to make pheromones glow brightly—something like the spray used by *CSI* forensic detectives that makes blood stains visible—so that all these hidden details would be clear. Regardless, it seemed unlikely that the workers of either species were acting as a mass-foraging group.

The slow-motion foraging of *affinis* and the Nigerian sub suggests possibilities for how army ant–style raids originated in the army ants, and among the marauder ants as well. Here's what scientists call a thought experiment: Consider what incremental changes in behavior over time could transform a species that searches for food alone into a mass forager. It's not as if scouts can begin recruiting nestmates only *part* of the way to food. A plausible option might be to start with a trunk trail that guides ants to places where they spread out to look for food individually. A trunk trail reduces the time required to obtain assistance: a successful forager need only draw from the ants on the trail, not the ones back at the nest, for help. The faster the trail system grows, and the less far the workers depart from it in a search for food, the shorter the lag time between finding a meal and getting help to harvest it. Reduce this lag time sufficiently, and the ants begin to become proficient at overpowering quick and dangerous prey. If these changes continue, the trail-making process begins to take on the form of an army ant raid.[24]

The full sequence, from individual recruitment trails to the emergence of trunk trails and continuing with the development of a coherent raiding army, offers for me the same sense of wonder that other biologists have experienced in unraveling the evolution of the human eye, in all its complexity, by looking at the light receptors in the skins of simpler creatures. Such are the pleasures of studying life at the scale of a superorganism.

Every evening the primatologists would return to the field station sweaty and aching, mostly without having seen a chimpanzee. As I sat by the washbasin, I reflected that with ants, discoveries don't necessitate an arduous trek. Plop down anywhere, and wonders show up before your eyes. One good friend, Stefan Cover, found a new ant species floating, half drowned, in a swimming pool in Arizona. Bill Brown, a maverick of ant research, was renowned for collecting dozens of kinds of ants from a single rotten log.

When I was about to leave Gashaka for good, I poured some honey bait among the subs in the basin as a gesture of farewell. The cook, wearing a long, bright-colored shirt locally known as a buba, watched me from a shady corner of his kitchen hut. He looked relieved that I, who had sat in the heat-addling sun for days staring at offal, would be out of his way. As I departed, glancing over my shoulder at the station, I saw him dash out of the hut, get down on one knee, and stare into the basin before scratching his head and returning to the shade.

weaver ant

empress of the air

Oecophylla, tropical Asia, Africa, and Australia

9

canopy empires

Cross River National Park, with its rich yield of driver ants, was the highlight of my visit to Nigeria with Caspar Schöning. But the highlight of that highlight was a dramatic *Mission: Impossible*–style assault I saw occur between driver ants and weaver ants.

It was our only night inside the park. Caspar and I had pitched our tents at the edge of the rainforest, near the guard post. All was silent. I awoke early, my back aching from the hard ground, and got up to find a line of bright orange weaver ants hanging by their long legs from nearby tree foliage. Marching swiftly below them was a column of the darkly pigmented driver ant *Dorylus sjöstedti,* returning from a night of raiding.

Lowering herself carefully, suspended by her hind legs, one of the weaver ant workers reached into the mass of driver ants and, without drawing the attention of the other ants in the surging column, grabbed a worker by the waist. Then, with the assistance of other weaver ants on the same leaf, she lifted the driver ant from the column and pulled it into the tree, where the group tore it to pieces. A minute later another weaver ant did the same thing.

I called out to Caspar and reached for my camera: this was a moment worth recording. *Oecophylla longinoda* weaver ants dominate Africa's treetops, creating territorial empires sharply different from the army ant's roving bands on the ground. (The weavers have an equally rapacious cousin in tropical Asia, *Oecophylla smaragdina,* a species that tends toward green in the

Previous page: Weaver ant workers in peninsular Malaysia pulling together the foliage that will form their nest.

A weaver ant nest of bound leaves at Ta Prohm temple, Angkor Wat, in Cambodia, where the ants are a local delicacy.

easternmost extreme of its range in Australia and the Solomons.)[1] Their initiative in executing driver ants is well documented.[2] The move I saw seemed a kind of Russian roulette, though: surely the heavily carapaced driver ants would demolish the slim weaver ants the instant they noticed their depredations.

I saw a way to test this. Up in the tree above the driver ant column was a leafy sphere held together with white silk: a weaver ant nest. I clipped off the lightweight bundle and dropped it onto the driver ant column, expecting to see a full-bore retaliation resulting in the death of all the weaver ants. Instead, the normally unflappable *Dorylus* driver ants withdrew entirely, with every one of the workers doing an abrupt about-face.

So many tens of thousands were in retreat that the traffic artery swelled to the width of my clasped hands; it then grew arms as ancillary columns shot out from it at almost the full ant-running speed of 3 meters per minute. When I scratched the dust in front of one arm, the workers pooled there, confused. That told me the panicked ants were making their escape by backtracking along the base trail of that day's raid as well as all its abandoned subsidiary trails that the ants could still detect by scent. Thick ribbons of ants flowed on a dozen routes spread over an area the size of a one-bedroom apartment. The army ants were obeying their life's simple rule "follow a trail." In applying this rule now, there was a danger: when the path to their nest is cut off, backtracking on old pheromone trails can eventually lead army ants in a circle many meters across. The throngs starve to death as they follow each other round and round, stuck in the social equivalent of the endless loop of a defective computer program or unbalanced brain. This colony was fortunate: by chance one ribbon extended on a path that linked back to the driver ants' original trail on the other side of the weaver ant nest, enabling the driver ants to retreat to their own nest while giving the weaver ants a wide berth. Soon, half a million ants had drained from the system of emergency trails.

It's common for ants to react explosively to serious threats. Marauder ants carrying food flee when their trail is interrupted, and carpenter ants rush up grass blades during a driver ant raid. But the scale of this driver ant retreat made "overkill" an understatement, especially for an army ant, whose mode of life is to eat other ants.[3] Strangely, though, not a single worker on either side had come into conflict, let alone been killed. Clearly the withdrawal began with the detection not of the weaver ants themselves but of their nest. What about *Oecophylla* merited such an extreme response?

Each weaver ant nest can hold thousands of workers. A colony can occupy hundreds of these nests in one or two dozen full-grown trees spread over 1,600 square meters of ground, with a total population of half a million, the same size as many driver ant colonies. Weaver ants don't have stingers; like driver ants, they overwhelm their enemies with their superior numbers and sheer mandible power.

And quite a bite they have, too. Javanese children are warned that they will be tied to trees, to be overrun by biting weavers, if they misbehave. People around the world have learned from experience to back off when they notice the spastic, open-jawed motions of alert weaver ants on leaf and twig outposts in their nest trees. Though driver ants are more powerful, weaver ant bites are intensified by acidic secretions, which feel like lemon juice rubbed into abraded skin. (That said, some cultures have learned how to benefit from these ants' defensive chemistry. The best chicken dish I ever ate was served in Cambodia with a tangy weaver tapenade. And in Australia I joined an Aborigine suffering from nasal congestion in sniffing a cake of fresh-crushed weaver ants; the fumes had the sinus-clearing effect of the mentholated gel Vicks VapoRub.)

I retrieved the nest I had dropped on the trail. It tore like paper and revealed a mere fifty workers. From the driver ants' reaction to this little nest, I predict that if there ever were a battle between weaver and driver ants, the ordinarily unstoppable driver ants, normally the terrors of the jungle, would be unlikely to come out on top.[4]

One basic difference between army ants and weaver ants is this: army ants strip away the workers' autonomy to the point where few signals are needed to coordinate their troops, whereas weaver ants move with more freedom—even getting lost can be an asset if a stray ant blunders into an overlooked meal.[5] But this freedom comes at the cost of added logistics, because a worker must persuade others to come together to help her perform tasks as varied as making a leaf nest and killing driver ants. Perhaps the endless looping that can happen to an army ant column shows that their colonies can at times be too much like an organism for their own good—no such catastrophic failures are known to befall the weaver ant.

BUILDING AN AERIAL NEST

Weaver ant nests are most common in the outer, often uppermost, sunlit branches of trees. The site of energy influx and photosynthesis, this shell of greenery is where most biological action in forests takes place and thus where the majority of resources sought by the ants accumulate. There the ants bind adjacent living leaves into a kind of arboreal tent. Ranging from the size of a baseball to the size of a volleyball but weighing not much more than an inflated balloon, the nests look frail, but they shelter the ants from wind, rain, and rivals. Transpiration from their leaf walls creates a built-in HVAC system, providing relatively stable temperature and humidity.

To begin building a nest, a worker pulls at the edge of a leaf, and if she's successful in bending it, nearby ants join her. The workers may stand side by side while gripping the leaf margins, but if the leaves are too far apart, they climb on top of one another and, seizing each other by the waist, form leaf-to-leaf chains that are strong enough to drag the foliage together.[6] Within hours, the nearby leaves are drawn tight and aligned in a nest configuration.

The name *weaver ant* comes from the next step, which involves a kind of child labor. In many ant species, the larvae spin silk cocoons in which they transform into adults. But a weaver ant larva does not make a cocoon. Instead, it produces silk at a young age, when still small enough to be held and manipulated by an adult worker. After bearing the larva to the construction site, the worker locates a leaf edge through palpations of her antennae, then lowers the larva's head to it. The larva attaches a silk line to the edge, and the worker then shuttles it back and forth between the leaves, like a weaver working a loom, until the foliage is bound by woven sheets. As a finishing touch, the nest is detailed with tidy entries and internal walls and galleries.[7] The nests, I suspect, can last for years: when the leaves wither, workers bind fresh ones into the structure to replace them.

Weaver ants avoid the inconveniences endured by most ants, which, as central-place foragers, spend considerable time commuting from one central nest. This is evident in driver ant raids, where hundreds of thousands of ants regularly travel dozens of meters or more. Weaver ants minimize the amount of moving around they do by spacing leaf nests throughout their territory, erecting them wherever their workers are needed and foliage is available for construction. This also makes it easy for them to handle unforeseen events quickly: a worker seeking assistance need only communicate with the ant reserves in the nearest nest.[8]

Inside the tent, among the brood piles, are smaller workers with shorter limbs. In most polymorphic ants, the major workers are scarce and specialized, but with *Oecophylla* the opposite is true, with the majors doing the foraging and nest construction, serving as the workaday ants rather than "soldiers," in the sense of a specialized defensive caste. The minors are less numerous and tend the eggs and small larvae. The physical differences between minors and majors are more modest than in the marauder and driver ants, but the two are relatively distinct, with only occasional intermediates. Typically, the queen is in a nest toward the center of the territory near the top of a crown, though she moves from time to time.[9] Her eggs are distributed among the nests by her workers.

Because weaving a nest requires an assembly of workers and larvae, one wonders how weaver ant colonies get started. What does the first nest look like? Once, in the Australian outback, I

A weaver ant grasping a larva that is dispensing a silk thread to bind leaves for a nest in Queensland, Australia.

peeled apart two small leaves sewn together at chest level to find four queens and forty workers, the latter each the size of a small major worker in a mature colony, cohabiting in a space the size of a change purse. Making such a tiny nest need not be difficult. Before their first workers are old enough to do the job, the neophyte queens are likely to join forces to hold larvae and weave the nest together.[10]

MAPPING AND DEFENDING AN EMPIRE

Weaver ants are excellent nest builders, but they excel equally in transportation and communication. Their colonies employ a flexible network of routes between population centers and valued resources, much as human civilizations have done since ancient times.[11] The ants travel via trunk trails, marked by pheromones produced by the rectal gland, an organ unique to these ants. Trails are made more durable by droplets of worker excrement, which they deposit in a wide swath along the thoroughfares. The droplets harden like shellac when dry, creating a sort of "blacktop" that renders the path extremely persistent.[12] The workers can rediscover lost routes even after months of torrential rain or drought.

The workers use different forms of communication to convey different messages.[13] Before a trail is reinforced through time and usage to become a highway, it is likely to be ignored unless the ants producing it provide additional signals. As a worker deposits trail markings from her rectal gland, she employs a combination of signals—jerky moves, gentle touches, regurgitation—to identify whether her chemical path goes to some distant food, a leaf construction site, a newly finished nest, fresh territorial space, or an enemy confrontation. An ant returning from an unexplored part of the hinterlands, for example, lays a trail while shaking her body at passersby. Recruited ants, following the trail, enter the unoccupied area, then do what any self-respecting dog would do: they mark the space by relieving themselves. In this situation the fecal droplets are the ant version of urine and serve to identify the society, rather than the individual (as is the case with a dog), and claim the area in the colony's name. Given that the marks persist for months, their use along trails and at borders most likely serves a long-term strategic function: the colony that first marks a site has the edge in later conflicts.

When an ant contacts an enemy worker, she rushes back to familiar territory, encouraging others to follow her by conducting mock fights in which she stands tall and jabs at her fellow workers as if to bite them. Like a dog's bared teeth, this serves a warning function, in this case communicating with nestmates about the battle to come or already afoot. Her movements are one of several symbolic ant behaviors, actions normally associated with a practical activity, such as fighting, that have been ritualized into communication signals about that activity.[14]

Among weaver ants in Australia, combatants are usually recruited from "barrack" nests at the territory perimeter near the outer crowns of trees or at the base of their trunks. These guard posts look like normal nests but rarely hold brood, being occupied instead by expendable elderly ants. "It can be said," write Edward O. Wilson and Bert Hölldobler, "that while human societies send their young men to war, weaver-ant societies send their old ladies."[15]

Once the battle is joined, the ensuing conflict involves thousands of workers circling one another, high on their legs, with raised abdomen and open jaws. When a worker seizes one of the foe, she releases a substance from a second gland unique to weaver ants. This "short-range" pheromone secreted from the sternal gland diffuses and dissipates quickly, like an emergency flare, inciting nearby ants to rush to her aid. (She uses the gland to similar effect when seizing prey.) Mortal conflicts do not continue indefinitely; eventually, a "no-ant's land" emerges between colonies.[16] Defined by the fecal droplets, these perimeters can remain in place for a long time. Similar buffer zones have been common among both humans and aggressive species such as chimpanzees in hotly contested areas with large populations. Other social species don't defend a specific property. The honeypot ants of the American Southwest fight not so much over land as over shifting patches of food.[17] This lower-cost defense stratagem is not unlike the one adopted by small bands of human hunter-gatherers.

While most major human civilizations have undertaken full-time defense of large tracts of land, until modern times borderlands were often fluid, reflecting not wars but decisions about current utility. For example, in fallow months the Mongols would abandon the pastures on which they grazed livestock during fruitful seasons.[18] Our modern fixed territories are similar to the "absolute territories" of weaver ants, where space is defended all the time, and reflect the close packing of their populous societies. In a rainforest filled with ants, no colony can afford to relinquish its territory. "Free space is the enemy of true warfare," writes the military theorist Robert O'Connell, and the stranglehold weaver ants maintain over their crowded canopy dominions gives credence to this view.[19] The only territorial changes typical of a weaver ant society are shifts in battle lines with a neighbor or changing levels of worker activity on the forest floor—which weaver ants treat as a less essential, and often seasonal, part of their home range. They abandon it, much as the Mongols did their marginal grazing lands, when conditions are too wet or too dry. Their territoriality reflects this: when certain competing ant species move on the ground beneath weaver ant–occupied trees, for example, the weaver ants merely avoid them, but when the same ants dare to ascend the weavers' tree, the invasion elicits a massive fighting response.[20]

The versatility of weaver ant communication systems is without parallel among the ants, but that's no surprise given the nature of their operations. There are parallels here between the size of a superorganism and the size of an organism. To handle logistical issues within their bodies, big creatures often require more-elaborate organs, including brains and hormone-secreting endocrine glands, than do tiny ones, which sometimes get by with no neurons or hormones at all. Large body size can also mean a capacity for behavioral innovations, which are most common in vertebrates with big brains, such as chimpanzees, who use sticks to catch ants.[21] The massed workers in a large ant weaver society may be similarly adept at solving problems or achieving goals, including keeping track of territorial space, enemies, prey, and good sites for leaf nests.

For humans, it's thought that when communities and their institutions, from the government to the marketplace, evolve beyond a threshold size, the potential arises for more complex social mechanisms.[22] Weaver ants, with their intricate transport and communication systems, conform to this expectation of emergent complexity with greater social size.[23]

HUNTING, GATHERING, AND ANIMAL HUSBANDRY

Weaver ants forage everywhere, while incessantly protecting every leaf, twig, and branch. With ready access to other weaver ant foragers in the canopy and to a surfeit of ants in nearby nests, the workers are able to handle unpredictably scattered prey and enemy incursions. If a dense army ant raid is like a powerful net trawling a limited area in a narrow swath, the workers spread across a weaver ant territory act as an immense, and only slightly weaker, fixed net. The space that weaver ants occupy is so huge that the influx of prey into their territories is enormous. Indeed, a mature colony processes millions of victims each year.[24] Each worker typically stands at one spot, using something like the sit-and-wait strategy of the bumpy *Proatta* ants I watched in Singapore, but with greater effectiveness. They keep their jaws open and body erect, pivoting occasionally. Unlike the sightless driver ants, weavers are so visually acute that they can follow the flight paths of tiny fruit flies and snatch them out of the air before they alight.

Even when a weaver ant is capable of subduing an insect on her own, capture of all but the smallest prey is almost always a group enterprise.[25] Attracted in part by the struggle, in part by the sternal gland pheromone, weavers use the spread-eagle technique I saw them apply to driver

A scorpion being carried into the canopy by weaver ants at Kirirom National Park, Cambodia.

ants. This is how many belligerent ants make their kills, and this technique may have been in the repertoire of the ancestors of humans as well; two or more chimpanzees, for instance, will pin a male from a competing group while others bite, tear, and beat him with their fists.[26]

One day during my stay in Cambodia, I was overturning rocks in the forest when out from under one scrambled a 5-centimeter-long scorpion. Though fierce enough that I couldn't grasp it with my forceps, the scorpion was immediately seized and held in place by a single weaver ant—doubtless one-thousandth its weight—until backups arrived to pin its limbs. The expanded soft pads on weaver ant feet allow them to maintain their grip on their substrate, even when their bodies seem stretched to the breaking point.[27] According to one report, a lone weaver ant was able to support the full weight of a 7-gram baby bird that hung below her.[28]

Two dozen workers ganged together to haul the scorpion 6 meters up a tree in about the same number of minutes. When I extracted the leviathan from their grip, he was groggy but alive. Weaver ants don't gnaw off limbs to immobilize their quarry, as marauders do. Death is thought to come from being pulled with enough force to dislocate appendages, though few limbs break free. But I suspect the ant bites, with their acidic secretions that leave me feeling dizzy after a day in the field with these ants, deliver toxins as well.

Against powerful invertebrates like scorpions, short-range recruitment and drawing-and-quartering techniques are very effective, and such prey could represent a larger part of the food intake of weaver ants than it does of army ants. Though small arthropods are the foundation of their diet, weaver ants may even target vertebrates: one colony in Cameroon contained the remains of two lizards, a snake, a bat, and three birds. Though we don't know if the ants slew the animals or found their corpses, the workers can kill a bird by pulling it taut, their dozens of bodies linked in chains like those formed when they build nests. Perhaps to avoid theft from even larger animals, weaver ants conceal hefty prey under leaf litter while they subdue it and organize transport on the ground.[29] When the carcass has reached its final canopy destination—as Conservation International biologist David Emmett told me happened in Phnom Penh with a 15-centimeter-long dogfaced fruit bat—workers often construct a leaf encasement around food, similar to the soil barricades made by marauder ants. This permits both species to eat in privacy—in the case of the dogfaced bat, down to and including many of the bones.

No adult ant can swallow the prey she kills because solids can't pass through her impossibly narrow waist to reach her stomach. Instead ants drink the fluids oozing from its body, perhaps after some chewing. Many species use child labor—just as weavers do to build their nests—to transform prey into a form the workers and queen can ingest. Workers of Arizona's *Pheidole spadonia* "big-headed" ants place chunks of prey in a bowl-like depression on the bellies of their larvae; the larvae spit out digestive enzymes that dissolve the meat into something like a protein shake for the adults, which slurp up what they want and feed the rest back to the larvae.[30] The *Adetomyrma* "Dracula ant" of Madagascar takes a more gruesome approach. The workers immobilize prey such as centipedes with their virulent stings, then move their larvae to the food, which chew it up. After a larva has eaten, the workers pierce its thin skin to drink the hemolymph that leaks out, leaving their young literally scarred for life.[31] Larvae of Asian *Leptanilla* avoid such blood-letting: they are fitted with spigots from which the adult ants can obtain a drink.[32]

Weaver ant workers have other sources of liquid sustenance. Not nearly as meat reliant as army ants, they are partial vegetarians, which has advantages, as they are always walking around on an excellent source of nutrition. Why look farther than the plant underfoot? While workers can't digest the cellulose and have no taste for seeds and fruit, they eagerly lap up sap.

Although plant sap is low in nutrients, it offers energy and sufficient nourishment, provided the ant drinks enough of it. The most successful canopy ants are therefore built to tank up.[33] The workers of some species transport droplets between their jaws to other ants, which drink the sap as if from a bucket. Other species carry the liquid in a thin-walled internal sac called a crop, from which they take food into their stomachs as needed or regurgitate it for their sisters. This makes a convenient storage place, and it leaves the mandibles free for other work.[34] However it's done, fluid meals are transferred from ant to ant so that each receives a sampling of the nutrients passing through the colony at the time and is aware of the colony's general food needs.[35] The ant superorganism has, in effect, a social stomach—an approach that even Napoleon, whose army traveled on its stomach, never imagined.

Nutrients aren't everything, however; that stomach needs to hydrate, too, and drinking water can be in short supply. I discovered that the nest entrances of a certain south Indian ant resemble dead birds because the workers decorate them with feathers, which collect dew each morning—a sort of proto-tool for harvesting moisture.[36] Even in a wet rainforest, rain quickly drips out of trees, leaving the canopy parched much of the time. For weaver ants and other species, sap is a prime water source. Workers often drink from wounds on vegetation. Typically, they prefer the watery sap from leaves and twigs over the unpleasantly viscous fluid from the bark that prevents infection of the trunk and can harden and fossilize as amber. (Ants can get caught in this sap, which is bad for them but sometimes useful for us. In one deposit in Kenya, a population of thirty-million-year-old weavers, preserved in crystalline form, revealed that the division of labor, based on similar minor and major workers, was probably organized then much as it is today.)[37]

When it comes to nutrition, though, much more desirable than everyday plant sap are the enriched, sweet fluids secreted by glands on the plant surface. Even more than the ground-strewn bonanzas of fruit and meat frequented by the marauder ant, these nectaries tend to be both persistent and distributed in patches.

Traveling around tropical Asia, I noticed blemishes on the leaves of many dipterocarps, the local tree giants. Later I learned that these "green spots" exude nectar.[38] Unlike the nectar-producing organs inside flowers, which attract bees, butterflies, and other pollinators, so-called extrafloral nectaries like the green spots occur on leaves and stems of diverse plants and are tailored to ant cravings. *Oecophylla* are among the many ant species that guard such nectaries, and in so doing they protect the plants by making lunch out of any foliage-eating insects nearby. Some nectaries even develop close to flowers or fruit, thereby ensuring protection for the plant's reproductive parts during its breeding season.[39]

Weaver ants don't just seek out food sources such as fruit flies, bats, and green spots. They also, fascinatingly, farm them—in the form of such insects as mealybugs, scale insects, and plant hoppers, which many tropical ants care for the way ants tend aphids in the temperate zones.[40] Classified by biologists as Homoptera, these sap-sucking species excrete excess plant fluids in a condensed form known as honeydew—often directly into the mouths of ants.[41] Their excrement, which is more nutritious than nectar, is considered delectable even by some humans: in the Bible, it's called manna.

Weaver ants tend many kinds of Homoptera, as well as certain caterpillars that produce similar sweet secretions. These "cattle" range from species that do fine without the ants to a few that are found only where weaver ants thrive. The ants are as protective of their livestock as any cowboy is of his herd, keeping them from harm and even moving them to fresh pastures when the plant sap runs dry—preferably to a site with young foliage, which is more easily penetrated by their mosquito-like mouthparts. Often, the ants construct a "pavilion" nest, a kind of holding pen, over the insects, as they do also with nectaries, before settling in to exploit them over the long haul.[42] Such leaf enclosures protect the Homoptera from predators, parasites, and weather and may even be essential to the ants' claim on their herd, since workers are less active outside their nests at night, when other species could steal from resources not under heavy guard.

In short, their continuous social dialogue enables weaver ant workers to exploit resources no matter how they are distributed. Throughout their absolute territories, the workers scatter widely to hunt prey on foliage or cluster densely to feed at nectaries or the sites of their homopteran herds. Colonies grow big and strong with a balanced meat and vegetable diet, so they are most vigorous where all of these sources of food are bountiful, such as in the young and succulent vegetation along forest edges.[43]

Among ants such as the weaver, the flow of food and other goods is likely to be regulated by what's available and what's needed, a supply-and-demand market strategy.[44] This is best observed in the workers of the red imported fire ant, which monitor the nutritional needs of the other adult ants and of the larvae and change their actions as necessary.[45] When scouts and their recruits converge on the nest laden with a variety of foods, they hawk their merchandise by regurgitating samples into the mouths of "buyers" in the nest chambers, who in turn roam through the nest to distribute the meals to the larvae and queen. If the buyers find their "customers" have become sated on meat, they peruse the marketplace for other commodities, until they find, perhaps, a seller offering nectar. When the market becomes glutted and the sellers can no longer peddle their wares, both buyers and sellers wander off to engage in other jobs, or take a nap. This is an excellent way to run a superorganism: if only our digestive systems served us this well, rejecting any excess fats arriving in our meals![46]

The diet choices of weaver ants affect their anatomy. Compared to the meat-sustained driver ants, the largely vegetarian adult *Oecophylla* are thin-skinned, with no special armaments. This may be because nectar and honeydew derived from plants are poor in the nitrogen needed to build proteins. In the economics of weaver ant existence, the carbohydrates in these readily available liquids are the fuel that the adults burn on their labor-intensive hunts for prey, while the prey themselves provide the bulk of the protein the larvae need to complete development and keep the superorganism growing. Even with their feeble armor, few adult weaver ants are killed in encounters with these prey; they are so fleet-footed that even army ants succumb to them.

Both the weaver ants and the army ants are predatory titans, but the two approach their lives differently. The army ants' narrowly concentrated raids comb wide areas to gain enough of the protein they require, especially from large and aggressive prey.[47] But weaver ants remain settled in one area and minimize travel within it by harvesting a steady and more varied local supply of plant foods and honeydew in addition to small and large prey. These differing tactics have allowed both ants to flourish with colony populations reaching into the hundreds of thousands.

10 fortified forests

Oecophylla weaver ants swarm through the tropical rainforests of Africa, Asia, and Australia, but because life in the trees has so many advantages, the New World has its own hyperaggressive canopy-dwelling ants.

One morning in late spring 1990, I found myself slung by ropes a dozen meters above the jungle floor rummaging for beetles in clumps of litter on tree branches. I was in Peru, on assignment for *National Geographic* magazine, to document the rainforest canopy, one of my research specialties.[1] Since finishing my thesis four years earlier, I had served as the curator of Harvard's ant collection (where I'd first seen the marauder ant and decided to make it my quest) at the Museum of Comparative Zoology under my former adviser, Ed Wilson. But now I had gone freelance, planning to support my life and research with writing and photography.

That morning at sunrise I had been on the ground with Terry Erwin, a Smithsonian beetle expert who inventories the species diversity of canopy insects. To get close to our targets I had sent a fishing line into the tree with a slingshot and used it to pull a climbing rope over a branch; I then got into my climbing harness, clipped two ascenders to the harness and the rope, and wriggled skyward.

But as I rose, my support rope shifted; I abruptly fell several inches and began to spin in space. Plant bits shaken loose from the branch above whirled into my eyes and blinded me. My hands were full of cameras and entomology gear. To stabilize myself, I threw my legs around a branch high to one side.

Big mistake! Swinging through the air, my foot smashed a mass of canopy-rooted plants, or epiphytes, that concealed a well-defended ants' nest. In an instant, workers covered my legs and then dropped like dive bombers onto the rest of my body. As they gashed my skin with their mandibles and sprayed formic acid into my wounds, I recognized not only the species, *Camponotus femoratus,* but also the fact that I'd found my first "ant garden"—albeit the hard way.

Regaining my balance while slapping at this vicious species of carpenter ant, I noticed the presence of a second ant on my skin—the smaller *Crematogaster levior,* a shy species of acrobat ant that does not bite. The ant garden is a result of their collaboration and represents an infrequent instance of harmony between ant species. Nestled in this mass of epiphytes, a confederation of these two ants had constructed a quarter-meter-wide treetop house of carton, papery sheets they produced by masticating plant matter and soil. The workers then collected seeds and embedded them in the carton. There the seeds grew into cacti, bromeliads, figs, orchids, philodendrons, and anthuriums, creating a bounteous garden.

The plants and ants depend on one another. The plant roots strengthen the carton, keeping it from disintegrating in rain and giving the ants a stable home.[2] The ants, in turn, seem to be necessary for the plants' survival, since these particular species of flora never occur on their own.[3] (Though we can't say yet if the seeds die if the ants don't find them, or if the ants are so thorough at snapping them up that these plants have no opportunity to germinate elsewhere.) In any case, the ants were clearly protecting both nest and garden with zeal.

In this striking example of mutualism, the *Camponotus* and *Crematogaster* jointly created the nest and protected the epiphytes. They shared trails, helped each other find prey (though *Camponotus* can be a bit of a thief), and tended the same sap-sucking insect "cattle" as an additional source of fuel. The acrobat ants then drank the honeydew excreted by the smaller Homoptera, or aphid relatives, and reared them to a size suitable for milking by the carpenter ants, which played the more important role in finding and planting the seeds that developed into fresh garden plants.[4]

Nauseated by an overdose of ant toxins in my bloodstream, I pushed myself away from the ant garden to another tree trunk. The garden was truly elegant, I could see, though for the moment the ants on it seethed. As I recovered my balance, I pondered what it was about life in the canopy that fostered both mutualism and belligerence.

BIOLOGICAL SUCCESS

Success in nature is often described in terms of the number of species in a group. By this measure, the ant-garden ants I had stumbled upon belong to two of the three most successful genera of ants (the third is *Pheidole,* or big-headed ant), with hundreds of species each. But success isn't always associated with a proliferation of species; the number of individuals and their effects on nature can matter more. Indeed, ants are the prime example.[5]

The tropical forest canopy, with its multiple levels of foliage and branches, can have ten times more habitable real estate than the ground, a much higher ratio than in the temperate zones. With all that elbow room, it may be no surprise that a study in the Amazon basin found eighty-two ant species in a single tree, almost twice the number of ants in the entire British Isles.[6] Though that sounds like a lot, compared to other insects in tropical canopies, ants have an almost negligible diversity. As Terry Erwin points out, a single tree in Peru can contain thousands of species of beetles alone. Still, arboreal ants more than make up for their relatively few species with an astonishing bounty of individuals. Workers, in particular, often make up 20 to 40 percent of the organisms in trees, microbes aside. Measured by weight rather than numbers, all of the ant species in combination account for 10 to 50 percent of the mass of arthropods living in tropical trees. Ants also weigh more than all the vertebrates in the same area, from frogs and lizards to parrots, monkeys, and leopards. With so many ants and so little else, canopy ants sustain their populations through heavy reliance on plant matter, as we saw for weaver ants.[7] The same is likely to be true for tropical ants living on and in the ground, where they also roam in overabundance.[8]

Pervading the tropics of three continents with just two species, weaver ants are a particularly good model of success without diversity. In this regard, weavers and humans have a similar history. Our ancestors adapted better than Neanderthals and earlier branches of our evolutionary tree, which stopped producing offshoots in the wake of *Homo sapiens'* aggressive dominance of the Earth—with six billion members now and counting. Weaver ants seem to have followed a similar course, controlling the environment to such a degree that they are often able to push out or mow down the competition. Along with South America's *Camponotus femoratus,* they are among the most militant ants on Earth, capable of eliminating all adversaries except the most fierce. This they accomplish by being numerically and behaviorally, and therefore ecologically, dominant, using their force of numbers and tactical skills to suppress or conquer territorial competitors and thereby control the environment.

Is numerousness essential to weaver ants' success in fighting, or is it their belligerence that allows them to expand their population? The two conditions seem to go hand in hand, making it

Weaver ants tearing apart a driver ant captured in Ghana.

difficult to distinguish cause and effect. Although marauder and army ants at times use strength of numbers and battle skills to overpower the competition, the goal of most violence in these mass-foraging ants is the practical one of securing food supplies. In contrast, weaver ant societies, much like Peru's ant-garden ants, fight other colonies to control the surfaces on which they live.

This difference in goals has parallels in human groups. Most early hunter-gatherers moved often in pursuit of foods that offered immediate large payoffs, such as big game. After the Pleistocene, human population pressure caused these slowly replenishing foodstuffs to become depleted and eventually forced people to settle down in areas chosen for the availability of fast-breeding foods such as grain and small game, which required more time and labor but could be harvested sustainably. This shift in turn necessitated vigorous defense of these territories against would-be usurpers.[9] In the insect version of this "broad-spectrum revolution" (as anthropologists refer to this shift in human diet), each densely packed ant garden or weaver settlement, with its foraging centered on a broad range of such quickly renewable resources as insect prey and nectaries, has come to approximate a warfare state. Among animals, all-out war against their fellows occurs only among the largest societies of humans and ants.[10]

TAILOR-MADE ANT ACCOMMODATIONS

Where do so many ants find homes in the trees? Many nest in hollowed twigs or galleries in bark, or the litter that accumulates among the roots of orchids and between the leaves of tree ferns.[11] Cavities capacious enough to hold large colonies are rare, though, and the success of such colonies often depends on constructing nests, such as the weaver ant's tents, with materials they find in the canopy or produce themselves. There are other ants that use silk, usually combined with leaves, to build their nests—mostly larval silk, though in one African species the adults synthesize silk of their own from a gland near their mouths, and an Australian ant steals its silk from spider webs.[12] The ant-garden ants are among many species that use carton.

There are ants that nest terrestrially and forage in the trees, giving them a toehold in both environments. This is more common than the reverse, a canopy-nesting species that primarily forages on the ground. It's a classic suburban commuter's compromise between the best housing and the best income: the forest floor provides more roomy nesting opportunities, and the food and other resources found in the canopy make the transit worthwhile.

While the canopy species mentioned thus far live on any plant that offers a suitable nesting cranny, certain trees, epiphytes, and vines provide custom-fitted ant accommodations. Some of these so-called ant plants cater to a specific ant, providing food and board suited to no one else.[13] Why? These residents are proficient at eating herbivores, and they kill anything that sets foot on their host. As a boy, I read how *Cecropia* trees house *Azteca* ants in spacious compartments in their trunk joints, feeding them pale, glistening "food bodies" more nutritionally balanced than nectar or honeydew, which exude from the base of each palmate leaf. On my first trip to the tropics as a college student I ran into one such tree—literally—and learned that *Azteca* don't just pick on creatures their own size.

Weaver ants, though similarly aggressive, do not occupy specialized ant plants. They can live in any tree by creating their own nests and finding their own food (if the plant has nectaries, so much the better). Childhood experience guides the choice of plant homestead: workers and queens prefer to nest and forage on the tree species they grew up on as larvae and young adults, and, like humans, they become more set in their ways as they get older. Still, the ants show a special affinity for mango and citrus, a fact that encouraged the Chinese to use *Oecophylla* to control citrus pests as far back as A.D. 304.[14] In parts of Africa and Asia, their use in biocontrol continues—though pity the laborers who climb those trees to pick the fruit!

Is the weaver ant's presence good for the trees? To answer such a question, ecologists conduct a cost/benefit analysis. In some ways the benefits clearly outweigh the costs: weaver ants cull leaf-munching insects, and tree foliage lasts longer where the ants reside. One type of beetle, though fond of foliage, flies away from a tree the moment it senses weaver ant pheromones.[15] Another benefit to trees might come from weaver ant hygiene, or rather the lack of it. Nutrition is a problem in tropical forests, where soils are thin—but a tree can absorb nutrients through its foliage as well as its roots.[16] The workers in some ant-plant mutualisms use leaves and stems as toilets and trash chambers, thereby feeding their plant. The fecal droplets that weaver ants scatter over leaves to mark their territorial claims might serve as fertilizer, too.

Known as the "dinosaur ant" for her primitive appearance, this *Nothomyrmecia macrops* worker from Poochera, South Australia, is guarding scale insects that have exuded so much honeydew that they appear to be covered in sugar.

On the negative side, the leaves that the weaver ants incorporate into their nests may be lost prematurely to wear and tear, as the ants pull them from their ideal alignment for photosynthesizing.[17] However, since only a tiny percentage of vegetation is tied up in nests, these costs to the tree should be low.

A bigger deficit item may be the ants' nurturing of their homopteran cattle. In the United States, you can locate *Formica propinqua* ants by the dead cottonwood trees around their nests, which have been sucked dry by the aphids the ants raise.[18] In addition, some sap-sucking insects carry infections, making them the plant version of the malaria mosquito. In most situations, though, the cost of Homoptera to trees is not so severe. *Azteca,* for example, raise sap-sucking insects in moderation on their *Cecropia* hosts to no evident ill effect. In fact, some trees may have evolved to be tasty to such insects *because* they attract protective ants, as an alternative to producing nectaries.[19]

Some of the weaver ant cattle reside not on the trees themselves, however, but on vines in their crowns, which have wide vessels ideal for feeding by Homoptera. Heavy infestations of "plant lice" raised by *Oecophylla* may in this case inhibit vines from shading a tree or weighing it down and breaking its crown, thus working to the tree's advantage.[20]

Overall, weavers are thought to benefit most trees. Could the relation of the ants to choice tree species such as mango and citrus be a rudimentary mutualism, as with *Azteca* ants and their *Cecropia* trees, though less precise and obligatory? Researchers have noted of mango and citrus trees that the "odors of the plants may have evolved to attract ants for protection."[21] And the tree wouldn't be the only one to profit from this arrangement: anything that increases its vigor should benefit the ant colony it houses, by yielding more durable homes and sweet and savory foods—honeydew and prey.[22]

SPECTACULAR DEFENSES

A few years after my trip to Peru, Dinah Davidson of the University of Utah offered to show me another dominant ant species and impressive adversary of the weaver ant in Brunei, a small, oil-rich country in northern Borneo. After touching down to an evening view of the Sultan Omar Ali Saifuddin Mosque, I arose the next morning for a forest river journey on a canopied boat. Kuala Belalong Field Studies Center was just as I remembered it, solidly built at the base of thickly wooded hills. Dinah, a compact woman with hair cut short for the field, took me up a steep path while pointing out weaver ant territories, which alternated with trees occupied by any of sixteen species of *Camponotus* carpenter ants belonging to what's known as the *cylindricus* group.

The *cylindricus* ants have dramatic methods of defense. The major worker's head, for example, is flattened into a disc, enabling her to serve as a living door to nests in hollow branches. She allows her nestmates inside only after they identify themselves by tapping the blockading disc with their antennae (a technique also seen in other ants). Dinah took me to the territory of one of the more unusual *cylindricus* species and told me to grab a minor worker that was climbing the trunk. I did, and the ant's leg fell away in my hand, in much the way that a lizard will lose its tail.

Still other *cylindricus* species exhibit the most extreme behavior of all, employing the "suicide bomber" response to its enemies that I had come to Brunei to see. Wishing me luck, Dinah left me at the base of a tree occupied by one of these colonies and departed. I pulled out my camera, adjusted my flashes, dripped some honey next to the tree from a vial in my pocket, and waited. After an hour, weaver ants along with another species of carpenter ant located the bait and started arriving at the *cylindricus*-occupied tree. One of them started up the trunk, but then came down again. That one would live another day. Another climbed a bit higher and attempted to walk by a *cylindricus* minor worker. Just as I clicked the shutter there was a splash of yellow, and both ants were immobilized in a sticky, grotesque tableau.

That picture made my journey halfway around the world worthwhile. Photography is, for me, a tool for storytelling, and this ant's story left my heart pounding. Approached by an adversary, the *cylindricus* had blown herself up, her body rupturing with a muscular convulsion that spewed forth a toxic, lemon-colored glue that pinned her foe to the ground, killing both of them straightaway.[23]

A Brazilian ant I've yet to see, *Forelius pusillus*, has an equally fatal approach to protecting the nest. Up to eight sacrificial individuals stay outside at night to seal the entrance with sand, kicking the final grains in place until no trace of the hole is visible. Walled off from their sisters, by

A *Camponotus cylindricus*–group "exploding ant" has ruptured her body to spew a sticky yellow glue, which has killed both her and the larger worker of another species of carpenter ant in Brunei, Borneo.

dawn almost all are dead, for reasons unknown—perhaps the squad consists of the old or sick. The ants in the nest then clear the passage to begin the day's foraging. That night, more victims seal the door.[24] No one can say what prompts this preemptive defense, though dangerous army ants would be one safe bet.

In northern Borneo, *cylindricus* often jointly control their canopy territories with certain *Polyrhachis,* which have their own self-destructive defense. The first time I saw a gleaming gold specimen of one big, attractive species of this genus, I couldn't resist touching her—and immediately had the worker embedded in my finger and unable to remove herself, thanks to the fishhook-shaped spines on her back. Birds and lizards must learn to avoid these pincushions.

DOMINANCE AND SUBORDINATION

If these colonies are viewed as organisms, a worker's death is of no more consequence than a man cutting his finger. The larger the colony, the less consequential the casualty. Extremist defenses, then, are a manifestation of a large labor force. Such extremism in handling risk is an example of how death without reproduction can be of service to queen and colony, and a reminder that anything humans concoct—even suicide missions and terrorism—probably has a parallel in nature.

Just as trained armies and impersonalized warfare came into being among people as populations exploded with the development of city-states, inexorable, large-scale offensive and defensive conflicts between rival ant nests usually involve the numerically dominant species, with their huge colonies. One likely reason is that the necessary communications are best orchestrated in large societies, whether they involve written languages in humans or pheromones in ants. Another reason is simple efficiency. Larger human settlements have higher per capita productivity, with fewer resources required to feed and house each individual.[25] This pattern, if similar among ants—which remains controversial[26]—may enable large societies to more easily accrue the spare time, energy, and resources that can be invested in creative endeavors (by people) and armies (by both ants and people). As a result, not only are the ants of large societies more expendable individually, but the group as a whole may also be able to take more large-scale risks, given that losing 10 percent of an army will be more devastating for a society of ten than for one of a million.[27]

Yet another advantage of community size is that populous societies control large spaces, and large spaces have relatively small perimeters (a large circle, for example, has a smaller circumference relative to its area than a small one). Thus, the bigger a colony, the smaller the proportion of its population that needs to be employed in border surveillance, and the more troops it has free to commit to offensive battles.

Ecologically dominant species are usually too competitive to coexist. On occasion there is a détente between two of them based on different nutrient and housing needs, as with the ant-garden ants, which eat different-sized prey, or the *Polyrhachis* species that share territories with the exploding ants. *Polyrhachis* nest on the ground and eat honeydew and insects, while *cylindricus* nest in trees and specialize in licking the microbial film growing on foliage.[28] Even with a large labor force, it pays to be selective in targeting competitors. The *Polyrhachis* and exploding ants have largely put aside their differences in a coalition against the weaver ant. Weavers in turn show enemy specification, picking out intruders of other dominant species and excluding them from their territories as if they were competing weaver ant colonies. Weavers don't win all their battles, and sometimes they choose to retreat. The exploding ants can keep them at bay, while the workers of the Asian *Technomyrmex albipes,* a species known for noxious chemical warfare, may force them to aggregate into fist-sized balls of thousands of brood and workers that rain out of the trees to the safety of the ground.[29]

Yet other, less quarrelsome canopy dwellers manage to survive in the territories of dominant ants by being overlooked or ignored. They creep out of view, run from trouble, or blend with the environment. Commonly referred to as nondominants, these ants may even compete for the same resources as the dominant or "extirpator" species. They may be opportunists, to use the entomological term, subject to attack by the dominant species but able to harvest food before the bullies drive them off. Others are insinuators, who rob meals from under the dominants' noses by virtue of speed, stealth, or tiny size, in some cases even parasitically sneaking along the dominant ants' recruitment trails. In other cases, the insinuator is active at times of day when the dominant ant is not, or it may simply forage in a nonthreatening manner, for food the dominant species does not want or at sites that it cannot reach, such as the narrow furrows in bark that the speck-sized workers of the ant *Carebara* explore under the feet of weaver ants.

Most nondominant ants have societies of a few thousand or less, and often much less, as colonies of mere dozens thrive in any ready-made spaces they can find. The relative timidity of these small colonies parallels the behavior of small bands of human hunter-gatherers, which similarly lack basic infrastructure, with no entrenched dwelling places, territorial land, elaborate trail constructions, or stockpiled resources to protect. Full-bore warfare is unnecessarily risky for groups of this size: for human hunting bands, for example, most conflicts are small in scale and arise, as they do for many animals, over issues of power or reproduction.[30] Given their mobility and lack of rivalry over land and resources, small groups are otherwise more likely to choose flight over fight—making nondominant ant species easy targets for domination.

Wherever weaver ants occur, they rule over the best sources of nectar and honeydew—those with the most amino acids and sugars. The subordinates remove whatever's left over, at times sharing the inferior spoils among themselves equally or accessing them in a pecking order.[31] Every once in a while one of the more tenacious of these subservient species has its moment of glory, taking over a swatch of the canopy when dominant species are absent.

The weavers' control of the canopy is so extreme that in times of food scarcity, they will raid the nondominant species nesting within their territories to eat their larvae, in a sense using the contents of these colonies as if they were reserves of food.[32] Subordinates could move away, but they may prefer to live with the enemy they know. They might also rely on weaver ants as a homeland security system, scaring off their usual competitors.

Canopy conditions encourage aggressive dominance by species like the weaver ant. For all their complexity, the living spaces in forest canopies are easier to control than areas of equivalent size on flat ground. Hill forts made it easier for people to fight approaching armies; ants similarly take advantage of the height and geometry of plants, which results in chokepoints that limit access to a territory, simplifying its defense. The borderlands that canopy ants guard are restricted; they consist mainly of vines, tree limbs, and the boundaries between tree trunks and the earth. If it were possible to squash flat all the trees occupied by a weaver ant colony, the surface area would be a pancake with a border hundreds of meters long, a frontier as imposing to patrol as the Mexican border is for the U.S. government. To control the equivalent area in the trees against even large armed forces, the ants need only employ a few expert fighters at a few bridgeheads.[33] Their payoff in terms of arboreal land per military expenditure is thus vastly increased.

In 1993, I spent an afternoon at Guanacaste National Park, Costa Rica, watching the thin, wasp-like workers of *Pseudomyrmex* ants that resided in the hollow thorns of their ant plant, an acacia tree. The action was ongoing as ants killed or drove off caterpillars and other insects. But what especially irritated the workers was a vine touching one of the acacia twigs. They examined its looping tendril, then spent an hour pulling and shredding its tissues, at which point the vine fell away. Why so much attention to what for this predatory species must be inedible vegetable matter? The answer is simple: a vine can become an access point for invasion by neighboring colonies.

Jorge, a Matsigenka Indian, standing in a spirit garden in Manu National Park, Peru. The undergrowth has been cleared away by *Myrmelachista schumanni* ants, whose aggressive attacks have warped the bases of the surviving trees.

This specialized form of clearing, which was so thorough that the ground around the base of their tree had been denuded of all other plants, served the acacia as well, if only coincidentally.[34]

It's common for ants to clear the area around their ant plants, and those that do it the best are found in Peru's forest glades, called spirit gardens by the local Matsigenka Indians, who believe that spirits clear the underbrush. I visited a half-acre spirit garden with an Indian named Jorge and Doug Yu, at the time a graduate student at Harvard studying the coevolution of ants and plants. Sweat bees, wasps, and killer bees, hungry for the salt in our perspiration, landed in such droves that we had to shout over the buzz and hold our arms out stiffly like Frankenstein's monster to keep from being stung. The trunks of the few big trees were swollen like potbellies with malformed bark, which the Matsigenka ascribe to fires set by the spirits. Slicing a trunk with his knife, Jorge showed us the true cause of the deformity: tunnels eaten out by the minute *Myrmelachista schumanni* ant, which were killing the trees. Doug pointed out the ant plants nearby, small trees, easy to miss in the clearing, that shelter the brood and queens of this ant in their stems. The trees were doing well under the unobstructed, illuminated conditions provided

Opposite: A *Pseudomyrmex* worker in Guanacaste, Costa Rica, tearing the tendril of a vine that has touched her nesting acacia tree. If it were permitted to grow, the vine might overwhelm the tree.

A cross-sectional view of rainforest along the Amazon River near Iquitos, Peru, showing the canopy layers, including a tall emergent tree at left, as well as vines and epiphytes. Different ants confine themselves to particular strata.

damp, and at night the temperature drops sharply. Between the extremes at top and bottom, an elevation change of a few meters up or down a tree can be equivalent to the environmental transformation we experience while traveling kilometers over the ground, say from inland mountain slopes to the seaside.

Even ecosystems with less height than a forest can offer similar variety. The overarching grasses and wildflowers in a meadow, for example, form an upper canopy. A suburban lawn can have layers of vegetation as well defined as those of a forest 90 meters high—from ground-huggers like the procumbent pearlwort and fairy flax, to midlevel scramblers such as sweet vernal, to the upright stalks of the lawn canopy giants: white clover and any of a variety of grasses.[3] For a human on the ground, forest interiors provide a mild climate relative to the sun-roasted air above the trees, and a lawn's interior offers similar conditions—for an ant, anyway.

Ants that stay within one canopy stratum, perhaps nesting and foraging their entire lives on the same tree branch, need not climb any more than ants on terra firma. However, nearly all ant species are natural climbers. *Camponotus gigas* carpenter ants, the biggest of all ant workers at a length of 2.8 centimeters, walk all over trees in the same Malaysian forests that weaver ants occupy. I remember huffing and puffing up one immense tree, assisted by ropes and gear, and looking over at the trunk to see a chunky *Camponotus gigas* major worker race ahead of me to the crown from a barrack nest at the base of the tree. Why was her ascent so effortless? When walking on flat ground, an ant burns a lot of energy relative to her mass because she has to move her little legs quickly to get anywhere; for her the added cost of a climb, compared to moving horizontally, is almost nil.[4]

For many people, going up a small tree can be a pleasure—at least for those of us who've kept a little bit of our kid selves inside. But in the tropics, even small trees harbor risks. At the Tiputini Research Station in Ecuador, I free-climbed a slim tree near the residence cabins during a much-needed break from rainy hours photographing the falling behavior of turtle ants—an effort that required scaring one worker after another off a little ledge and entailed over 6,300 clicks of my camera. I had gone up the tree because I surmised there was a turtle ant nest in it, which would provide me with a fresh supply of workers. What I found instead were workers of the giant *Paraponera clavata*, a species called the bullet ant for its fearsome sting. Looking at the ants lumber past my fingers on a branch next to my face, I recalled a story about one of Terry Erwin's assistants who had overheard Terry explaining that if stung by this ant, a person should cauterize the wound with a cigarette; stung by three, the assistant had been in such pain that he'd been unable to stop burning holes in his leg. Wondering about my own pain threshold, I glanced down to see more of the rugged black workers mounting the trunk below me. Now I recalled that bullet ants nest at tree bases but forage in the canopy, where they deposit trails that guide nestmates to meals. In some cases, they arrive by the hundreds.

As a tree-climbing biologist, my anxiety that the branch bearing my weight could break is matched only by my fear of confronting bullet ants while high on a rope and unable to flee. Luckily, on that day I was neither on a rope nor high. Letting go of the trunk, I thrust out with my feet and fell the 2 meters to the ground, landing safely away from the bullet ant nest.

For humans, height is significant because we fear a fall. But falling is not the same for every being, as biologist John Haldane describes: "You can drop a mouse down a thousand-yard mine shaft and, on arriving at the bottom, it gets a slight shock and walks away. A rat is killed, a man is broken, and a horse splashes."[5] For animals larger than a mouse, there is a height above which a fall can cause harm; call it the critical-injury height. People come away from the majority of short falls with no more than a bruise, while scampering squirrels plunge meters without injury. But an ant can, in theory, fall forever without being bruised.

The distance covered, however, can have serious implications for ants. The farther a worker drops, the longer she takes to return home, and the more likely it is that she will get lost or die in enemy hands. Perhaps for this reason canopy ants are particularly good at hanging on tight. I made the acquaintance of one such species, *Daceton armigerum,* while exploring the Orinoco basin of Venezuela. *Daceton* are unmistakable: big and spiny, with lobes at the back of their head to accommodate muscles that power snap-action mandibles. These ants really knew their place within the forest strata: this colony was nesting 6 meters up a small tree, but the few that descended the trunk avoided the ground altogether, retreating the instant they touched the earth. These same ants had no aversion to walking on a humus-covered branch laden with epiphytes

A turtle ant, *Cephalotes atratus*, gliding backward, toward the left in this image.

I placed before them. I hadn't a clue how the *Daceton* workers recognized the ground, or why they found it so alarming, but for them the forest floor seemed like truly foreign soil. Given this distaste for touching down, it came as no surprise to us that after removing the tree and bisecting its hollow trunk at the village hotel, over the protestations of our maid, we found that each of the 2,342 workers could grip anything—nest, bathroom tile, ceiling—as if her feet were glued in place. I realized then how important clingy feet are to foraging and living in the treetops.

Weaver ants show no antipathy for the earth, ranging freely from treetop to ground. For them, a fall has little cost: a plummeting ant will land within her colony borders, with almost no chance of going astray. They still must avoid losing their grip when struggling with enemies and prey, which is why they resemble *Daceton* in having strong gripping feet. To avoid coming to physical blows, which might lead to tumbles, many arboreal ant species fight at a distance with noxious sprays.[6]

One way or another, ants do fall, and in such numbers that they can amount to an ant rain. Ground-dwelling European wood ants, *Formica aquilonia,* which forage in the trees, plummet from the canopy by the millions each day. *Formica* rain harder, so to speak, when near foraging birds. Some are knocked off branches, but others jump to avoid bird pecks. Still, the ant rain continues even when there are no birds. A portion of these ants lose their grip by accident, but some might fall simply to save time on the commute back to their nests.[7]

Whereas ants who don't live in the trees tend to tumble willy-nilly and hit the ground blindly, canopy species frequently seem to be able to control their falls. Jack Longino of Evergreen State College and I have spent some time dangling from ropes, where we've often contributed to ant rain by knocking various species of ants off branches, for the most part unintentionally. More often than not, it seemed to us, the ants would land back on the tree, as if they could control their flight path and hit a target. We couldn't understand how they did it.

Stephen Yanoviak, then at the University of Oklahoma, noticed the same thing and set out to prove that certain ants from Peru and Panama indeed can glide.[8] The species he concentrated on was *Cephalotes atratus,* a slate-black "turtle ant" with a flattened body. High-speed videos proved that when dislodged from a tree, a turtle ant stretches out her body and limbs and aligns herself with respect to the ground so that she doesn't turn head over heels. Detecting a tree trunk by its relative brightness against the dark greenery, she twists in the air to point her abdomen in that direction, glides backward at a steep angle—a behavior that I was eventually able to capture for this species with my camera in Tiputini, Ecuador—and grabs the trunk on impact.[9]

Other ant species make a tight spiral as they fall, directing their bodies with apparent intention, much like a parachutist who aims well enough to strike the earth at a good spot and on his feet. In a rainforest, numerous leaves lie between a plummeting ant and the ground. I believe that if she can slow her descent while keeping her clingy feet oriented downward, a worker can greatly improve her chances of landing securely on one of these leaves rather than bouncing off, as she'd surely do if she were tumbling head over tarsi.

The method employed by a falling worker depends on where she lives and the dangers she faces. Weaver ants neither glide nor spiral down, but rather plummet head over heels, a reflection of how little a fall matters to them, whereas *Daceton* are virtuoso gliders. Both the turtle ant and *Daceton* nest in tree trunks or thick branches. When one of these ants falls, a trunk is likely to be in range, and gliding to it is the obvious choice. For species that nest farther out among the twigs it makes little sense to aim for a trunk that may be too far away to see, let alone reach by gliding. Foliage is a sensible target, and parachuting in a tight spiral is the way to make a firm landing.

TRAVELING IN THE CANOPY

What convoluted territories arboreal ants inhabit! The navigational problem faced by a small ant is that a tree for her is a highly warped surface, one that is much more complicated than the surface of the earth is for us. She can usually monitor her movements up and down by gauging the influence of gravity on her body segments, but these gravitational effects can be masked by the movement of a plant in a breeze.[10] Because she may often have no idea which way is up, a worker in the canopy does not experience the geometry of the world the way we do. She can walk in one direction and find herself back at her starting point (she circled a trunk or branch). If she makes a ninety-degree turn, she may either reach the end of the world (a branch tip) or be lost forever (having walked down the trunk and onto the ground).

It helps that individual trees have some common features, such as a limited number of branching patterns—compare the alternate branching rhythms of an oak to the terminal leaf clumps of palm, for instance.[11] Unlike a rat forced to navigate a psychologist's maze that has been constructed with no thought to the geometry of nature, arboreal animals can use a tree's predictable structure as a navigation aid.

Ants exploit many aspects of plant architecture. On trunks, columns of workers often follow grooves or edges to orient themselves upward. When a worker reaches a horizontal leaf splay, she tends to stay on its upper surface, where she is less likely to be knocked off. She can survey a leaf by moving along its edge to trace its outline, deviating if she chooses to explore the leaf's center or underside before going back to the margin. When she departs a leaf and encounters a branching point among the twigs, her best bet is to always turn onto the next stem in a consistent direction—say, to her left. By tracing leaf outlines and being consistent in her choices at forks, she is able to move through a leaf spray without revisiting the same leaf twice and without ever having to mark a route or memorize the terrain.[12] This technique enables an ant to examine a plant more efficiently than is possible when she explores the ground. That is true even when the ground is flat and bare, which is rare: ground-dwelling ants navigate through jumbled decaying matter and plant parts, geometries far more haphazard than trees. For weaver ants, the solution is to commute on the ground as they do in the canopy, following crestlines offered by exposed roots or fallen sticks.

Once an ant begins to range widely from one tree to the next, the messiness of canopy topography forces her to rely on a variety of orientation cues. Leafcutter ants, for example, measure thermal radiation to locate the sun-warmed foliage they prefer to cut.[13] Visual cues may also be valuable. A weaver ant can choose a course at a particular angle relative to the sun or moon; if the sky is shaded, she uses a less accurate, internal magnetic compass.[14] Some ants create maps by taking mental snapshots of the greenery against the sky.[15] Although it may be relatively easy to use these snapshots to navigate on flat ground, using them within the trees must require an overwhelming feat of insect memory.

A good memory can be essential to canopy survival. The weaver ants Ed Wilson raised on a small citrus tree in his office when I was his graduate student became excited when a novel object such as another tree was placed nearby, gathering on a twig in an attempt to reach it. Apparently they remembered enough of their surroundings to recognize this change. We might expect the workers in a large colony to keep to the small portion of the territory that they know well, and in fact weaver ants on border patrol don't move around a lot.[16]

Age and experience can play a big role when ants explore farther afield. The chemical trails of bullet ants, for example, often cross one another among the interdigitating branches and vines of the canopy, and the workers can distinguish the scent routes laid by different nestmates to

In Queensland, weaver ants—known in Australia as green tree ants because of their coloration—form a chain to connect branches during a foraging expedition. In most situations the chain is many ants thick.

different destinations.[17] These ants also have exceptionally good eyesight and keep track of their whereabouts by eventually memorizing the location of such landmarks as branches.[18] Novice workers follow the trails, while their more experienced compatriots come to navigate by the landmarks almost exclusively.[19]

It's not easy even for arboreal species to range beyond one tree. Rainforest crowns are separated by open spaces and seldom intermix. Vines offer shortcuts as well as an abundance of honeydew-producing insects, explaining why weaver ants prosper at forest margins where the canopy is cluttered with such connections. Adjacent tall trees lacking vines tend not to be occupied by the same weaver ant colony. However, in these situations weaver ants can create their own shortcuts. I saw this ten years ago during a stay on the north coast of Papua New Guinea. I was drying myself off after snorkeling when I noticed, stretched between the branches of two citrus trees a few meters above my head, a chain of weaver ants 6 centimeters long. I broke the chain with a finger to see what would happen. Ants accumulated at the site, climbing on one another in the direction of the neighboring tree to form a fingerlike mass jutting into the air. After an hour, the wind rocked two branches close enough for the workers at the end of the

A *Camponotus schmitzi* worker free-diving into the digestive fluids of a pitcher plant in Brunei, Borneo, where it will retrieve the corpse of a cricket.

Camponotus gigas of Malaysia, swimming, like climbing, is no big deal; instead of detouring, workers paddle across any puddles in their path.[29] Indeed, a *Camponotus gigas* should find it easier to swim than a marauder ant, because water offers more resistance to a smaller individual. However, the marauder will carry more air down with her, proportional to her size, which she can use for breathing and to make herself more buoyant for her slow haul back to shore.

In the mangrove swamps of northern Australia lives an ant that swims as a matter of course. Nests of the spiny ant *Polyrhachis sokolova* can remain underwater for several hours at high tide. As the waters begin to rise, the workers swim on the ocean surface to reach the raised entrance cones in the mud, rowing with their front two pairs of legs and using the back pair as a rudder. Once inundated, the sandy cones collapse, plugging the colony safely inside. If an ant doesn't get back to the nest in time, she awaits the return of low tide on the trunk of a nearby mangrove tree. When the waters recede, the nests are opened by the ants sealed underground, which then walk out to hunt small crustaceans on the mud flats.[30]

One plant-dwelling carpenter ant species has incorporated swimming into its foraging routine, in a behavior so extraordinary I had to see it to believe it.[31] To do so, I returned to Brunei, home of the exploding *cylindricus* carpenter ants, and drove an hour west of the capital, crossing most of the breadth of this tiny country to a remnant patch of red meranti, an endangered timber tree with a long, pale trunk. There I found *Camponotus schmitzi* workers crawling on pitcher plants that grow as vines at the base of the meranti.

A pitcher plant is not ordinarily a healthy place for an ant, since these plants are carnivorous. A cup that grows from the twisted tendril at the end of each of their modified leaves holds a liquid into which insects tumble and drown after "aquaplaning" over the pitcher's slick rim much as a man slips on a banana peel.[32] The plant secretes digestive enzymes into this liquid that break down the corpses and help the pitcher absorb their nutrients. Ants are the plant's most common meal, except for the resident species of carpenter ant, *Camponotus schmitzi*, which nests in the pitchers' tendrils and takes dips in the liquid, emerging alive and well.

That afternoon, I watched ants dive into the cups for a swim, staying underwater for up to thirty seconds. At the floor of one pool, two workers tugged at the corpse of a cricket, dragging it up through the water meniscus—a feat in itself, given how difficult it is for a small being to break the surface tension in a body of water. Then the pair carried the body up the pitcher walls, an equally tough job because the surface is slippery, thanks to a flaky wax that helps the pitchers entrap their prey. Slowly, the ants dragged the cricket to the underside of the pitcher rim, where a dozen other workers gathered to eat it.

What looks like theft turns out to serve the plant. By working in twos or threes, the little divers retrieve insect corpses several times their weight. These bulk items can't be tidily digested by the plant and so tend to putrefy. Liquid fouled with ammonia and sullied by organic matter gives the pitcher plant the equivalent of acid indigestion and causes the pitcher to rot. The ants therefore aid the plant by removing large prey, but they also feed it: as my ant workers ate their cricket at the plant's rim, small chunks of the insect dropped back into the liquid below, to be absorbed by the pitcher plant. For *Camponotus schmitzi* the pitcher is a first-rate "ant plant," providing for its residents' every need: housing and meat, and even sweets, in the form of the nectar at the rim of the pitcher that also attracts its hapless prey.

DOES SIZE MATTER?

Whether walking, swimming, climbing, or falling, an ant's diminutive size influences how she travels through her world. Although we think of all ants as small, they vary in size several thousandfold. The average species has workers a little less than 3 millimeters (an eighth of an inch) long—smaller than a weaver ant and about the size of a marauder ant minor worker. But ants at the small end of the spectrum, such as *Carebara atoma*, the "atom ant," are truly Lilliputian. I once dislodged a flake of bark from a tree in Singapore, only to expose four hundred yellow specks: an entire colony of its close cousin *Carebara overbecki*. The minor workers were almost the size of an *atoma*, their oval heads about as small as a single-celled paramecium. The slightly larger soldiers have elongated heads with two little horns.[33]

The ant worker at the other end of the scale, the major of the carpenter ant *Camponotus gigas*, is nowhere near the car size of the ants that terrorized Los Angeles in the 1950s cult film *Them!* At a little more than an inch long, she is indeed only fair-to-middling in size among insects and falls far below the world record holder for an adult insect, a female giant weta cricket I collected on Little Barrier Island in New Zealand (it weighed 71 grams, three times as much as a lab mouse). One scientist observed that ant species with bigger workers tend to show a greater number of behaviors, and he proposed this might be because of their larger brains.[34] Still, *Camponotus gigas* workers don't strike me as being especially quick-witted, and indeed there are many physical advantages to staying relatively small. Although the ant's little body loses heat and water more easily than yours or mine and overheats more swiftly in the sun, it also circulates nutrients without as complex a cardiovascular system. As we've seen, an ant's size enables her to climb almost effortlessly and fall without the possibility of breaking a leg. Ants can float or swim when caught in a downpour, and survive long periods of immersion thanks to their sluggish metabolisms. (Yet weaver ants drown inexplicably fast, making it easy for Cambodians to collect them in water and pull out the bodies later to add as spice to a meal.)

But looked at another way, ants aren't small at all. A leopard may impress with her bulk and power, but compared to the ant, she is a minor part of the forest in which she lives—measured in terms of both her ecological impact and her size. Ants have, in effect, two body sizes: the individual's and the colony's. To understand this basic truth, I use a mental exercise I learned as a graduate student studying the marauder ant. First, I follow an individual ant. Then I take in several ants collectively, a group of workers busy at a task. Finally, I liberate my imagination from what is directly before me, emulating German chemist August Kekulé, who discovered the beautiful structure of the benzene molecule in a dream. Allowing my reverie to expand beyond what is visible, I contemplate the functioning of the whole: all the ants, in the nest and out, with the workers integrated like the cells of a human body into a superorganism.

This is more than mental gymnastics. By living socially, ants break through the glass ceiling imposed by their exoskeleton. At nearly 40 kilograms, a large driver ant colony is the size of an eleven-year-old boy. However, this particular young man would be a kind of superhero, one who can disassemble, such that his hands can stop a crime while his head commutes to the office to write up the news report—both Superman and Clark Kent at the same moment, an analogy that is particularly apt for ant species whose workers spread out widely. Even an atom ant colony, which might fit within the head of a *Camponotus gigas* worker, is a superhero in miniature.

A colony is an organism divided, with no loss of integrity. Its body spreads over space in pieces that give it a multitude of eyes and brains with which to glean nutrients, energy, and information. It does this with a microscopic attention to detail that no unified body can match. It's more flexible than an organism, too: the superorganism counterparts to tissues and organs range from transport teams to nurseries and are easily assembled or taken apart or shifted to a different function. Whereas a human vascular system has well-established roles, its colony analogue, the trail system, is flexible; it can serve as a snare for food and later be co-opted for migrations or a fight.

This fragmentation helps a collective of ants to succeed when a single big vertebrate would fail. The workers in a weaver ant colony, with a combined weight, at 14 kilograms, of a young leopard, can disperse among leaf nests on many frail branches—or, in other species with big colonies, fill up cracks, crevices, and galleries in wood—and thereby live where no hefty vertebrate could. Furthermore, most of the food available in nature is present in packets too small for a large animal to glean. A young leopard or a man would starve trying to gather the tidbits that make up the diet of a large ant colony, and neither is muscular enough to carry as much food as all those ants can move collectively. A whale trawling for zooplankton is the only vertebrate creature that scoops up as many bits of prey; indeed, a baleen whale is the marine equivalent of an ant colony.[35] To accomplish this task as a group, bulky individual ants (even ones of weta cricket size) would be at a disadvantage. That's why so many plants have evolved to support ants, but not aggressive mammals or birds, as their guardians. Only ants can scour a plant's surface relentlessly enough to weed out its enemies, large and small.[36]

Their scouring behavior illustrates the repetition and fast tempo that are the hallmarks of large ant societies. We saw this with the swarms of marauder ants that crisscross the terrain within a raid, rooting out prey (and weaver ants do much the same in much looser bands). An advantage to having many do the same thing at once is that if one individual fails to finish a task, whether subduing prey or building a trail, another will do it.[37] Also, the frenetic workers in large societies often make mistakes; close inspection may reveal one ant going the wrong way or leaving building material at an inappropriate site. Such a blunder might be lethal for a solitary creature that has but one chance to do a job right; the same may be true for ant nests with few individuals, in which workers carry out every move with meticulous care. But in a large society, differences in performance assure that often enough a task is done to perfection. Even if lapses or errors occur, they are quickly corrected by another individual. In fact, with sufficient redundancy, variations in performance can lead to useful novelty and innovation, as when an ant on a busy recruitment trail overshoots the intended prey and, while she is lost, happens on another; or when a friendly competition for goods and services brings a kind of market economy to a nest.

The redundancy of worker actions gives a superorganism other survival advantages as well. While a human life ends if a wound destroys the brain or heart, the functions of brain and heart are spread throughout a colony, making it harder to damage. To bring the comparison to the level of the society, we humans have erected increasingly elaborate top-down hierarchies and centralized systems of control to deal with the disasters, from plagues to terrorism, that so easily disrupt our modern nations.[38] The lesson from nature, however, is that the war on terror will never end: all living things fight back against enemies (parasites, predators, and competitors) in a continual arms race in which new defenses emerge but the dangers never disappear because the adversary always evolves a counterstrategy. Under such circumstances, it doesn't pay to consolidate power; better to have redundant operations with few or no established chains of command, as ants do.[39] Because they have no central command center that can malfunction or be crippled or manipulated by outside forces—indeed, no established leaders or unique individuals—the destruction of part of the population will never bring down the whole.[40] Weaver ants apply this redundancy even to their nests by having multiple leaf tents instead of a vulnerable central domicile. With the exception of reproduction (most colonies succumb to the death of the queen), this safety net permeates all aspects of ant social existence.[41]

Is there an ideal size for individual workers within the superorganism? The answer is unclear—as it is for the size of the cells in the body, for that matter. One weaver ant major worker can stretch across the nail of a pinkie finger. Among ants, that's pretty large. Weaver ants compete with other dominant ants that have much smaller workers. But imagine a mature weaver ant colony that, instead of being polymorphic, contained only minor workers, and instead of half a million ants, held several million. The colony consisting only of minors would burn more energy at rest than the original colony or a colony of only large workers of the same total weight—perhaps an economic disadvantage; and no minor worker could hold prey as large as one major can restrain for long enough for reinforcements to arrive. But the foragers would exchange information at a higher rate and be more effective at rooting around in locations previously hidden to them; and the greater number of individuals would be more effective at ganging up against the competition.

Variation in worker size is associated with a division of labor, and the redundancy afforded by large ant societies helps make specialization pay off, as it does for large human groups and complex organisms. With humans in, say, a small military squadron, the loss of the one person who knows how to radio headquarters could be devastating; in small groups it therefore pays for everyone to be a generalist, with overlapping knowledge and skills. Larger divisions can afford to include more specialists, among them helicopter pilots, tank drivers, and snipers. Large ant societies can similarly produce more specialists. Consider the outrageous polymorphism of the marauder ant: by having workers that span the size distribution of many of their competitors combined, they may be able to outperform them all the more effectively.

We don't know why, given a certain outlay of resources, one ant species produces only large workers, another produces only small ones, and a third—the weaver ant—produces a mixed population, biased toward the majors. Yet so omnipresent are the major workers, so complete their concerted action when they sense a person, that I have often felt, on walks in Africa or Asia, as if some predator were spying on me from the trees—only then to hear on all sides a muted drumming of alarm, similar to the sound of peas dropping onto a plate, as one study puts it.[42] That is the collective sound of a superorganism, generated by the crowds of agitated weaver ants striking their leaf roosts.

amazon ant

the slavemaker

Polyergus, northern temperate zones

For the average Amazon ant, the royal treatment continues after she arrives home. Entering the nest, she lounges around, at most grooming herself or her nestmates, while the *Formica* slaves tend to her needs. Her daily efforts last for a couple of hours at most. Yet Alex told me only about half the slavemakers are likely to go on a raid; the rest stay behind, doing nothing all day. For a booming society of several thousand ants the tempo of life is abnormally laid-back for *Polyergus breviceps*. (To our way of thinking, the average ant worker's life sounds like slavery even when she is in her birth nest, though in that case she is at least toiling for the benefit of her mother, the queen.)

That afternoon's confrontation at the *Formica* nest was a soundless blur lasting twenty minutes, more or less the normal length of such a slave raid attack. The timing was typical too: conducting their raids late in the day, *Polyergus* is forced to get the job done and head home before the sun wanes, since they don't stay out after dark. It is unclear why the raiders don't give themselves more time by beginning earlier. Perhaps the delicate pupae would cook if trundled away in the midday sun. Or perhaps as the afternoon cools the *Formica* ants fetch brood from deep in the nest to warm nursery chambers near the surface, and so the slavemakers don't have to invade as far underground, thus minimizing their exertions even further.[2]

A mature Amazon nest houses five thousand ants, comprising both the slavemaker workers and their more numerous slaves. Inside the nest, the slaves tend the stolen pupae until they transform into adults. Like hatchling birds that imprint on their parents, a young ant quickly learns to recognize the individuals around her and thereafter treat them as family. This imprinting is based on the scent of other pismires, an archaic term for an ant that derives from a colony's pungent odor. Whenever ants meet, they sweep their antennae over each other to confirm the presence of the blend of compounds that identifies their nestmates.[3] If the odor matches expectations, they treat each other as sisters-in-arms. If an individual smells wrong, the workers will either run away from each other or fight.

In most ant species, this imprinting is infallible, because the youngsters are surrounded by sisters in the nest of their mother, the queen. But when slaves-to-be mature in an Amazon nest, they imprint on their captors. Assimilated into the wrong society, the ants are duped into a life of servitude, doing all the drudge work their masters won't: building nests, foraging for prey, harvesting honeydew, slaying free-living *Formica* that enter their territory, and taking care of the brood. The Amazon slavers' only job is to go on raids, replenishing the store of *Formica* pupae as their enslaved workers age and die.

The slavemakers do so little for themselves that when I pulled a sandwich from my backpack and dropped a bit of turkey in front of an Amazon worker, she walked right past it. Incapable even of recognizing a meal, she is unable to feed herself. Eventually one of the slaves found the stash of poultry and retrieved it to the nest.

At once more brawny than a slave and yet as helpless as a baby, the Amazon worker gets her sustenance only after her servants find food and, like birds with their nestlings, regurgitate it to her. She can neither excavate tunnels nor raise the queen's young. She is a fighting machine, nothing more. The curved daggers she bears as jaws are useless for any chore except assaulting free-living *Formica*, but they deliver the ultimate in all-purpose tools: a new stash of slaves. Even

Ant slaves harvesting a dead grasshopper as one of their Amazon "masters" walks by in the foreground. The Amazon ants would starve if their slaves didn't feed them.

with their superior armaments, though, the Amazons are so outnumbered they would be massacred if it weren't for a chemical known as a propaganda substance that they wield as a social weapon, released from a gland associated with their mandibles, which throws the bombarded colony into mayhem and flight.[4]

The tolerance of *Formica argentea* to frequent raids is a sign that this species shares a long history with *Polyergus breviceps*.[5] After countless generations of attack and counterattack at Sagehen Creek, the *Formica* have apparently come to treat their losses as a cost of doing business. "Resistance is futile," declared the species-enslaving *Star Trek* creatures known as the Borg, who make decisions collectively, like ants.

LIFE IN A NUTSHELL

A year later, far from California, I found myself studying a second, unrelated species of slavemaker ant—*Protomognathus americanus*—in the hope of experiencing some of the variety of ant slavemaking behavior. For the second day in a row, I stooped over the little *Protomognathus* ants, trying not to disturb the action. One of the dark-pigmented slavemakers had been about to find

An acorn from Ohio containing a colony of dark brown *Protomognathus americanus* slavemaker ants and their orange *Temnothorax* slaves.

an acorn housing a nest of *Temnothorax* when two *Temnothorax* workers managed to sting the slightly larger ant to death. That had been the last of several close calls for the free-living *Temnothorax,* two of which lay nearby, killed in earlier confrontations. Another hour crept by, as my arms cramped under the weight of the camera, before a slavemaker with better luck found her way into the acorn through a split in its side. Dozens of the *Temnothorax* within immediately fled, each grabbing a larva or pupa in the stampede, with the queen beside them. Left alone in the empty shell, the slavemaker stood on a heap of abandoned brood. After a few idle moments she picked up a pupa and returned to her own acorn. She was soon back with reinforcements, who helped her abscond with most of the remaining stock of future slaves.

Glancing around, I was struck by the unreality of the situation: all this time I had been so caught up in the action that I had forgotten I was in a laboratory, surrounded by Petri dishes and Bunsen burners. To learn about slavery in a nutshell, I had come to the Mecca of acorn-ant research: Ohio State University, base of Joan Herbers. Joan specializes in *Protomognathus americanus,* which enslaves three species of *Temnothorax* in the temperate deciduous forests and yards of eastern North America.[6] It is difficult to observe raids of these pygmies in nature. Before I flew to Columbus, Joan had been kind enough to sort out some colonies for me, collected by her students and encased in Ziploc bags. Each bag contained either a slavemaker colony or a colony of free-living *Temnothorax,* housed in an acorn. All I had to do was put a mix of these acorns in a plastic arena, settle down in front of it, and wait.

Fifteen hours later, I had finished documenting my first *Protomognathus* slave raid. Over the course of an hour, the slavemakers had taken part of the *Temnothorax* brood to their old acorn, while expanding their colony into the new one as well. Having multiple nests like this is called polydomy, and it is common among ants that live in acorns and other small, convenient places. Meanwhile, the *Temnothorax* adults were still scattered over the ground, having lost both progeny and home.

To bring troops to a *Temnothorax* colony, these slavemakers employ a variation on something called tandem running, a follow-the-leader approach to recruitment in which an ant tracks the successful scout to a site by touching her repeatedly or, if they lose contact, by orienting to a short-range pheromone released by the first ant. Because the leader is responsive to the follower, stopping at intervals to wait for her touch, the relation has been likened to that between teacher and student.[7] With *Protomognathus,* the "teacher" brings along a whole class, for a conga line of several nestmates follows the successful scout.[8]

The Amazon ants, *Polyergus,* belong to the Formicinae, a group of ants that includes the carpenter ants and their relations. The Formicinae evolved slavemaking several times independently in different species in different locations. *Protomognathus* belongs to the Myrmicinae, a second large group in which slavemaking is common.[9] The *Temnothorax* species it enslaves commonly reside in fallen acorns that have been opened up by one of two acorn specialists, the acorn moth or the acorn weevil. The adult females of both these insects lay eggs on or in the nuts; the larva then eats part of the meat before chewing a hole in the nutshell, from which it emerges. The exit hole becomes the entryway for a succession of motley residents, often culminating in *Temnothorax.*

I described this array of relationships for *National Geographic* magazine while I was a graduate student looking for cool projects in my neighborhood.[10] In researching that article, I spent a lot of time gathering acorns and dropping them in water. Those with residents float because of the eaten-out cavities. Cracking them open—and exercising patience—I eventually uncovered a whole society: several dozen *Temnothorax* with a queen and pale brood, occupying hollows carved in the nut. After hundreds more acorns, I came upon a mixed colony with *two* forms of worker—much scarcer. One kind (with a bigger head, stronger jaws, and a groove along each side of the head into which she withdraws her antennae for battle) was the *Protomognathus;* the other was a *Temnothorax* worker—in this circumstance, a slave.

Slavemakers like the Amazon and the acorn ants are known as social parasites. They acquire nutrients not by tapping into an organism's tissues, in the manner of a tapeworm, but by exploiting the selfless, cooperative behavior of a host animal or, in the ants' case, host society, as one superorganism exploiting another.[11] Social parasites escape the burden of foraging by letting their captives collect food. Slavery is just one means to this end. In some ant species that share a nest, the two groups of ants benefit equally from the housing arrangement; as we have seen, the acrobat ant *Crematogaster levior* and the carpenter ant *Camponotus femoratus* raise plants together in canopy gardens. At another extreme are colonies that occupy adjoining chambers in a nest, with one kind of ant soliciting food from the other or surviving on the other's rubbish. The minute and stealthy thief ant nests in the walls of the chambers of larger ants, infiltrating to steal food and brood. The British banker and naturalist Sir John Lubbock found this social parasite appalling. In 1883 he wrote, "It is as if we had small dwarfs about eighteen inches to two feet long, harbouring in the walls of our houses, and every now and then carrying off some of our children into their horrid dens."[12]

In Ohio, while I was watching ants come and go from acorns, Joan Herbers and I talked about ants and people. Naturalists have referred to abducted ants as "slaves" ever since Swiss entomologist Pierre Huber first used the term to describe the behavior in 1810.[13] Darwin devoted several pages of his *Origin of Species* chapter on instinct to a discussion of what he characterized as the "remarkable" "slavemaking" activities of certain species of ants.[14] Although the analogy is not perfect, it has become established in the literature.

Ant slavery has notable differences from human slavery. Ants, lacking commerce between societies, don't buy and sell or trade slaves from colony to colony. Ant species such as the Amazon ant are more dependent on their slaves than humans have been, aside from a few "slave societies" such as the Roman Empire at the time of Augustus.[15] Ant slaves can't breed (but then, they fare no better in their birth society: in ant colonies, usually only the queen has the privilege of reproducing). Ant slaves also seem remarkably acquiescent about their subjugation. Only two species exhibit signs of mutiny: some acorn-dweller slaves will, if not outright revolt, at least undermine the colony by killing their masters' pupae.[16] (Though this could simply be their normal response to finding something strange in the brood pile—some brood just doesn't smell right.) And slaves of the Amazon ant in Europe will make a getaway at times, with some running off to form satellite nests, even adopting a nest-founding queen of their own species should one pass by. Independence is usually short-lived, however, because the Amazons retaliate with periodic raids to retrieve the escapees, which engage in restrained fighting that quickly leads to acquiescence.[17] Other than in these situations, ant slaves seldom try to thwart their captors or attempt to escape them; they die in the defense of their masters exactly as they would have done in their birth nest.[18]

Kidnapped before they have formulated an identity, the victims imprint unconditionally on their captors through ignorance, not brainwashing. They are similar to the human working class described by Karl Marx—a whole population whose efforts are misdirected to benefit an oppressor.[19] As slaves, the ants have lost not freedom (which they never had) but the biological imperative to raise the offspring from their own genetic family.

The indolence of ant slavemakers relies on their captives being programmed to slog through the day without objection, which they do. And just as ant slaves seem oblivious to the fact that they are in a slavemaker colony, slavemakers might not be able to distinguish slaves from their own kin.

For some ants, this is hardly surprising. In the red imported fire ant and some other species, one colony will raid a smaller colony of their own kind and rear their brood as slaves.[20] But Amazon ants and certain other social parasites enslave not their own kind but related species that have similar diets, nests, and communication systems. In these cases, the activities that come instinctively to a slave are of continuing use, allowing the adjustment to slave life to occur rapidly.[21] No domestication through breeding is needed, as occurred with humans and dogs, nor is training or coercion necessary, as it is, for instance, with human prisoners in a chain gang.

Instead of taking slaves outright, human civilizations bent on expansion have often usurped villages and exacted tribute and labor from them while expanding their dominion to encompass the vanquished people's land. The losers are often allowed to remain with their families and communities; unlike with slaves, their former identities are not completely lost. With time and luck, they may even be incorporated into the victorious society as full citizens. This middle ground of empire building, which requires a large population of victors to quell rebellions, is unknown for ants, for whom surrender followed by a midlife switch in social allegiance is not possible. As part of the spoils of war, ants either take slaves or kill the losers (in which case cannibalism is frequent, as it was in the early stages of human warfare).[22] Though the victors commonly reduce the defeated colony in size, they seldom destroy it—as we saw also for army ants, which raid a nest to the point of diminishing returns, then leave the remnants. Unless a colony is weak or its queen is killed, it will likely see another day.

Are ants and humans the only animals that have slavery? Female primates may capture or enforce the adoption of an infant. In Old World monkeys with hierarchies of female dominance, such as the Lowe's guenon of West Africa and the Bonnet macaque of India, a female may take a baby from a low-ranking mother, possibly to interfere with a competitor or simply because she is attracted to the infant. Female langur monkeys share their young, but an inexperienced juvenile impatient to get her hands on a baby may abduct one from another troop. In no primate, though, is the abducted individual a source of forced labor.[23]

Much more analogous to the ant model is the activity of a large Australian bird, the white-winged chough. During the four years it needs to reach maturity, a chough stays with its parents and helps them raise its siblings. Without enough assistants, the parents will be unable to adequately build their elaborate mud nest, incubate the eggs, and feed the chicks. In some situations, the parents or their immature helpers will bully a neighboring family until one or more of its fledglings can be shepherded to their own nest, where the new youngsters are raised until they can serve as helpers. Just like ant slaves, fledglings don't recognize that they've been abducted.[24]

Given our differences from animals, is it reasonable to apply the word *slavery* to ant practices? Most words usefully encompass a variety of phenomena, and *slavery*, like many of our names for things, was used first and foremost to define human relationships, before being applied by analogy to the natural world. But just as slavery as practiced in Augustan Rome doesn't correspond exactly to the behavior of an ant, neither is it the same as slavery in other human societies (consider today's largely secret slave trade); similarly, the behavior of one ant species does not match that of any other. The attributes of slavemakers vary even from place to place in a single species of ant.[25]

Joan offers *piracy* as a more neutral word for this activity in ants.[26] Yet piracy, though it involves robbing or plundering, inaccurately describes the ants' behavior. After all, if pirates forced a person into a life of servitude, that person would be described as a slave. Retired Hunter

College professor and Amazon ant expert Howard Topoff quips that *adoption* may be a better word, but this term isn't used for situations so insidious as to entail theft followed by a life of travail for another's benefit.[27]

I don't have any problem with using the term *slavery* for ant behavior unless people commit the naturalistic fallacy—assuming that if a behavior exists in nature, it is somehow "natural" and therefore morally acceptable in human society. The earliest humans were hunter-gatherers organized around equality, not dominance, and had little use for slaves and limited capacity to keep them, which suggests that slavery is not a part of our own heritage.[28] Thus if, as Edward O. Wilson and Bert Hölldobler propose, Marxism is better suited to ants than to humans, then, by orders of magnitude, slavery is even less well suited to the human social order.[29] While apes and some other vertebrates have been known to express empathy and to act in accordance with rudimentary moral standards, ants do not.[30] Regardless of the power of Aesop's fables, among animals other than ourselves, actions are neither right nor wrong. They just are.

PROPAGATING A SLAVE COLONY

Given the slavemaker penchant for avoiding work, how does an Amazon queen go about founding a new colony?[31] I caught a glimpse of the first step during the raid at Sagehen Creek. Among the onrushing workers ran ten queens, their cellophane wings glittering. En route they left the mass to scramble up tufts of grass. At such raised locations unmated queens attract males, using pheromone secretions from their mandibular glands.

After mating, the queen has a choice: she can continue with the raiding party and establish her colony in what remains of the *Formica* nest after the slavemaker workers have plundered it; or she can strike out on her own to locate a different *Formica* nest and found her colony there. The first choice has its risks. If she moves into the conquered nest, ants from her original colony might come back later to raid the site again. Retaining no memory that she is a relative, they would kill her burgeoning family.

In either case, the Amazon queen rushes the nest with savage fury, shoving aside any *Formica* workers that get in her way. She is protected from their attacks by both a tough exoskeleton and repellent secretions.[32] Leaping on the *Formica* queen, the Amazon punctures her counterpart repeatedly with her dagger jaws and then licks the fluids draining from the dying queen's body. The transformation of the colony is almost instantaneous; mere moments after their queen's death, her workers undergo what Howard Topoff has called brainwashing: "The *Formica* workers behave as if sedated. They calmly approach the *Polyergus* queen and start grooming her—just as they did their own queen. The *Polyergus* queen, in turn, assembles the scattered *Formica* pupae into a neat pile and stands triumphantly on top of it. At this point, colony takeover is a done deal."[33]

Because the ants in a colony imprint on each other's scent, licking the dying queen is the slavemaker queen's macabre way of applying the home-grown perfume. Once she has the colony's identifying odor, the invader is accepted as one of its own. The opposite is true as well: if the mother queen is removed from a *Formica* colony prior to the arrival of an Amazon queen, the impostor has no way to appropriate the local scent, and the workers will bite her until she dies.[34]

The coup d'état does not end with the first queen's murder. *Formica argentea* colonies have multiple egg-laying queens, which the slavemaker queen roots out from their safe havens in the nest and executes one after another. Why this is done is not known. Having procured the correct odor by killing one queen, she could commute the death sentences of the others and leave them to produce more workers, which is the resource she will need from the colony. The Amazon queen would thrive without her own workers, and her species would eventually evolve to lose the worker caste entirely, enabling her to concentrate on laying eggs that will grow into future parasitic queens.

Some ants employ that very strategy. *Teleutomyrmex schneideri* is a European species that produces no workers. The queen infiltrates a colony of a *Tetramorium* pavement ant, and her concave undersurface enables her to attach herself to the back of the resident queen, where she clings for the rest of her life.[35]

Polyergus queens, however, have retained their workers. I assume there must be a downside to keeping a breeding stock of slave queens around. Perhaps Amazon nests that retain a slave queen are subject to insurrection, in which case it wouldn't pay for the slavemaker to share reproductive control, the ultimate source of power in an ant society.

13 abduction in the afternoon

A week into my stay at California's Sagehen Creek, I was lounging among the grass tufts next to a *Polyergus breviceps* ant nest with Faerthen Felix, the station's assistant manager. I was trying to determine exactly how their raids transpire, and we had shown up early to catch the moment when the action began. But for most of the day not a single Amazon had shown her face. When the first slavemaker appeared at 4 P.M., she showed a supreme indifference to the slaves that were industriously collecting bits of my jelly sandwich. By 4:45, a hundred Amazons were milling about. While the foraging slaves went far afield, the Amazons stayed within 2 to 3 meters of their nest entrance.

That had been the habit of every colony I'd observed: no Amazon ants most of the day, then a slow buildup of milling workers near the nest late in the afternoon. Faerthen and I tried to figure out what all the wandering Amazons were doing. In time, the raid would depart, and they would join it, but until then they occupied their time examining every cranny near the nest. Were they foraging? Faerthen asked. No: finding meals is slaves' work. The milling ants never picked up a thing, nor was there any indication that they laid trails or took notice of one another. Was their exploration part of colony defense? Perhaps. I had heard that when raiding workers encounter another Amazon colony, they attempt to destroy it and carry off its brood as a source of slaves

This Amazon worker is positioning her daggerlike mandibles to pierce the head of an ant from an alien nest. Her own head has already been punctured behind the eye.

of their own kind.[1] So I dropped an alien *Polyergus* in front of the milling workers to see if they would respond—and was not surprised, given the threat she represented to their colony, that they immediately punctured her to death.

But as this was the extent of their safeguarding activities, there had to be a better explanation for the slavemakers' desultory bustle. I would find that Amazon colonies raid almost simultaneously at the end of each day, with virtually every wanderer joining in. By the time the raiders from one colony have gone far enough to encounter a competing nest, the raiders from that second colony have probably departed along their own path. Protection against other Amazon colonies must depend chiefly on the surfeit of workers that never leave the nest.

If not foraging or supplying defense, what do the milling Amazons accomplish? Watching their incessant scrambling within this staging area, I proposed to Faerthen that they might be energizing themselves for the upcoming battle. That would make them a bit like human troops performing drills—sharing movements and chants, a practice known to deepen identification with the regiment.[2] Wolves and wild dogs engage in similar rallies, which assure that everyone is "awake, alert, and ready" before a hunt.[3] In any case, after this seeming display of bravado, which typically ended between 5 and 6 P.M., the raid started fast and moved quickly.

The first indication of an incipient raid is a surge of Amazon workers in one direction, with the milling ants in that quadrant joining an outpouring from the nest. The exodus bears some similarity to the way a raid explodes from an army ant bivouac, except that with Amazon ants, in the few minutes after most of the milling ants join the raid, the number of ants lagging behind declines to zero. A stream of workers a few meters long moves away from the nest like a swift snake, albeit one that is straight as a stick.

For a distance of up to 140 meters, the procession glides at speeds amounting to nearly 200 meters an hour—or a foot every few seconds. That's ten times faster than an army ant raid. Though I've become pretty good at stalking animals, on two afternoons I staked out a colony and missed the whole raid: both times I looked away from the nest for a moment, and all the ants were gone. Even the ants' orange color and the open landscape didn't help me find the raid once it slipped out of sight.

What triggers the ants to leave the nest, and how are they able to travel so quickly and efficiently? Most of the details of the subjugation of *Formica* by *Polyergus breviceps* have been gathered by Howard Topoff. Working in Arizona, Howard discovered that, unlike raiding army ants, Amazon ants target a single location and are brought there by a leader. Earlier in the day, in fact while her nestmates are ambling haphazardly around their nest, this motivated individual has been scouting for victims, and when the raid begins, she is in charge of the group.[4]

Researchers studying other populations and species of *Polyergus,* however, have denied the existence of scouts.[5] Watching the Amazons at Sagehen Creek, I understood their doubts. I had been unable to pin down the leader of the raid or to discover her earlier in the day, while she was seeking a target colony. But animal behavior can vary. Even the species of ant enslaved by the Amazon ant differs between locations. Perhaps the California and Arizona Amazons had different scouting behaviors as well. I took it as a personal challenge to confirm the existence of scouts at Sagehen Creek.

And I failed. Hour after hour, day after day, I could not identify a single scout.

Not only was I unable to pick out any leader within a raid, but I was puzzled by certain things about the raids themselves. I could scrape away the soil in front of the column and the ants would continue, which told me that, like foraging army or marauder ants, the Amazons were not tracking a scent. But if the foremost individuals weren't following an obvious leader, they also didn't replace each other the way those at the front line of an army or marauder ant raid do, by advancing incrementally before retreating. Instead, all the Amazon workers kept pace with each other while moving forward, with only occasional meanders and a few abrupt shifts in course. They had to be acting on some kind of guidance. That would explain why Amazon raids advance more quickly than army ant raids, despite the fact that the army ant workers are much faster runners.

More than anything else, the Amazon ant raids brought to mind the termite-hungry gangs of *Pachycondyla* I'd seen while studying driver ants in Nigeria. Instead of taking victims en route, as army ants do, the *Pachycondyla* raids target a clump of termites, make their kills, and then retreat. Compared to the Amazon raids, though, the *Pachycondyla* groups were slow and even more tightly compact—one or two ants wide—making it easy to pick out the group leader who had scouted out the termites earlier in the day.[6]

Amazon ants at the advancing end of a slave raid rush long distances at Sagehen Creek, California, led by a single worker.

polarization signals were obscured, so the guide ant could no longer shepherd her nestmates. After some fruitless snooping around, the workers returned to their residence.

Rain, like clouds, may force a raid to be canceled. In another experiment suggested by Howard, I was able to get a raid to turn back in a panic by showering the ants with a pinch of water. Fortunately, most afternoons during the raiding season are warm and sunny, the kind of weather the Amazon ants prefer.

Once the procession is under way, the guide simply halts when she reaches the site where she found the *Formica* nest, presumably ceasing to release her pheromone signal and thereby allowing the battalion to spread over the nearby terrain. At this juncture, her troops seem to switch from "follow the leader" mode to "search for nest" mode. (The change is remarkable and abrupt, mirroring the change in the scout's behavior from "run ahead single-mindedly" to "search for nest" on her initial, outbound trek.) Insofar as searching for slaves is the closest the slavemaker workers come to foraging, foraging has begun.

As I watched the Amazons rooting around for a passage into the besieged colony, a simple explanation emerged for the wandering they did at their own nest prior to a raid. They were acting no differently than they were now. Perhaps the milling ants had been looking for *Formica* from the moment they emerged from their nest. The leader's pheromones (I hypothesized) could have a hypnotic effect on her nestmates, diverting the overly eager raiding workers from their ferreting behavior long enough to take them to a location more sensible for their search.

A worker leading her companions with only the help of the sun is unlikely to arrive at the precise location of the *Formica* nest entrance she found earlier in the day. No matter: colonies have multiple entry holes. Brought to the outskirts of the targeted colony, her comrades need only spread over a few square meters to find a way in. Each raider seems to emit a pheromone as soon as she detects an entrance, for reinforcements come fast—with luck, before the *Formica* can erect a blockade.

Whether the raid is a success or a failure, the return trip is a cinch. The pack is no longer dependent on its leader. The pheromones the workers released to keep themselves together on their outbound journey persist long enough for them to follow home.[8] The retreating ants make use of polarized light as well, though. Taking on a Pied Piper role, Howard has managed to delude the ants into turning around and returning the brood to the nest they just raided by shading the advancing column and using a mirror to redirect the sun.

HOW DID SLAVEMAKING BEGIN?

In *The Origin of Species,* Darwin proposed that the ancestors of slavemakers may have eaten the brood of other ants, much as army ants do today. He suggested that when the booty wasn't consumed fast enough, some transformed into adult workers that imprinted on their captors and ipso facto became slaves. At first the nascent slavemakers would have relied on their slaves most heavily for jobs that were costly or dangerous, such as foraging and defense. But with their slaves constantly performing all duties out of habit, the slavemakers would have gradually lost their other domestic skills as well, culminating in the modern Amazon ants that will starve to death if there are no slaves to hand them food. Over many generations, the addition to the supply of laborers would have become so valuable, even essential, that predatory raids metamorphosed into slave raids. Darwin saw the raids in effect as a circuitous kind of food search: what was once a foraging enterprise became a quest for individuals that would do the foraging—that is, a fresh batch of slaves.

Today, the most widely accepted conjecture about the origin of slavery in ants holds that the slavemaker ancestors were competitive species that took brood as part of their war booty and then ate it, even though ant brood wasn't an everyday part of their diet, and some of these occasionally survived to become slaves.[9] This idea has gained favor because the groups of ants that most often evolve slavery have been species that fight over territories or other resources rather than the specialist cannibals of other ants. Despite their predatory raids, for example, no army ant species has become a slavemaker.[10] Remember the acorn-dwelling *Temnothorax* that are subject to enslavement by *Protomognathus americanus*? In territorial battles between their colonies, *Temnothorax* eat any immatures they seize.[11] But on rare occasions, and seemingly by mistake, they will rear one to adulthood. Such an individual will serve as an accidental slave—unless her confused nestmates kill her first. Indeed, the evolution of slavery must have included a suppression of the worker impulse to kill foreigners, thereby allowing would-be slaves to coexist with them.

The honeypot ants of the American Southwest are a territorial species with an aptitude for slavery. They show fascinating ritualized behaviors designed to avoid lethal conflict, engaging in mock fights called tournaments in which ants from different colonies circle one another on tiptoe to try to make themselves look larger. In this so-called stilt walking, they will sometimes climb on pebbles to stand taller than their neighbors, a ploy that anthropologists call tactical deception, which is associated in primates with keen intelligence.[12] For species with a modest labor force, ritualized conflict is a reasonable strategy to avoid casualties, though if one side determines that their workers are larger and outnumber the others, they raid the weaker nest, gorge on its brood, and drag back its honey-filled replete workers as slaves.

Slavery evolved in both ant and human societies out of the need for a compliant and manageable labor force to drive their extensive systems of collective production. Ants are concentrated on producing the brood that will be the source of the next generation, and the labor force geared to this task is normally made up of the ant colony's own workers. But why work? If we consider the slavemaker colony as a superorganism, it's as if a creature obtained its arms not by growing them but by grafting on fully formed limbs that it pulled off someone else and can replace whenever it needs to, something like the horticulturalist who grafts a branch of one apple tree onto the trunk of another. Apart from the effort required to make this graft—or steal the slaves—in the first place, the outcome is cost free and pays off handsomely as long as each slave brings in more food than the meat she would have provided had the slavemakers eaten her instead of letting her live. By this measure, a slave probably nets the colony a profit in a few days, with an hour-long raid yielding many thousands of hours of services by the ants enslaved.

The graft has the most value when the initial cost to the colony of procuring it is least.

When honeypot ants engage in ritualized combat over food, the colony that draws fewer big workers retreats, often without injury. By standing on a pebble, the worker at right is "cheating" to appear larger than she is, driving off a larger opponent near Portal, Arizona.

Capturing pupae is the key. Although nursing the larvae is labor intensive, pupae—ants at the nonfeeding stage between larva and adult—yield a new slave crop quickly and with almost no effort. In fact, the slaves raise only pupae to become new slaves; the few eggs and larvae sacked by their masters become their food.[13] These snacks reduce the number of mouths the colony has to feed. There's no explanation for why the slaves make this decision about what to eat. But the slaves even consume the pupae if they are hungry enough, as always regurgitating a portion of each meal to the Amazon workers.

These observations are a clue that sustenance and slave raiding are linked: after all, raids bring in the slaves that will in turn bring in the food (or, at the whim of other slaves, end up as food themselves). Amazons raid less often when there is a food glut, which makes sense, given that a well-fed colony probably already has an adequate supply of enslaved foragers.[14] Future research should resolve whether raids are prompted by hungry slavemakers and slaves. If they are, eating more of the booty may be a means of sustaining a colony during the time between procuring brood and turning it into foraging slaves.

SEASONAL STRIKES

Slavemakers are commonplace in the temperate zones, even in suburban backyards. But their absence in tropical climes is an enigma. There are theories, of course. One has to do with numbers of ants. Temperate-zone ants frequently subject to enslavement, such as *Formica* and *Temnothorax,* tend to be superabundant and yet easy to attack. Despite the large number of ants in the tropics, few tropical species are as plentiful as *Formica* and *Temnothorax,* and the ones that are, like the weaver and marauder ants, are extraordinarily well defended.

I once traveled a few weeks on the Paria Peninsula of Venezuela and in the Arima Valley of Trinidad with Robin Stuart, an expert on the North American slavemaker ants that reside in acorns. We spent our time watching little brown ants called *Nesomyrmex* come and go from nests in the hollow twigs of roadside shrubs—while trying to ignore the roaring logging trucks at our backs. These weren't attractive locations, but Robin and I had decided *Nesomyrmex* were promising candidates for enslavement, and it would be a coup for us to find ant slavery in the tropics.

Nesomyrmex interested us not only because they are common and innocuous but also because they are cousins to the temperate *Temnothorax,* or acorn ant. By occupying a number of acorns, a single acorn ant colony is fragmented into isolated housing units (polydomy), each of which may have one or more egg-laying queens (polygyny). Both these attributes may increase the openness of the colony to invasion by outsiders, including slave raiders. Robin and I found that *Nesomyrmex* colonies are likewise characterized by polydomy and polygyny, and yet we found no social parasites in their nests.[15]

Perhaps it is not the differences in colony organization but rather the seasonality of the temperate zones that is conducive to ant slavery. The production of brood in temperate-zone ants has an annual cycle, and slave raids of the Amazon ant take place during the few summer weeks when the free-living *Formica* nests contain workers in the coveted pupae stage of development. Earlier in the year, while the *Formica* colonies are still rearing their larvae, the Amazons are full-time stay-at-home loafers, their slaves tending the Amazon queen's brood that will mature into a fresh cadre of warriors by the time they are needed, in the raiding season. Seasonal production of brood is less pronounced in the tropics. It's also been proposed that seasonally cool temperatures might dull the ability of ants in the temperate regions to recognize the alien queens when they infiltrate their colonies, a necessary step in the evolution of many slavery species.[16]

I believe there is another possibility. Slavemakers show a rare forbearance in giving up the immediate gain of eating food (stolen brood) for the potentially greater long-term benefit of having slaves.[17] Such delayed gratification can be especially advantageous in temperate environments due to the hardships arising from predictable changes with season, as well as more extreme and unexpected cold or warm spells.[18] In essence, slavemakers such as the Amazon ant have chosen to hoard not meat but slaves, whose efforts help tide them over in lean times. Hoarding in the tropics is less often a life-and-death matter, because animals are more likely to procure a steady-enough food supply that they can eat meals when they find them or shortly thereafter.[19] For Amazons, the payoff is a comfortable lifestyle in which workers avoid work outside of the raiding season, leaving their slaves to toil, come rain or shine.

14

a fungus farmer's life

In a pasture near Botucatu, Brazil, two cows were staring at me with heat-addled eyes, when all hell broke loose. Luiz, a laborer we had hired to help us, gave a shout of pain as part of the trench he was digging in collapsed. In front of him was a gash wide enough to hold a treasure chest, from which spilled not gold but a porridge-like material. Despite Luiz's screams, my heart started beating with the same excitement a prospector must have felt when he struck a vein of gold. I ran over and lowered myself into the trench.

We were digging for leaf-cutting ants. These fall into two genera, *Acromyrmex* and *Atta,* with a total of thirty-nine species, all from the New World.[1] *Atta,* the most impressive and ecologically important of these ants and my focus in this chapter, are most prevalent in the tropics, though *Atta texana* range through east Texas and west Louisiana, and *Atta mexicana* cross into Arizona. Their popular name derives from the medium-sized members' habit of cutting foliage. They hold the pieces aloft like little green parasols, then stream across the ground to their immense lairs, where they use the plant material as a substrate for underground farming.

In four ten-hour days, our ten-person team—led by Virgilio Pereira da Silva of São Paulo State University—had so far only scratched the surface of a labyrinth of chambers that seemed to go on forever. It was thrilling to think we might finally be reaching the heart of the nest. I now stood in one of two 7-meter-long corridors we had dug, both deeper than we were tall. Everywhere along their walls were bisected galleries. A leafcutter nest can extend 7 meters into the earth

Previous page: A leafcutter ant, *Acromyrmex octospinosus,* slicing a leaf in Guadeloupe.

and contain nearly eight thousand chambers. The biggest hold the ant's trash, buried—with greater thoroughness than humans use in handling nuclear waste—as deep as the ants can go, sometimes as far as the water table. In this colony the largest cavities were still below us. But we were getting close. The gash Luiz had just opened revealed several kilograms of this refuse. Passing my hand through the loose stuff, I detected first the heat of decay, then motion. In my palm wriggled browsing beetle and fly larvae, a moth pupa, and several cockroaches—species that speed the breakdown of the material.

My excitement wasn't much consolation for Luiz. His back was covered with leafcutter soldiers the size of horseflies, which had latched onto his flesh with scalpel-sharp mandibles, and his shoulder was bleeding where a soldier was busy slicing an especially deep curved groove.[2] I indicated to Luiz that he should protect himself by wearing a shirt, but my own defenseless hands were marked with similar crescent moons, painful as paper cuts, as if I'd been clawing through brambles. More soldiers poured out of a hole in the floor of the trench and swarmed up my legs as I helped two of Luiz's friends pull the ants from his back. Then everyone picked up their picks and shovels and went back to work.

Leafcutter soldiers have few duties. They have been known to use their offensive skills to dice up tough fruit, but mostly they protect the nest and its immediate environs from army ants, hungry armadillos (whose powerful forearms help them tunnel easily into the heart of a nest), and curious entomologists and their unfortunate assistants.[3] The damage they had done to Luiz's back and my hands reminded me of *Lophomyrmex bedoti,* an otherwise unremarkable ant with jaws almost as fine-toothed as saw blades. In Indonesia, I had seen three workers of this species shear off both antennae, four legs, and the tips of two body spines of a big-headed ant with grisly ease, in less than thirty seconds.[4]

I would have guessed that leafcutter jaws work with an equal effortlessness, but leaf carving is an arduous activity for these ants, equivalent to the cost of flying in other animals.[5] Follow the leafcutters to the source of the ants' parasols, and you will find the ants on vegetation, cutting arcs like the ones chiseled into my fingers. Rather than using her mandibles like saws or scissors, a leafcutter sticks the terminal tooth of one jaw into the leaf to fix its position as she pushes her other jaw against the leaf edge. She then forces the second jaw into the tissue with a rocking motion, the way people use a lever-type can opener. Leafcutter jaws are also like a can opener in that they get some of their strength and rigidity from metal, having a zinc content of 30 to 40 percent.[6]

While cutting, the worker herself acts like a geometer's compass: she anchors her back legs at the leaf margin and moves in an arc around that point, adjusting the size of each fragment by flexing her legs or head to different degrees. By such fine-tuning, she slices off fragments that she (or other ants) can carry: smaller pieces from thicker leaves and larger pieces from thinner ones.[7] She can also be precise in her choice of leaf: workers tend to end up on the foliage they cut best. This is often based on their size, which is quite diverse in leafcutter workers, even among those who specialize in cutting leaves. The smallest cutters abandon foliage too tough for them to handle, while larger ones tend to depart from soft foliage after being pushed aside by bustling smaller ants. There's even some evidence that larger cutters are drawn to recruitment trails that lead to the sturdier foliage.[8]

Unlike marauders and army ants, leafcutters always move their burdens individually and never need help.[9] Because leaves are flat, large pieces have a small mass that a single ant can easily heft, and larger individuals slice and haul larger loads.[10] But leafcutters take fragments just a few times their body weight at most. They are capable of hefting heavier fragments, so this may seem inefficient, but small pieces reduce congestion on the trails and inside the nest, speeding up the processing of foliage overall.[11] Additionally, because workers carry the fragments vertically and high above their heads, on steep slopes the ants can become unbalanced, causing them to slow down, flip over, and even fall. The ants seem to anticipate when the route home will be uphill, and cut smaller parcels.[12]

HOW DOES YOUR GARDEN GROW?

Leafcutters are almost unique among ants in their total dependence on vegetable nutrition. True, many other species rely to some degree on plant matter, notably in the canopy, but vegetation contains meager protein. Hence, treetop ants like the weaver ant are omnivores, the workers using the energy they get from carbohydrate-rich plant food to procure animal flesh for their protein-greedy larvae. (Only the dwellers of a few ant plants can afford to be strict vegetarians; their plants secrete protein-rich food bodies in payment for the ants' protective services.) Leafcutters have a similar strategy.[13] The adult workers drink sap from leaf fragments (their primary energy source), energizing themselves to process the foliage that serves as compost for their larvae's sole food: a protein-rich fungus that is related to the button mushroom sold in supermarkets. Leafcutter colonies have come to depend on the fungus so much that even the adult ants' digestive tracts lack key enzymes for breaking down some proteins.[14] The fungus does that for them, and also removes from the plant tissues any insecticides that would stop most herbivores in their tracks.[15]

Most people know fungi as mushrooms, but mushrooms are the substantial yet fleeting reproductive outgrowths of microscopic strands called hyphae that spread in a latticework to infiltrate and ingest soil, decomposing matter, even rock. In the leafcutter nest, this latticework spreads through the foliage that the workers have mulched in order to free its contents for absorption, a necessary step given that the fungus likely can't digest cellulose and only takes nutrients in solution.[16] The hyphae and its vegetable substrate fill the majority of leafcutter nest chambers with what is called a fungus garden, a mass of featherweight, fissured gray matter that looks like a human brain and can reach a similar volume. A garden is given its cerebral shape by workers that continually add fresh leaf matter to its top and sides while dismantling and disposing of the bottom, older half.

Leafcutters grow and harvest their fungi using farming techniques no less complex than ours. Along with the fungus-growing termites I saw attacked by driver ants in Africa, they are among the only animals besides humans that can be considered agricultural.[17] The invention of agriculture has enabled societies of humans and leafcutters, which were farming long before people, to support massive populations. These massive populations then give rise to massive

A small fungus garden of *Atta colombica* in Paraguay: the white fuzz is fungus, which is tended by the workers. Hiding in the nooks are winged queens.

structures—for shelter, food rearing, and so on. The most enormous nest I have come across was in the dense forests of the Kaw Mountains of French Guiana, with soil mounds rising chest high over an area 14 meters wide and about 160 square meters total. Trails initiating from the nest's far corners led into the forest in each cardinal direction. Scaled to human size, the space occupied by such a nest would exceed the dimensions of the Empire State Building. Such a colony might easily contain several million workers.

A leafcutter colony's chambers and tunnels can require the excavation of 40 tons of soil, as they must house not only queen, brood, and workers, which even in the millions occupy only a tiny fraction of the space, but also fungus gardens in the hundreds or even thousands. The garden chambers are distributed along tunnels in a pattern that can resemble grapes on a stem—with the garden-containing "grapes" the size of soccer balls and "stems" as wide as a child's arm, which must give the jostling leaf-bearing ants elbow room aplenty.[18]

Cramming gardens and millions of ants together underground produces air pollution, and too much heat or too little oxygen will slow garden growth. In the Kaw Mountains, I took time off from swatting mosquitoes to peer into a nest entrance that thrust from the earth like a volcanic cone. The metabolic heat of fungus and ants struck my face like the exhalation of a great bull. Illuminated by my headlight, the smooth throat, over 7 centimeters wide, gracefully curved out of view a meter or so down. This vent was near the center of the nest, where the population is densest and the heat of metabolism therefore highest; humid air escapes through such openings, to be replaced by fresh, cooler air drawn through perimeter entrances. In open habitats, wind striking these turrets could be the principal source of air conditioning.[19] For these reasons colonies can have a thousand entryways; those not being used as ant thoroughfares can be opened and closed to regulate the conditions below.

A colony this big is comparable in many ways to a cow or a deer. The ant population weighs 15 to 20 kilograms, as much as a newborn calf—or an adult red brocket, a Latin American deer that lives in the same forests as many leafcutters. Leafcutter nests consume as much vegetable matter in a year as does one red brocket: up to 280 kilograms, enough leaves to blanket a soccer field.[20] As we shall see, the assemblage of ants in a colony processes forage in much the same way as a cow does, from chewing the raw material to excreting the remnants. The gardens are the equivalent of the cow's rumen—but whereas the rumen makes use of a slurry of bacteria, protozoa, and fungi to extract the proteins and fatty acids a cow needs, a garden requires only one fungus species to process foliage into a complete ant chow.[21] Both ant and cow find and prepare plant matter for their microbes, which they house under ideal conditions; the ants will relocate gardens if the temperature or humidity falls outside a suitable range.[22] Just as algae and fungi have combined to form an organism known as a lichen, and the gut flora have become an essential part of a cow or a deer, the garden fungus has been integrated into the leafcutter superorganism.

INDUSTRIAL FARMING AND TRANSPORT

Farming requires a diverse skill set in ants, as it does in humans. Today, humans farming on a large scale use tools and machinery to handle different steps in the process, but in ants, different skills reside in different workers, and as we have seen, polymorphism plays an important role in this. The biggest leafcutter colonies are extraordinarily polymorphic, with the largest soldier having two hundred times the mass of a small worker.[23]

Ant colonies have been likened to a factory in a fortress.[24] I find the metaphor particularly apt for leafcutters. Their multiple-step procedure, in which all get involved, dwarfs the two-step process by which a marauder-ant media worker extracts a seed from a grass stalk and a minor carries it away. The leafcutter workforce is self-directed, adjusting to the local requirements of colony and fungi without the oversight of any foremen. In business terms, it has the flat organizational structure adopted by corporations from Hewlett-Packard to IKEA, with the absence of middle management enhancing cost effectiveness and the organization's responsiveness to rapid shifts in needs.[25] Most leafcutter activities are accomplished with little communication: as is done in any well-run assembly line, the gardeners simply do the task that comes before them.

A leafcutter factory might have been the envy of Henry Ford: different workers collect, transport, and mince foliage, apply it to a garden, and eject its decayed remnants in an orchestrated

This small leafcutter worker from a Panamanian colony of *Atta sexdens* is taking a tuft of white fungus to "plant" in fresh leaf mulch.

flow of material from environment to nest and back out again. Many steps are managed by ants in a narrow range of sizes.[26] Mid-sized workers cut the foliage, carry it into the nest, and drop it onto the garden surface, where, as the production line unfolds, ever smaller ants accomplish more delicate tasks. Workers with heads about 1.6 millimeters wide shred the greens into scraps. Slightly smaller ants further masticate the chunks, now discolored from abuse, into a moist pulp. Still smaller ants, using their forelegs, implant the pulp into the garden. Tiny ants with heads a millimeter wide lick the pulp and seed it with tufts of fungus from established parts of the garden, like horticulturists using cuttings from a vine to begin a new crop of grapes. The smallest workers of all, with head widths of 0.8 millimeter, reach into the garden's recesses to remove weedy species and contaminants that include bacteria, yeasts, and spores.[27]

Much like vintners trimming back the branches of grape vines to maximize their yield, minor workers prune the garden surface, stimulating more edible fungus growth.[28] The gardens' brainlike fissures dramatically increase the surface area from which minors can harvest meals.[29] The small workers seem to take on the primary role in distributing food. They drink sap while processing leaves and nibble on the fungus while tending the garden recesses. Then they feed their nestmates, either by regurgitation or by handing them edible wads, which they also give to the larvae scattered over the garden.

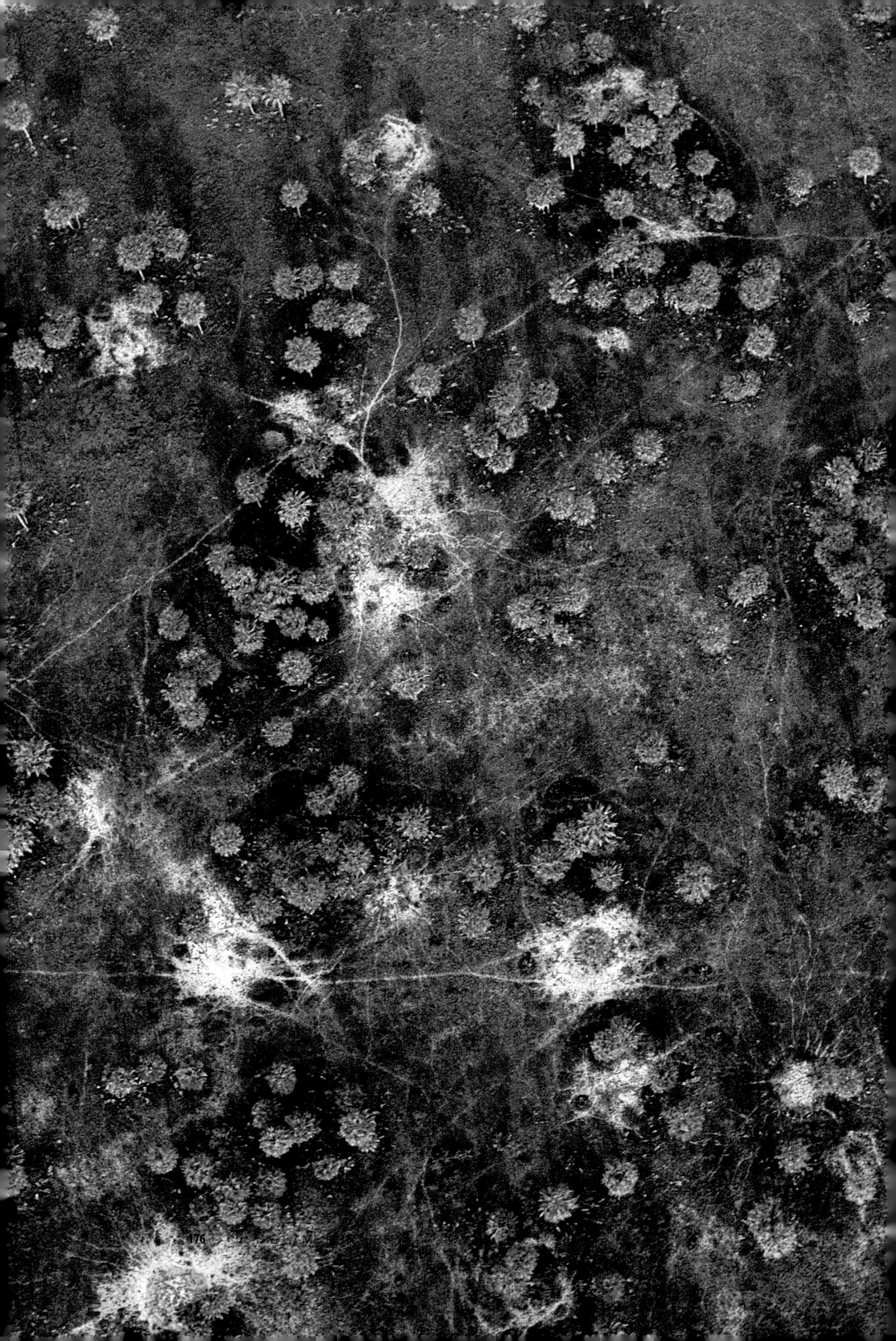

All along this conveyor belt, the workers defecate on the leaves. Their feces, like the manure we use in our gardens, contain ammonia and amino acids that promote garden growth. Leafcutter excrement also includes enzymes from the fungi they have eaten, which pass through their digestive systems intact and speed the breakdown of fresh substrate, helping each new tuft get better settled. Coddled and cared for, in a day the fresh fungus has sprouted what look like microbe-sized masses of cotton candy. The masses consist of swollen hyphae tips, configured to lie in easy reach of hungry ants. As seen under the microscope, they are arranged in grapelike clusters, much as the fungus garden chambers are along their underground runways, but at one-thousandth the size. Found in no other fungi and with no purpose other than to be eaten, the swellings reveal that the fungus has been selected over long periods of time by the ants to serve as food, just as plump grapes and rosy apples show generations of cultivation by human hands.

As in human industries, such large-scale operations require the support of extensive transport and distribution networks. On average, each leafcutter colony maintains a trail system 267 meters long at any given time, which requires the completion of 2.7 kilometers of roads over the course of a year. That much construction requires 11,000 ant-days of labor, during which the workers expend the energy attained from eight thousand leaf burdens. That sounds like a lot, but because a leafcutter workforce is larger than the human population that was employed to build the pyramids of Egypt, it takes a colony less than a day to fetch enough foliage to fuel a year's worth of road building.[30]

I came across some prime leafcutter trail systems in Paraguay in the early 1990s while driving with a companion through the Gran Chaco, an alternatively swampy and scorchingly dry savanna region. The thermometer read 125 degrees Fahrenheit, and the heat was overpowering—literally: on our second day, my friend was catatonic from exposure, staring blankly into space for half an hour before I could rouse him. We had come in search of *Atta vollenweideri,* which can unearth more cubic footage than that boasted by an average New York apartment.[31] The nests, with their excavated discs of pale sand, several meters across and visible to the horizon, each radiated trunk trails as wide as a human foot and more than 70 meters long, leading to the grass this species prefers, which it cuts into long, linear segments. Days later, I had a chance to view the area from a low-flying airplane. The leafcutter communities resembled highway maps of human urban centers; in fact, some leafcutter species reportedly have beltways encircling their metropolises.

The leafcutter transportation system comprises durable trunk routes with weaker side trails near their far end from which the ants spread out to forage.[32] In canopy-foraging species, each side route lasts for a few days, during which time it grows longer because, as ants harvest foliage, a lack of space for late arrivals forces them to move out to explore other, interconnected vines and branches.[33] These temporary outgrowths often cause the layout of paths

Opposite: The leafcutter ant *Atta vollenweideri* plays an important ecological role in Paraguay, where the Caranday palms and mesquites of Chaco savannas sprout from the fertile soils of dying nests.

for a colony to resemble a two-dimensional sketch of a shrub or tree, with its stable trunk and more ephemeral branches. This branched design is a reflection of local resources—or their absence. Every twist in the stem of a vine, for example, is a record of its responses to changes in light and support brought about by the comings and goings of the plants around it. The location and architecture of ant trails offer a similar record.

The parallels between ant highway systems and plant architecture are easiest to detect in clonal plants such as ivy that spread over surfaces with the intelligence of an ant superorganism.[34] Sometimes ivy uses what is called a guerilla strategy, developing long, unbranched stems that carry it quickly (for a plant) through sectors poor in resources—such as shady spots—with a minimal investment of tissue and little searching around. Entering a sunny patch, the ivy shifts to a phalanx strategy, combing the ground by growing more branches and short stems. It can even sense nearby plants (by shifts in the wavelength of light reflected from their foliage), and so can grow away from rivals.

Many kinds of ants similarly orient their trails to the location of food while circumnavigating the competition, as we saw with marauder ants, which construct long direct trails to distant productive regions and short branching ones within food patches.[35] But unlike marauder ant colonies, which employ scoutless raids that are uncertain of what lies ahead, leafcutter colonies send lone foragers to gauge conditions over several feet and lay recruitment trails to choice vegetation. Still, because the distances they search are so short, the colony as a whole—the superorganism—is effectively nearsighted and makes choices based on distorted information. A tree next to a trunk trail may be discovered no matter how far it is from the nest, whereas a sought-after specimen away from a trail will go unnoticed, no matter how close it is to home. Even if a forager were to go far enough to detect such a plant, she'd probably be unsuccessful in mounting a return expedition. As a result, leafcutters haul foliage from long distances for what may appear to us to be no good reason.

Marauder ants and certain seed-harvesting ants have trail systems that shift every few weeks, apparently to track the location of food, but leafcutters are obstinate about retaining old trunk trails and seldom start new ones. Some ground-cover plants show a similar static pattern, staying in place like sit-and-wait predators to absorb nutrients as they become available, before their more mobile plant competitors, creeping over the ground, can show up.[36] Many leafcutter trunk trails may function this way, their exact location a historical artifact of their being laid out in early life while the maturing superorganism was in an adventurous state of mind, with its workers exploring farther and in novel directions. Thereafter the routes are maintained through a kind of inertia that locks a colony into certain sectors within reach of a nest, where leaf flushes can be quickly harvested as they become available. Over time, heavy use and trail maintenance crews make the routes wide, smooth, and deep, enhancing their durability. Even if such a well-made trail becomes inactive, it's likely to remain visible for months, if not years. This physical persistence pays off by enabling a colony to revisit sites that have shown a high productivity in the long term. The greater experience of older ants has nothing to do with these choices, because the trails last far longer than the life span of the individuals. Traces of a trail are a kind of long-term memory at the superorganism level.

To keep their trunk trails operating at capacity, leafcutter ants lay waste to anything that gets in their way. Camping in the Kaw Mountains, I was awakened by a rivulet of water next to my face. I had pitched my tent on a nocturnal leafcutter route. After nightfall, the workers had cut crescents from the floor, opening gaps in the waterproof material to allow their traffic to continue. Rain had come, and I turned on my flashlight to find my sleeping bag flecked with nylon discs and my belongings thoroughly soaked.

System maintenance, including trail-clearing operations, is as expansive as trail building. Removing a kilogram of debris from a trail takes 3,359 ant-hours of labor, equivalent to the energy content of four Snickers bars. For an ant, that's a lot of kilojoules. Trail-clearing workers, most of which are larger than leaf carriers but smaller than soldiers, are present in sufficient numbers that obstacles such as litter—or tents—are quickly removed. They haul off small objects and gnaw larger ones while smoothing and widening the trail's surface, until any traffic problem is alleviated. Minor workers have a separate role in trail maintenance: they loop back and forth under the feet of the larger, leaf-bearing ants to reinforce the trail as its pheromone markers dissipate.[37] This is an especially important task when a section of trail is damaged by a falling branch or a passing animal or washed away in a storm; until the chemical signals are reinstated, commerce halts.

When foragers depart from the trails, they prefer to search on plant limbs, including fallen branches, rather than on the ground surface—which makes sense, given that the leaves the ants seek grow on twigs. Incorporated into trails, roots and branches serve the ants well: clean and smooth, they are maintenance free and suited for speed. By following them rather than moving along the forest floor, the multitudes of ants reduce their transit time in total by many thousands of hours over the course of a day.[38]

Some leafcutter trails are so well etched in the earth that I have gotten lost while hiking in the South American tropics when I mistook an abandoned ant roadway for a tidy human path. A well-built trail increases ant walking speed four- to tenfold.[39] With scores of ants waving leaf banners the breadth of a thumbnail, their caravans can seem Olympian in pace and scope. "If we magnify the scene to human scale," writes Edward O. Wilson, "so that an ant's quarter-inch length grows to six feet, the forager runs along the trail for a distance of about ten miles. . . . picks up a burden of 750 pounds and speeds back toward the nest at 15 miles an hour—hence, four minute miles."[40]

Yet with so few superhighways and so many ants on them, congestion still can be a problem. Unlike outbound marauder and army ants, which avoid those returning home with bulky prey by taking to the trail's edges, forming one inbound and two outbound lanes, outbound leafcutters simply slip to the left or right of the homebound leaf carriers with their slim loads. Though full-stop head-on collisions are rare, each slight run-in jogs a leaf carrier off her path. Paradoxically, the best traffic flow occurs when outbound and returning workers pass one another in equal numbers, maximizing this interference, which spreads all the ants apart across the trail, forming no lanes at all.[41] If this scattering doesn't occur, the carriers end up too close together and the leaves bump together, impeding each ant's progress. The small but frequent diversions therefore result in the fastest foliage retrieval overall, even though the ants are slowed individually.[42]

CACHING AND LETTING GO

> The two that watched the garden . . . did not notice the ants who were robbing them . . . climbing the trees to cut the flowers, and gathering them from the ground at the foot of the trees. . . . Thus the ants carried, between their teeth, the flowers which they took down . . . [and] quickly they filled the four gourds with flowers.[43]

This passage from the Mayan creation myth *Popol Vuh* describes leafcutters stealing flowers from under the noses of two guards to aid the "hero twins," Hunahpú and Xbalanqué. It is a description of caching, in which workers deposit fragments where others have been left, a kind of positive feedback that leads to consolidated piles like those in the flower-filled gourds.[44] Sometimes leafcutters leave their pile until the next day, though there's a danger a competitor will steal it in the interim. The delay may reflect a preference for wilted foliage, which loses its chemical defenses, is lighter, and, like aged restaurant beef, is easier to chew.[45] Caching is also efficient because the fragments are more likely to be transported onward when they are part of a clump than when they are abandoned in isolation.[46]

Caches of up to a thousand pieces accumulate when ants cut foliage faster than it can be processed or when the carriers' progress has snarled. Workers might try for a time to enter a cramped nest entrance with their leaves, then give up and add their fragments to a stash outside. Or a worker on a subordinate traffic artery may drop her leaf when she reaches the main trunk trail, perhaps because, like a nest entrance, this juncture is a bottleneck, leading to traffic pileups at trail intersections. It's also a sensible place for a cache because the ant carrying the leaf is likely to be familiar with the neighborhood where she found it and, once she's deposited her burden, can return to cut again. If she continued to the nest, it would be hours before she made it back to that tree. By then, her knowledge of the local foraging situation would be long out of date.[47]

While leafcutters lack group transport teams, caching is an example of a different method of coordinating a workforce, one they excel at, called task partitioning: the subdivision of a job, such as the carrying and processing of leaves, into sequential stages. Task partitioning isn't always effective. When I renewed my driver's license at the DMV, I spent the first half hour in a line to get a number to wait in another line. But task partitioning makes sense if an overall savings in time or effort is the result. Among many kinds of ants, for example, it's common for workers to take burdens directly from clumsy carriers. In leafcutters, handoffs from worker to worker often lead to a better match between leaf size and worker size. (I suspect this is because it takes a worker larger and stronger than a leaf's carrier to get the clumsy ant to release her grasp.) The result of these transfers is speedier delivery.[48] Some corporations have become similarly proficient at this kind of task partitioning, avoiding logjams by setting rules that mandate that employees who move faster in one step of a complex procedure take over from colleagues slower at that step.[49] By contrast, an ant taking a leaf from a cache will likely move slower than the one who dropped it there because it is not easy for her to select a burden appropriate for her size from the pile.[50] But despite this seeming inferiority to handoffs, caches are still common. The difference in local and large-scale efficiency between direct handoffs and transfers at caches reflects the traffic problems these techniques solve: a handoff is an immediate response of one worker to the difficulties of another (perhaps after she picked up a too-big leaf at a cache), whereas most caches are stopgap solutions to wholesale gridlock in the processing line.

Caches aren't the only way ant colonies trade individual effort for a society-wide increase in efficiency. I once witnessed a remarkable sight deep in a rainforest near Manaus, Brazil: a rain of confetti spinning through the air beneath a tree. Peering overhead, I couldn't make out where it was coming from, but I did see a column of leafcutters climbing the trunk. They were apparently delivering their harvest to workers on the ground via airmail. Many of the pieces were larger than an ant could carry and so plummeted straight down, minimizing the loss that would have occurred if the pieces had been small and light and liable to drift over a wide region. As it was, the contingent on the ground located only about half the cuttings, slicing them up further for transport. A 50 percent yield may be acceptable, given that the ants saved themselves the trouble of hauling the foliage down from the treetops.[51]

In his classic 1874 book *The Naturalist in Nicaragua,* the British geologist and natural historian Thomas Belt described another instance of leafcutters using gravity to save time, in this case when transporting bits of fungus garden during a migration:

> I found them busily employed bringing up the ant-food from the old burrows, and carrying it to a new one a few yards distant; and here I first noticed a wonderful instance of their reasoning powers. Between the old burrows and the new one was a steep slope. Instead of descending this with their burdens, they cast them down on the top of the slope, whence they rolled down to the bottom, where another relay of labourers picked them up and carried them to the new burrow. It was amusing to watch the ants hurrying out with bundles of food, dropping them over the slope, and rushing back immediately for more.[52]

HUNTER AND PREY

The fact that leafcutters live on foliage and fungus doesn't mean they aren't as picky about their meals as meat-eating ants. Foragers may largely keep the interests of the fungus in mind—in a sense they are shopping for someone else—but because the adult workers are sustained largely by sap, some of the plants they harvest could reflect personal taste rather than the needs of the gardens. Still, the ants don't drink the sap while they cut and carry leaves; that typically happens in the nest, where the small ants lick the fragments and regurgitate the liquid to their larger sisters.

While the colony as a whole consumes varied foliage, individual ants become specialists on certain plants growing at sites they get to know intimately.[53] Workers are prompt at recruiting assistance to a plant species they know well; conversely, they recruit to an unfamiliar plant only after they assess its quality.[54] In this approach they resemble a bumblebee, which, after sampling a variety of flowers, comes to specialize, or major, in a single plant species.[55] It's unclear whether majoring makes a leafcutter in any way better at her job. In any case, tender leaves come and go, and, like bumblebees (and many college students), a leafcutter worker has to change her major now and then.

There are several aspects to leaf desirability—for the ant, the fungus, and the colony. Leafcutters prefer vegetation that is easy to slice. Also, they gravitate toward foliage in direct sunlight, which is the most nutrient rich. Red leaf flushes indicate chemicals toxic to fungi,[56] and leafcutters avoid them in favor of older leaves or, ideally, soft, defenseless young leaves with less of the cellulose their fungi can't assimilate.[57] Such foliage is particularly abundant in pioneer trees, which are species that spring up in early-successional habitats—relatively open places where the mature trees of heavily shaded, old growth forest have been felled by storms or old age. Where there are many pioneers, leafcutters have the luxury of selecting the few most desirable plants, whereas in older forests, colonies are forced to constantly sample from dozens of less-choice trees.[58]

Human land-clearing practices keep vegetation in an early-successional stage, which is why cultivated land is the leafcutters' favorite grocery store. Many human cultivars are of Old World origin and have no native defenses against leafcutters, or they have had the toxins bred out of them for human consumption, turning them into perfect fungus garden fodder and allowing the ants to strip them bare.[59] For these reasons, leafcutter populations have thrived along with human populations to a degree that can be as crippling as a biblical plague of locusts, resulting in hundreds of billions of dollars in damage annually.[60]

It's curious that plants don't do a better job of fighting leafcutter incursions. The munching of caterpillars or beetles can induce plants to produce chemical deterrents that make their tissues unpalatable or even deadly, in much the way our bodies fight a viral infection by producing antibodies.[61] Plants damaged by leafcutters don't escalate their defenses in this way, an unexplained oversight that allows colonies to harvest from the same tree again and again over the years.[62] But despite the inadequacies of plant deterrents, the ants seldom completely denude full-sized native trees. Some researchers have suggested that leafcutters are "prudent pruners," taking only a portion of each plant so it can recover for future exploitation.[63] Still, it's doubtful that ants are more sensible than humans when it comes to resources. They likely extract foliage as fast as they can, at times removing enough leaves—20 percent or more—to adversely affect a tree's survival and reproduction.[64] Yet they depart after taking the best leaves. In fact, like a kid who's eaten enough chocolate to make herself sick, the ants can tire of certain plants, shunning a once-choice species for weeks.[65] Foliage-cutting workers particularly avoid a plant species when their small comrades on the gardens detect signs of fungal ill health, suggesting that the fungus may be informing the ants of its needs.[66]

Foliage isn't the only thing leafcutters take from trees. Fruit, seeds, and flowers make up the bulk of their collections during tropical dry seasons, when fresh leaves are scarce.[67] Rich in calories and containing few noxious chemicals, these plant parts, which workers remove directly from plants or snatch after they fall in near-mint condition, can be more sought after than foliage. The ants also collect the pulp of fruit discarded by birds or mammals and extract seeds from animal droppings, adding any attached pulp to the fungus gardens after discarding the seeds in their garbage heaps.[68]

With their sweet tooth for fruit and sap, it's a surprise that *Atta* workers have never been seen drinking from the sugary nectaries on plants that other ants visit so readily.[69] Nectaries encourage predatory ants to protect foliage and flower rather than cutting them up, as leafcutters do; but how do the same nectaries keep leafcutters away? Perhaps their fluids contain fungicides that discourage leafcutters. No one has investigated this possibility.

Leafcutters are concerned with more than hunting down plant parts—they must guard against becoming prey themselves. Once in the early 1990s, I squatted for three days straight in the narrow space between strangler fig roots near one of the Mayan ruins at Copán where priests had once performed ritual beheadings. I was looking for a phorid fly, and I knew I had found one when a leafcutter worker threw herself back to make a quick jab at a tiny speck that appeared suddenly over her head. A phorid floats around leafcutter trails like a dust mote until it swoops down on a worker's head, inserting an egg through the ant's neck or mouth. Some flies even find the carried leaf fragment a convenient site to cling to while they insert an egg.[70] The hatched maggot then consumes brain and muscle until finally the ant's head falls off—hence one common name for the phorid: the decapitating fly. The worker I watched warded off her pursuer, but if she had been carrying a leaf—making her unable to move fast or defend herself—she would have been an ideal target.[71]

Because decapitating flies need to see the ants in order to aim for their heads, some leafcutters forage only at night, and only the workers too small to be parasitized venture out during the day. This strategy compromises productivity, however, because the smaller ants are less effective at cutting and carrying most kinds of foliage than the bigger ants that emerge after dark.[72]

It appears that the best strategy for dealing with these pests, like most leafcutter strategies, involves a specialized labor force. As workers cut foliage, they stridulate, producing a vibration that travels through the ant's body to her mandibles. This causes the leaf to stiffen, but unlike with an electric carving knife, this doesn't improve the speed and efficiency of the cut; rather, it sets the leaf itself vibrating. The better the leaf and the hungrier the cutter, the more often she chirps, which suggests the vibrations communicate the location of a good leaf to nearby workers, encouraging them to follow recruitment trails in her direction.[73] Stridulation is a modulatory signal, meaning chirp intensity motivates workers to respond, much as we take cues from how feverishly a dog wags its tail.[74] The vibrations become especially urgent during the worker's final moments of sawing and her initial maneuvers to carry the leaf, at which point small workers react to her signals by climbing on the fragment as she carries it away.[75] The function of these feisty hitchhikers has been the subject of much research, but primarily they serve as shields to keep the leaf carriers from losing their heads.[76] The shotgun riders thrust their legs or snap their mandibles when a fly comes close. Too small for decapitating flies to target, the bodyguards become more numerous whenever the flies are abundant.

Even when a leaf carrier lacks a protective force, she can summon one quickly by stridulating.[77] That's because the small workers that ride on leaves are the same ones that reinforce trails with pheromones, during which time they patrol for threats and respond to any sign of trouble. They are also the workers that, in preparing the leaf fragments inside the nest, lick them to

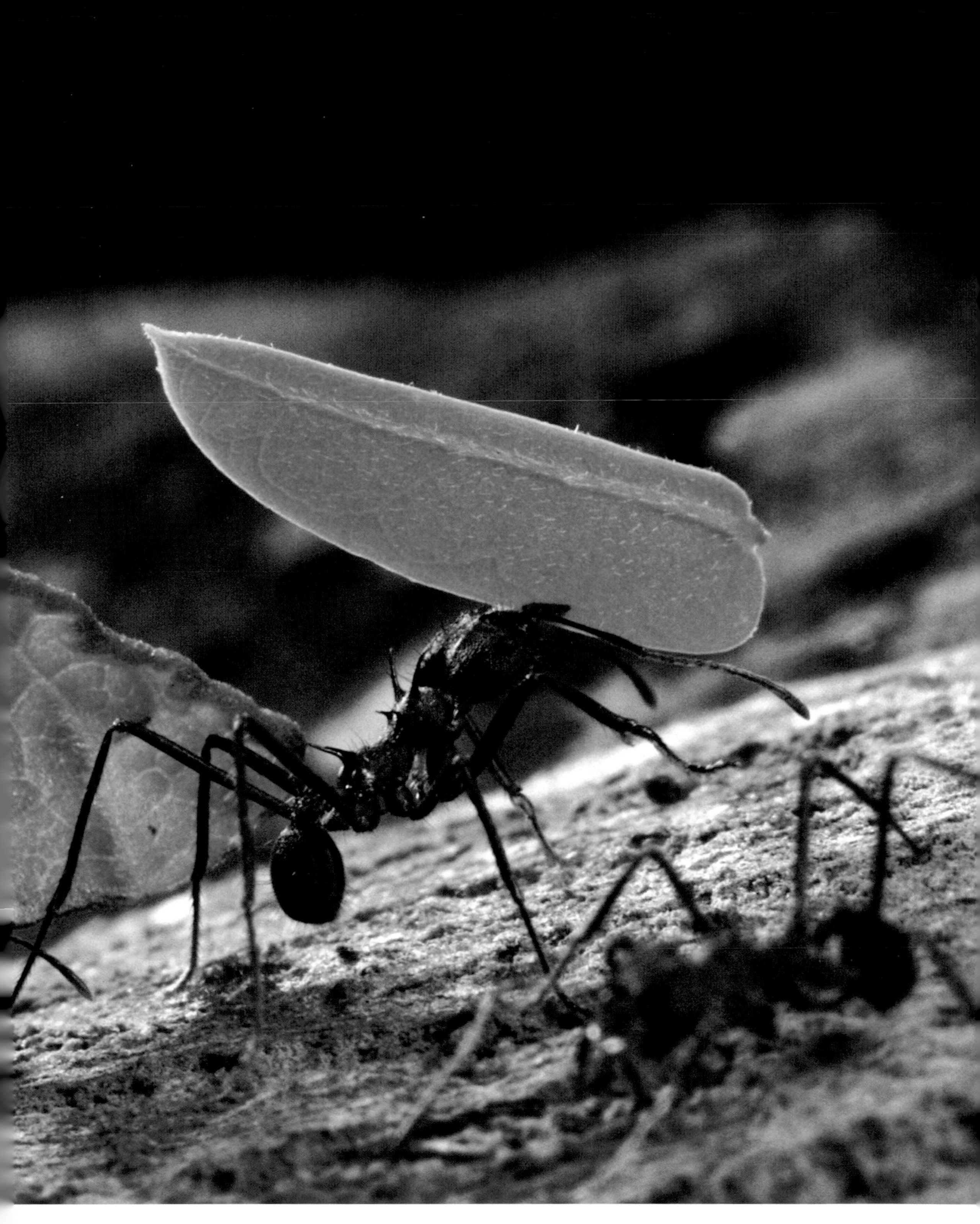

Workers of the leafcutter ant *Atta cephalotes* toting leaf fragments on Barro Colorado Island, Panama.

ingest a meal of oozing sap and, more important, to scrub off any contaminating microorganisms. It makes sense to get started on this essential task before the foliage reaches the delicate gardens—and in wide-open spaces rather than the cramped quarters of the nest.

The flies, and the occasional raid by predatory army ants, are among the biggest problems the leafcutters face. Gardening frees them from competition over food with other kinds of ants, though leafcutter colonies have been known to fight with each other along contact zones that shift back and forth like the battle lines at Gettysburg.[78] Otherwise their chief competitors are solitary plant-feeding insects, few species of which have anywhere near the ant aptitude for search and seizure. Little has been written about whether leafcutters ignore, scare off, or kill caterpillars, bugs, and beetles, but in all likelihood these leaf eaters are inconsequential to them.

THE EMBRYONIC EMPIRE

Perhaps the biggest challenge any colony faces is getting started. The process is much the same for all ants. Leafcutters add a wrinkle to the story with their fungus, which is an essential part from the beginning. New queens, larger than the workers of any ant species and each the size of an unshelled peanut, tuck a wad of it into their mouths when they leave their birth nest. To pursue the superorganism metaphor, they are like so many eggs cast out by a fish, and are similarly fertilized as they disperse. (As in all ant species, they often have several midair consorts, who die immediately. Most queens don't last long, either, but soon become food for animals and humans. Eating fried queens is like eating crunchy nuts.) Afterward, each queen will search for a place to rear her new colony. Once she digs the first chamber of her new nest, which she will probably never leave, she spits out the fungus and defecates to fertilize it. This moment must be as delicate as blowing on sparks to start a fire, for if her fungus dies, she will, too.[79]

With luck, in a couple of days she will have a small, robust garden. Meanwhile, she lays two kinds of eggs: small ones that develop into larvae, and large, infertile ones that serve as food for her developing brood. (Later the queen will eat similar eggs, laid by a clustering retinue of young mid-sized workers born with rudimentary ovaries that shrivel away as they get older.)[80] From now on the queen serves as the ovaries for the whole—the superorganism she has created, which has, like a fish or a person, a life cycle of its own. Her workers take over the other bodily functions, equivalent to the animal organs that scientists call somatic: muscle and bone, for example. Like these parts of a body, the workers cannot reproduce, but they provide a safe environment in which the reproductive parts can create the next generation.

In its early stages, a colony undergoes a kind of embryonic development at a superorganism level.[81] By the time the queen's initial brood is old enough to leave the nest, the young society seems as precocious as a calf, with its wobbly ability to stand at birth. The queen's first twenty to sixty workers encompass a minimum range of middling worker sizes needed to form a simplified version of the processing line of mature colonies, just enough to tend the garden and the young and to cut and process leaves, and thereby get the colony going. In the months that follow, it will grow in complexity as smaller and larger ants appear.[82]

With luck, the queen found a nesting spot in an open habitat. Even the recent death of one tree within a forest may provide a good-enough start. A tall tree rends a hole in the canopy when it collapses. This treefall gap lets light into the understory, allowing herbs and pioneer trees, which do poorly in deep forest shade, to move in. Even a juvenile colony's modest labor force can process saplings of these plants, whose soft foliage is easily cut by small, less powerful workers.[83] During its first, exploratory years, a colony creates temporary trails straight to such small plants within a short distance of the nest.

A growing colony will add larger road-building and leaf-cutting workers as it expands its reach to the canopy. This is to be expected: work tends to be divvied up in large societies (and in the bodies of large organisms) as a result of the differing functions or duties that must be performed in the larger, more variable area they occupy; the first complex human cultures, for example, arose where populations were dense enough for trading to become practical, and local groups could develop and maintain particular skills involving goods, such as flints for tools, specific to where they lived.[84]

When the larger leaf-cutting workers begin to appear, they first test their mettle on tougher tree foliage that has fallen to the ground. Finally the colony begins to build its first durable trunk trails, which never lead directly to single bonanza plants like the colony's first feeble trails did, and the workers start to climb tall trees. It takes a couple years to add soldiers to the mix. By this time workers have begun harvesting in the high canopy and carving out giant midden chambers well before they are needed—which will be long after the crews that built them are dead, because the workers live only a few weeks.[85] By its fifth year the superorganism has reached sexual maturity, producing males and queens that depart on mating flights. The superorganism can survive twenty-five years, it appears: that's the record life span for a leafcutter queen, and there are no backups.

15 the origins of agriculture

In Ecuador I once had the good fortune to sit in view of the Napo, an immense tributary of the Amazon. To my right flowed what, for their size, seemed to be an equally mighty river of leafcutter ants hefting pink petals; to my left, a similar line of army ants carrying their slaughtered spiders. Looking from one species to the other, I recognized the apogees of two distinct lifestyles: the sedentary communities of vegetarian farmers and the migratory hordes of meat eaters.

How did leafcutters shift away from the hunter-gatherer habits of most other ants to a life of agriculture?[1] Could they have encountered some of the same fortuitous circumstances that set human beings on the road to farming about ten millennia ago?[2] When, for instance, early human hunter-gatherers harvested wild fruit and grain and then discarded or defecated seeds on waste heaps, these grew into a ready crop of fruiting plants. It might not have taken much skill for people to transform these useful "camp follower" species into sustainable gardens.

The leafcutter's fungus could first have arrived at ant nests by happenstance—as spores carried by the wind, for example. If it found its way onto the body or food of a browsing worker, all the better: it would likely end up in the oral pocket all ants use to hold the detritus they clean off each other or strain from a meal before ingestion. About once a day ant workers spit a pellet from this pocket onto the midden pile, like a cat spitting out a fur ball. These pellets include potentially edible trifles, bits of fungi among them.[3] Then, as with the fruiting plants in early human camps, the fungus may have begun to grow from excrement or castoff garbage near where they live.

Or maybe the ants began their relationship with fungi by eating mold sprouting on palatable but overripe food, in much the way Middle Eastern tribes are thought to have discovered cheese. The ancestor of fungus growers may have shared a characteristic with slovenly Malaysian *Proatta* ants, who will collect about anything. Lacking typical ant fastidiousness, they allow refuse to accumulate in the nest, which can resemble an insect version of a fraternity house. As happens with pizza under the sofa, the mess can get moldy.[4]

Sprouting from trash, excrement, or food, the fungus may initially have been a harmless interloper suited to nest environs but of no value to the ants, but any genetic change that made the fungus useful would have increased the odds of its survival among them. The camp followers could in this way have taken the first evolutionary steps in their own domestication. A fungus among ants or a plant among humans would be more successful, for example, if it were tasty, causing the ants or humans to change their behavior to nurture it. After that, its evolution would have in a sense been "domesticated." (By applying this domestication process across species on a global scale, people have shifted the very balance of evolution on our planet such that from now on, wild organisms will either accommodate to our existence or survive only as a result of our active conservation efforts.)[5]

The first apples near human dwellings were nowhere near as juicy as those of today. They have been unconsciously and incrementally improved by what Darwin called artificial selection, as people reached for the best available fruit and coddled the apple trees on their farms. Ants are just as picky. Workers selected the fungi that offered the greatest nourishment and gustatory satisfaction, which encouraged them to maintain the substrate for the fungus—garbage or old food—so they could feed off its blooms of hyphae.[6] And so, fifty million years ago, farming was born among the ants.

DOMESTICATION

Except for the edible tips of its microscopic hyphae, the garden fungus isn't very different from its wild relatives. The ants are the ones that have undergone dramatic changes during their evolution as agronomists, from developing leaf-processing assembly lines to gaining the skill to remove weedy microbes on the gardens to losing the ability to digest certain proteins. To keep the fungus cultures pure, what may have once been a fraternity-style slob species has been reconfigured over time into today's leafcutters, the ultimate neat freaks.

We suppose that in the history of humanity, the farmers controlled the domestication of plants and animals, but domestication and artificial selection, whether in humans or ants, are more often two-way processes. Each success story is a mutual accommodation, a symbiosis between parties. Domestication can turn being eaten into a good thing by providing benefits for some prey. In organisms that haven't been domesticated, the ratio of predators to prey can shift wildly, as has been shown for the Canada lynx and the arctic hare. When there are too many hungry lynx, the hare population takes a nosedive, which then causes the lynx to die from starvation. The low number of lynx in turn gives the hares the opportunity to reproduce and replenish their population. It's an endless, brutal cycle.[7] In a domesticated partnership, in contrast, the cow in a pasture or the fungus in an ant garden thrives when the species eating it is abundant, because the predator has come to tend to the prey's every need. By a biologist's yardstick of growth and breeding, the garden fungus is, like our apples and cattle, among today's most successful species.

Is there a downside to the garden fungus's life of privilege? The ants' constant pruning and eating of the fungus prevents the garden from forming mushrooms. The ants thereby obstruct the reproduction of their cultivar, which, like the many kinds of fruit and vegetables similarly pruned by humans, spreads only asexually within the gardens. This lack of sexual reproduction may help assure the continued compatibility of the cultivar with its host, by keeping the desirable traits of the cultivar intact.[8] Only after the death of the colony can the gardens seize the moment before they starve from a lack of mulch to sprout gilled mushrooms and interbreed.[9]

The mode of transmitting their gardening methods couldn't differ more between ant and human. Issues of lactose intolerance or allergies to wheat aside, most of the changes humans have undergone in their transformation to agriculturalists have been learned rather than genetic.[10] Though such social learning is known in ants, no worker has the brainpower or longevity for a complex education. Thus leafcutters have encoded gardening in their genes, with the exception of a few variables such as a forager's preference for the plants she knows best. In gaining the ability to farm, leafcutters have lost the ability to survive any other way. But agriculture provides the ants with a dependable supply of nutrition, buffering them from whatever disasters—floods, desiccation, disease—might befall the world outside the nest. When little vegetation is available to harvest, for example, the gardens shrink gradually but continue to provide food for weeks.

There are examples of farming and domestication among other ants as well: in a similar, two-way accommodation, certain ants have in a sense domesticated the plants that provide them room and board. Ant species that tend herbivores such as aphids and their sapsucking relatives come close to practicing a form of husbandry. That these "cattle" have been domesticated is shown by their adaptations to life with ants, whereby some of the husbanded insects forgo their ancestral defenses, such as chemical sprays or leaping, in favor of a dependency on ant protection services.[11] In humans, the pastoral way of life succeeded agriculture, about 6,000 years ago, with the domestication of goats, cattle, and sheep, whereas ant "dairy" farming probably

Herdsmen ants guarding their aphid "cattle" in Pasoh, Malaysia. The workers carry the aphids everywhere and protect them at night within bivouac nests.

preceded their cultivation of fungus, likely going back 100 million years or more.[12] Fungus gardening is a complex task for ants compared to the simplicity of tending honeydew-producing insects, which don't need to be fed or penned and aren't dependent on the ants for their survival.

In some mutualisms, however, ants and their herbivore "cattle" are completely codependent. The domesticated mealybugs of Asian herdsmen ants are never found without their ants, which carry them to fresh pasture—young foliage—in incessant migrations that recall the life of nomadic peoples. Like a leafcutter queen who brings a starter fungus with her to found a nest, a herdsman queen takes a mealybug on her nuptial flight to establish her own herd. In addition to drinking the honeydew, certain ant species cull their herds for meat. This makes such herders, much like the leafcutters with their fungus and leaf sap diets, virtually self-sustaining.[13]

HYGIENE

The sun was up and the flaming orange-red *Pogonomyrmex maricopa* harvester ants were running along well-trodden trunk trails to collect seeds, when a worker walked a couple of feet from the disk of white sand that marked her nest entrance and stood stiffly, high on her legs, with head and abdomen raised and jaws agape. Near her was a small hole—the nest entrance of a smaller, innocuous-looking cone ant, a species of *Dorymyrmex*. Within a few seconds, one of the cone ant workers dashed up onto the larger ant's body. Using my camera as a microscope, I watched, my heartbeat audible in my ears, as the cone ant climbed all over the harvester, licking here and there. At one point she scrubbed between the harvester ant's open jaws, like a fawn sticking her head in a lion's mouth. After about thirty seconds of this attention, the harvester ant flicked the cone ant off with a brush of a foreleg and lumbered back to her own nest. A few minutes later another harvester ant came to the same area and was given a similar treatment. What ecstasy to see such a novel, and apparently unrecorded, behavior!

I had been visiting the biologist Howard Topoff to learn about the slavery ants around Portal, Arizona, when I came upon this instance of one ant species cleaning another in the desert flats below town. The cone ants' boldness reminded me of the cleaner fish that nibble at the mouth and body of a larger fish that has stationed itself in a similarly rigid, open-mouthed posture, which encourages a cleaner to come aboard. As occurs among the fish, after a while the harvester ant tired of the attention or became irritated if her cleaner got carried away and gave her a nip. Sometimes too many—two or three—cone ants joined the first, at which point the harvester rolled on the ground to remove the cleaners before departing. However, when no cone ant came along, an ignored harvester ant sometimes backed into the cone ants' nest entrance, virtually begging for attention.

From their earliest evolution, ants have needed to control the microbes that find sanctuary on their bodies and in the stagnant, moist depths where most species nest. Group living in

Workers of *Dorymyrmex* near Portal, Arizona, climbing on a larger harvester ant, *Pogonomyrmex maricopa*, to lick her body. In this image, the harvester ant has begun to kick them away.

such places is a public-health challenge, requiring the rapid removal of the dead, among other essential tasks. The metapleural glands on the thoraxes of ants have been one of their primary weapons in germ warfare. Like humans scrubbing down with soap, workers use their legs to transfer antibiotics from the metapleural glands to other parts of their anatomy, spending as much time primping and grooming as the average supermodel. They are equally fastidious with each other, licking their sisters' hard-to-reach spots and spreading pharmacological secretions throughout their living spaces whenever an infection shows up.[14]

Why do harvester ants invite interspecific cleaning by cone ants, when harvester ants, like other species, can groom one another? Perhaps cone ants offer a special brand of antibiotic, or perhaps, being small, they reach spots that other harvester ants miss. Harvester ants are flecked with sugary flakes from the seeds they eat, which might make them prone to bacterial infection. The cone ants must get a reward, perhaps nourishment from such sweet substances.[15]

For leafcutters, issues of hygiene and nutrition extend beyond ordinary concerns of body and nest to the needs of the gardens. As with any growth medium, the leaf mulch becomes exhausted of nutrients by the produce growing on it. In ant gardens, this takes a few weeks. The hyphae then retreat, leaving behind nutritionally barren remnants filled with a concentration of the plant toxins rejected by the fungus. The inability of the fungus to break down cellulose means the gardens generate mountains of such trash, as I had seen in the colony excavation in Botucatu, Brazil. One *Atta sexdens* colony contained 475 kilograms (half a ton) of refuse in 296 underground chambers, while the nest of another species had a similar amount of waste in a single huge chamber.[16]

If the old substrate and foreign matter aren't removed promptly, they begin to spoil. Sanitizing goes into high gear when the policing minor ants discover an infected spot. The minors take contagion spores into the same pocket at the back of their mouths that they use for cleaning themselves and expel them a safe distance from the gardens.[17] They also pull out infected clumps of garden, even teaming up to extract bigger pieces, which their larger nestmates then ferry toward the dump.

The most noxious disease of the garden fungus is caused by another fungus, *Escovopsis,* whose web of filaments can slow growth or even overwhelm garden and nest. The garden fungus synthesizes an antibiotic against *Escovopsis,* but this chemical is of limited effectiveness. And so at least some leafcutter ants grow a filamentous bacterium on their bodies that they spread over the gardens, which in turn synthesizes a more potent fungicide that targets *Escovopsis.* A relative of the *Streptomyces* that we use to create most of the antibiotics in our own pharmacology, the ant-grown bacterium may also sterilize the harmful spores the workers gather in their oral pocket.[18] Genetic analysis reveals that the relation between the bacterium and the ant has extended back as long as the ants have cultivated their fungi.[19] The bacteria, like the garden fungus, has become a component of the ant superorganism.

Once in western Colombia, while waiting for the chief of an Embera Choco Indian tribe to give me permission to stay in their village, I spent an afternoon next to the mound of an *Atta colombica* colony. After two days getting to the village in a dugout canoe, I was glad to idly watch a stream of ant workers stagger out of the nest bearing bits of defunct fungus garden. I surmised that it's one thing to keep gardens tidy and disease free; it's quite another to handle the resulting trash. Most of the mid-sized leafcutter ants collect foliage and maintain trails; a few unfortunates are relegated to sanitation.

Not surprisingly, the garbage-transport and garbage-processor ants can die from handling hazardous waste.[20] Why take such lethal jobs? Perhaps contact with unhealthy or old gardens forces young workers onto this career path. To add insult to injury, the janitorial staff is treated as untouchables by the other ants, who evade infection by dropping off refuse at safely located waste-transfer centers and staying away from the trash heaps themselves, and by attacking any of the sanitation squad who approach them or their gardens. With no access to plant sap or fresh fungus meals, the waste-management ants that take the trash from the transfer centers onward to the midden must either scavenge from the offal or starve. Working herself into an early, diseased grave, a sanitation worker cannot change careers to become a forager (nor can a forager take the suicidal move of employment on the dump).

Such a permanent segregation is coldly logical. With waste management, in ants as in humans, it's important to separate the general population from its unhealthy by-products. This has often meant that human populations locate dumps in poor neighborhoods and arrange for socially ostracized laborers to deal with trash, such as the untouchables in India and the buraku in Japan. Among Americans today, sanitation workers have three times the on-the-job death rate of police and firefighters.[21] In leafcutter ants, the clean-up crews are an expendable part of this process.

Newly commissioned garbage collectors of *Atta colombica* first take on the job of shuttling refuse to the dump. When they grow older, they settle down on the exposed midden heap, where they tear up the defunct mulch. Because this species has aboveground waste disposal, contaminants can build up near the nest. Whenever possible, the midden ants take their refuse to an overhang, where it tumbles away, and the wind and rain carry it farther. I have also seen the ants heave trash down embankments into a river—a common, if ill-advised, strategy for people as well.

In species of *Atta* that store trash belowground, different individuals concentrate on removing the dead, hauling fresh or old waste, and rearranging all this trash.[22] By fragmenting and mixing the garbage, these last workers, along with the roaches and other insects that live in the refuse, speed the degradation of the compost into humus, which, when complete, is likely to render disease organisms impotent. More of the waste managers appear during *Escovopsis* outbreaks, when they quickly remove the dangerous material from the living quarters.[23]

A belowground dump may get its start as an old garden chamber, perhaps after the resident fungus goes bad. But other species, such as *Atta vollenweideri,* position their dumps deeper in the earth, in chambers created for this function. These isolated caverns can be the size of a human coffin, and just as pleasant. How many cumulative years of labor must be required of the workers to empty such volumes of hard earth, which can be 6 meters—the equivalent in human terms of 3 kilometers—underground! In this, as the Bible advises, we should consider the ant's ways and be wise: trained by the long march of evolutionary history, the leafcutters have come to invest far more in recycling, environmental safety, and public health than we do.

LEAFCUTTERS AND TREES: A LOVE-HATE RELATIONSHIP

I awoke to the gunshot sound of splitting timber, the creak of cable-thick vines stretched to the limit, the rumble and boom of a tree cleaving the canopy and striking earth, and the ensuing downpour of canopy plants, animals, and debris. I sprang out of my sleeping bag and unzipped my tent to the first glow of morning light in the forest, my heart beating hard as I tried to remember where I was. It came to me: Una Biological Reserve, in the Atlantic coastal forest of Brazil. I had heard such cataclysms many times before, but the sounds of a treefall always filled me with a sense of dread, even though I knew they carried for kilometers, and even though I knew such an event brought forth new life. A new space had opened in the rainforest for young pioneer tree species and for the leafcutter ant colonies that prefer their especially succulent foliage.

The evolution of leafcutters and their fungus is a story of an alliance between an animal and its food. While most research on leafcutters has focused on the damage they inflict on plants—and certainly, the ants cause harm to most of their victims—*Atta* develop a positive alliance with some of the trees they plunder, by providing, as they do for their fungus, conditions suited for the plants' reproduction, though in a form of species interdependence that is far harder to spot.

Treefall gaps are ordinarily filled in by vegetative regrowth, especially from pioneer trees. When a leafcutter colony matures in the gap, workers maintain the clearing by dismembering the foliage growing back at the site and hauling away the smallest bits of debris on the ground.[24] While the original treefall tears the canopy fabric from above, leafcutters clear from below, opening a space to human height above the nest. In addition to creating an exposed and well-illuminated environment, leafcutters—like other ants at their nests, but at a grander scale—loosen and aerate the soil, cycle earth from underground, and assist in production of humus through decomposition of trash, thereby adding to the soil's nitrogen and potassium content.[25] The mature trees around the perimeter of the nest extend their roots into the trash middens on the surface of the ground, where nutrient levels can be enhanced a hundredfold.[26] When they're around *Atta* species that bury their refuse, the trees tap into whatever chambers their roots can reach.[27]

Good light, rich soil, and a clean shot at the canopy—conditions at a leafcutter nest would be perfect for plant seedlings, too, except for the obvious fact that the ants strip away any vegetation growing on their mounds, killing smaller plants.[28] But just as the demise of a tree makes room for pioneer plants, when an *Atta* colony dies, it offers opportunities for new plant life. Over its lifetime, the tree *Miconia argentea*, for example, loses massive quantities of leaves to the leafcutters, but the ants retrieve its seeds to the nest, where they remove any clinging fruit for garden fodder.[29] Then they throw the seeds into their fertile middens, where they can survive being eaten by lying dormant until the colony dies. Leafcutter ants can hoist and carry only tiny seeds, which are the ones most often produced by pioneer trees. Thus the plants sprouting from a defunct nest are likely to be species suited to the next generation of *Atta*.

Saplings have less chance to sprout at the dying nests of *Atta* species that discard their trash—including seeds—far underground. Even so, I like to think these leafcutters still assist the reproduction of some of the vegetables they consume. During tropical downpours, a lot of booty, including seeds, is knocked from the mandibles of ants returning to the nest. On the opposite side of the earth from the leafcutter ants, on the island of Borneo, I have watched Dyak tribesmen walk along their traditional trails eating forest fruits and tossing away the pits. Like unintentional Johnny Appleseeds, both ants and people may sow the next generation of their favorite trees along their preferred paths.

Whether dropped along a trail or thrown out with the garbage, only a minute fraction of the seeds moved by leafcutters take root, whatever the plant species. But the constant loss of foliage, flower, fruit, and seed is a given among trees, for which on average one offspring in a centuries-long lifetime experiences a ripe old age. For plants such as *Miconia argentea,* then, a loss of foliage to leafcutters may be a small price to pay for reproduction. No relationship in nature and life is without its costs, after all: human love stories take time and energy, even couples counseling. In a similar way, "ant plants" and their resident ants form alliances beneficial to both, even when a plant's services of food and lodging come at a very high price for the plant.[30] As we saw for leafcutters and their fungi, in the diplomatic machinations of nature, there is no black and white, no good and evil. There are advantages to sleeping with the enemy.

THE FIRST FUNGUS GROWERS

Although the earliest steps of fungus growing by ants are lost to history, genetic studies have revealed many specifics about the evolution of their agriculture. The first ancestor of the leafcutters that developed a relationship with fungus was almost certainly a slovenly species living in the mold-prone litter of the tropical rainforest some fifty million years ago. Such ant Cro-Magnons were trash collectors, and their habits are retained today by inconspicuous relatives of the leafcutters known as the lower attines.[31] These New World ants—of which there are 180 species—raise fungi not on foliage but on such ordinary food sources of fungi as moist organic debris and the manure of other insects.[32] Their farming methods don't require a complex division of labor. Colonies of lower attines are small, simply organized, and so retiring that the ants curl up and play dead when threatened rather than fight; their fungi are virtually indistinguishable from wild species and can in fact go feral.

This mode of fungus farming was the ant world's only form of agriculture for thirty million years, until one species bred a fungus with improved edible tips, domesticating it to such an extent that its prospects for growing wild became less likely, if not almost nil. From this species arose further novelties. Some ants began to rear a spherical yeast version of the same fungus; another switched to a different kind of fungus entirely. The descendants of these two groups and their unique fungi are alive today.

Cutting fresh leaves was the third, most phenomenally successful of these evolutionary experiments, exploiting the tropics' unlimited supply of vegetable matter, a superabundant source of solid tissues generally unavailable to ants. Leafcutters arose eight million to twelve million years ago when South America was going through a dry spell.[33] Grasslands were spreading and rainforests were in retreat. In the insect version of the first hominids who left the forest to walk upright in the expanding savannas of Africa, leafcutters may have gotten their start in grasslands because of the newfound abundance of herbaceous plants these habitats offered. Like many human crops, the fungus, under the care of the ants, did especially well outside its place of origin, in situations where its forest-dwelling ancestors would not have survived.[34] Much later some leafcutters moved back into the forests, but today, even there, most prefer the relatively open and disturbed sites. By clearing forests for agriculture, humans have brought back the good old savanna days for many leafcutters.

OF ANTS AND MEN

Close examination of the evolution of leafcutters from lower attines suggests uncanny parallels with the history of human agriculture. Some early records of human subsistence farming document marginally domesticated species: like the lower attines, farmers frequently added wild specimens to their breeding stock, which in turn sometimes went wild, as cassava does in the Amazon. As a result of their limited inbreeding and frequent crossbreeding with free-range populations, the crops of lower attines and subsistence farmers were hearty, buffered from disease and environmental change by their genetic versatility. Since they weren't strongly modified for cultivation, however, yields were low, limiting the societies of lower attines and subsistence farmers to a few hundred or few thousand individuals.

The emergence of fully domesticated crops—modified for high yield in such a way that they are no longer able to revert to life in the wild—has permitted both leafcutters and modern humans to scale up their farming efforts to fully exploit novel resource opportunities such as those available in open habitats. Not that their lives have been made easier by it. Early on, human farming turned into backbreaking, multistep work that became the only viable option for survival when the press of high populations made a return to harvesting wild food impractical, a difficulty referred to as "the plant trap."[35] Leafcutters likewise invest massively in maintaining their gardens. Their specialized laborers, who, in addition to their other tasks, must still forage, but now to feed their fungus, toil at a tempo no less exhausting than that of their more predatory sisters, from army ant to weaver ant. But the efforts of both leafcutters and people have in turn supported (and to some extent required) their big populations. The agriculture-fed civilizations of both exploded in size, eventually into the millions.

Humans have been adept at cultivating many species, as reflected in our urban diets today. No ant has achieved this versatility (although those feeding on honeydew often tend several insect "cattle" species). The colonies of both lower attines and leafcutters subsist on a single cultivar, growing no more than one kind of fungus in a nest. All leafcutters, in fact, rear the same fungal species. That fungus comes in many subtle varieties, however, and each colony devotes itself to one strain, propagated by asexual cuttings in much the same way humans farm bananas. That strain is established by the tiny, and as a result probably genetically uniform, sample of fungus the queen uses to start her nest, which she carries on her mating flight in the same pouch in her cheek that her workers use to gather up disease organisms and debris.

The gardens' isolation underground must help keep a fungal strain pure. At the same time, ants will actively work to maintain genetic purity. If a scientist transfers a clone from one nest to another of the same species, for example, the ants will weed it out, even though it should serve them just as well as their own strain. Whether today's ants can detect and select improvements in their stock, judging differences in nutrition or yield, is unknown. Their predilections may amount less to selection for the ideal fungus meal than to a kind of imprinting on its chemistry—flavor or odor—much as ants imprint on their nestmates; like people who insist on particular childhood recipes, gardening ants seem to tolerate only what they are familiar with, relegating anything else to the trash heap.

Still, a practical reason for a colony to remain faithful to one variety is that the fungus clones from different nests hawkishly compete with each other, creating "no-fungus zones" between strains and lowering crop yields.[36] In an ant nest, the fungus lets the ants do this fighting for it: compounds from ingested hyphae pass through ant digestive systems unaltered but cause a chemical reaction if a worker should excrete them as fertilizer on another strain of fungus. Sensing this reaction, the ants weed out the alien strain, retaining the garden's genetic consistency over the life of the colony.[37]

Yet lower attines frequently shift to a new variety of fungus cultivar, and this happens from time to time with leafcutters also.[38] When a shift is forced upon a colony experimentally by replacing its gardens with a fungus strain the workers would normally reject, the ants, having no alternative, come to adopt the new variety within days.

Given the ants' diligence in keeping the family-fungus pedigree, how does such a shift come about in nature? On occasion, a nest must suffer a loss of fungus from famine, flood, contagion, or (with some lower attines) seizure by garden-eating ants that can't rear their own fungus.[39] Finding a replacement may be the only recourse. Robbery, a widespread practice among ants, is one option.[40] Gardening ants are known to pillage fungi from each other, thus spreading cultivars, much as people do through trade or sometimes theft.[41]

The result of the leafcutter ants' fussing over their gardens is that all but the subtlest genetic variations are bred out. The fungus flourishes only because the workers' efforts relieve their cultivar of the burden of living in the real world.[42] Without the ants' intimate supervision, the unvarying leafcutter fungus, in monocultures of hundreds of kilograms in a nest, is particularly vulnerable to catastrophic destruction by the *Escovopsis* fungus—a problem more deadly than anything faced by the hearty fungus breeds of the lower attines.

We share with leafcutters a refined sense of what we like. We settle on certain favorite foods and then use artificial selection, inbreeding, and clonal propagation to heighten and maintain the characteristics in them that we prefer—consider the flavor difference we've come to expect from varieties of apples such as a Macintosh or Granny Smith. As a result, our crops, like the garden fungus, have lost the genetic diversity that their ancestors could draw on to survive disease and environmental change. Along with the ants, we must actively protect our cultivars. Controlling plagues has been part of the raising of inbred crops since ancient times, and risks remain grave for foods grown in monoculture—as, nowadays, most are. On this issue, we might once again heed the biblical injunction to "go to the ant" and consider her ways. Human

In the Kaw Mountains of French Guiana, *Cyphomyrmex* rears fungus—in this case, a yeast—on caterpillar feces (alongside an incidental carapace). Unlike leafcutter ant gardens, those of lower attines are small, and their cultivar shows little sign of domestication.

agriculture began after the last ice age. Leafcutters have reared genetically uniform crops a thousand times as long. Why, then, haven't leafcutter societies collapsed from their own version of the Irish potato famine? The answer lies in the fact that the ants keep their crops immaculately clean. Through an elaborate division of labor unmatched by that of any living creature besides people, ants sow those crops, weed them, cull them, manage their wastes, treat them with pesticides, and divide them among many chambers so that, when diseases do appear, the stricken gardens can be quarantined and killed.

Pests of human crops have been swift to evolve a resistance to our pesticides. The ant's version of the potato blight, the *Escovopsis* fungus, must have done the same a staggering number of times over millions of years. Meanwhile, the ants must have countered each threat to their cultivar countless times over the eons with changes of their own. What can we learn from agriculture as practiced by ants? For them, the only effective response to threats from nature has been constant vigilance.

argentine ant

the global invader

Linepithema humile, worldwide

16

armies of the earth

Weaver ants attain colony sizes of a half million; certain driver ants, possibly twenty million or more. But we close this book with the Argentine ant, whose dominion is the granddaddy of them all—colonies that can span hundreds of square kilometers. Knowing that as colonies grow larger, their inhabitants tend to become more aggressive, I had anticipated that colonies this large would have an almost unlimited capacity for bloodshed, and I was on the brink of witnessing their battles.

It was the fall of 2007, and I was in southern California with David Holway, an energetic associate professor from the University of California, San Diego. With us was Melissa Wells, my partner in adventure since we had met nearly a year earlier, at the counter of Pearl Oyster Bar in Manhattan. A redhead with a swimmer's strong body, Melissa had a hunger for exploration (and for oysters) that matched my own. Though she proclaimed a preference for elephants, she thought chasing ants was pretty nifty, too. I had taken her to Laos and Cambodia, where she produced a video of me with the weaver ants, and now we were en route to what, I assured her, was the largest battlefield on Earth. Our visit would lead me to conceive of a new kind of superorganism, one not limited by the corporeal boundaries of time and space or the ordinary lines between individual, society, and species.

But at first Melissa and I must have looked doubtful as David drove past the suburbs along Del Dios Highway in San Diego County. He turned off in Escondido, a secluded coastal

Previous page: Argentine ants, shown here in Argentina, move continuously between nests within a colony, a fluid lifestyle that constantly mixes the population.

Dead bodies of Argentine ants piling up along the battle lines between the Lake Hodges Colony and the Very Large Colony in Escondido, California.

development north of Del Mar. Navigating through well-tended streets, he finally parked along a curb and proclaimed, "This is it!"

We were surrounded by tidy homes and hedges. Melissa shot me a quizzical look, but David urged us to our knees at the curb, and there, in a bare patch of dirt and continuing over the concrete, we saw a finger-wide, chocolate-brown belt of tiny dead bodies, piled up by the thousands. The heap of corpses continued out of view, hidden by the undergrowth. From what I'd read, the battlefront could extend for miles. David explained that thirty million ants die each year in border skirmishes between the Very Large Colony and the Lake Hodges Colony. That's a casualty every second.

"Last night's rain shower may have broken up the fighting," David warned us.

That may have been the case, but already the battle lines were reforming. Trails of ants, converging from all directions, led the troops over the remains of the dead. Scanning the action through my camera, I gave Melissa and David the blow-by-blow on dozens of fierce confrontations. Most started one-on-one, with a slow and meticulous approach followed by a thrust-and-grab. Atop the corpses, pairs of workers pulled on each other, indefatigably, for minutes (and for all I knew, hours) on end. Here and there, a third or fourth worker joined in. I focused my camera on a group of three ants pulling on another that was already missing an antenna. As

I watched, a hind limb tore free. The worker who wrenched it off stood for a moment as if surprised at her success, the leg hanging from her jaws, before dropping it and inspecting her adversary's stump.

This intimate view was reminiscent of the warfare of species such as the marauder ant and the weaver ant, where workers use a similar rack spreading technique. But while marauder ants fight only during the few hours when they happen to encounter another colony in a raid, and weaver ants engage in skirmishes that can go on intermittently for weeks, the Argentine ant battles ceaselessly. The ants actively police every centimeter of their territories, right up to a precise perimeter that constitutes a band of violence.

The quiet suburb we were visiting, it turned out, was just one front in a vast war between gigantic Argentine ant empires. Around Escondido and to the west lay the holdings of the Lake Hodges Colony, a kingdom spreading over almost 50 square kilometers. To the north was the dominion of the Very Large Colony, a single society whose territory, stretching almost 1,000 kilometers from the Mexican border to California's Central Valley and on past San Francisco, boggles the mind. Given that the average Escondido backyard can sustain a million ants, the total population of the Very Large Colony could approach a trillion individuals, with a cumulative weight approximating that of the human residents of Carmel, one of the California cities it occupies.

Little wonder, then, that these Argentine ant republics are called supercolonies.

Argentine ants are unicolonial, which means individuals mix freely among nests, or rather, untold millions of nests. Ants from other colonies or other species, though, are attacked. In disputed areas, combat rages. As the losses build on either side, the boundaries shift. This movement of borders can be as slow as the creep of a glacier.[1] The most extreme change recorded was 70 meters over the course of a month, when troops from the Very Large Colony overran the Lake Hodges Colony and seized a portion of scrubby Escondido land. In time, the Lake Hodges army took it back again.

Stepping away from the front lines, Melissa kicked a chunk of wood to reveal a mass of ants and brood. Three tea-colored queens ran to hide below a leaf. As a fourth queen rushed into view, David explained that the supercolonies owe their enormous populations in part to a myrmecological twist: they produce multiple queens that grow wings but travel only on foot, staying in their colony to give birth to still more ants. The California supercolonies contain queens by the millions. Nothing can stop a colony from thriving, growing, and expanding—except clashes with other supercolonies.

COLLIDING KINGDOMS

Until 1997, fighting among Argentine ants was unknown, and many thought the species was in effect one big happy family. That summer, UC San Diego undergraduate Jill Shanahand accidentally incited an Argentine ant skirmish—and so drew back the curtain on their internecine warfare.

Jill was assisting with a project on Argentine ants spearheaded by UCSD graduate student Andrew Suarez. The lab maintained a stock of the ants collected on campus. One day, Jill decided to augment their supply by gathering ants in her parents' yard in Escondido, where Argentine ants were plentiful. When she put the new ants into a plastic tub containing the ants from campus, she was shocked to see the two groups rip into each other. Were the fights a function of the ants' captivity, or were the workers clashing for some other reason, most likely because they originated from independent, hostile colonies?

The answer to this question came only in 2004. David had recently become a professor at UCSD, and another Melissa, Melissa Thomas, was his first postdoctoral student. For three months, day after day, she drove around San Diego County looking for a colony perimeter. She had batches of live ants stashed in the back seat, which she planned to mix with others from different sites like a chemist looking for a reaction. If her captive ants from UCSD were attacked by the local ants, it signified they were from a different colony; if they were not molested, it indicated they were from the same colony. Her goal: to use this test for aggression to find where the two colonies came into contact.

Melissa's task proved more laborious than she had anticipated, given how huge their territories turned out to be. Starting in Jill Shanahand's parents' neighborhood, she expanded her search street by street until she eventually found two different supercolonies, each occupying one side of a road—an asphalt no-ant's land that must greatly reduce the death toll, at least from combat. Then, early one day in April, eureka! She located a site where the nests were clashing, thus becoming the first biologist to witness a raging turf war that was hidden from view among suburban blades of grass.

Over the months it became clear to her how enormous and unique the colonies are. Even today, only four colonies are known to live in California: Escondido's Lake Hodges Colony, two other supercolonies in the southern part of the state, and the Very Large Colony, which controls not only the UCSD campus but also much of the rest of California.

With his stories of supercolony mayhem, described from the curb in Escondido, David had our full attention. Melissa asked how the Argentine ant had become invincible. "Why don't you visit Argentina with me?" David suggested. "That's where the Argentine ant comes from. What we've learned about *Linepithema humile* in its home range tells us a lot about how the species has conquered California. It's stunning."[2]

Melissa glanced around at the California ranch houses as we got back into David's car. Now that, she agreed, sounded more like a proper adventure. "Perhaps," said David. "But before you leave California, I must show you something."

BEWARE, LOCAL ECOSYSTEMS

David drove us to the edge of the housing development, where a potpourri of native chaparral plants spread off into the distance. As we hiked past buckwheat, sagebrush, black sage, and laurel sumac, David explained how Californians are accustomed to finding Argentine ants trespassing in their pantries, on their kitchen counters, and in their gardens, where they're regarded as a nuisance. However, the damage they do is far more insidious. The ants tend to immense numbers of Homoptera—aphids and scale insects—helping them to flourish on backyard roses and in California's fruit orchards, with severe economic consequences. But it is on the scale of natural ecosystems that their impact is most serious, for they are harbingers of death for many indigenous species. This is especially true for local ants and everything that depends on them.

Argentine ants are as tenacious in the wars they wage with other ant species as they are in battles with their own, annihilating even California ants with far bigger and meaner workers. Though the Argentines can't sting and are too small to bite humans, they use the energy-rich honeydew from their homopteran herds as fuel to quickly find and dominate every food resource they can reach, thereby leaving the competition hungry. But their depredations go further than that, for even when native species don't vie for the same resources and offer no physical threat, the Argentine ants plunder their brood for an easy meal.

Some native ant species mount a weak defense. David pointed out a large circular nest mound of seed-harvester ants. They had plugged their entrance in much the same way that *Formica* block slave raids by Amazon ants. To no avail: the passing columns of Argentine ants would whittle the nest away over time through starvation or repeated assault, giving this native species no chance to proliferate. Even when mature nests manage to hang on, they are, as David put it, "essentially the living dead."[3] The few indigenous ants that survive are able to do so only because they forage underground or come to the surface when it's too frigid for Argentine ants. All other species that live in the habitats taken over by the Argentine ants have disappeared, their defensive tactics no match for the persistence of the South American conquerors.

The cleansing of indigenous ants has occurred everywhere Argentine ants have spread, namely, to every continent except Antarctica—especially in regions with a Mediterranean or subtropical climate, which include some of the most coveted human real estate on earth, southern California included.[4] In each case, the ants form supercolonies, the largest of which extends from Italy to Spain's Atlantic coast, a distance approaching 2,000 kilometers.

The effects of the Argentine ants' conquest have only begun to be studied. On the mountain slopes of Hawaii, where the species invaded in the 1940s, the workers ravage populations of other ants, as well as predatory species like wolf spiders, herbivores such as caterpillars, detrivores such as snails, and pollinators like moths and bees.[5] These organisms may be far larger than the Argentine ants, but the workers wear them down over long periods of time until they succumb. When it has eliminated these prey, a supercolony continues to thrive by increasing its dependence on honeydew.[6]

Even some vertebrates have been disappearing from coastal southern California, thanks to the Argentines. Horned lizards, shaped like thorny pancakes, require a balanced diet of diverse native ants, whereas the Argentine ants are themselves too small and quick for the horned lizards to catch. Adding insult to injury, the Argentine workers are experts at harassment, literally driving the reptiles into the sand.[7]

Argentine ants feeding on a protea seed in the fynbos habitat of the Western Cape of South Africa.

The best-documented effects of Argentine ants on plants are in the fynbos, a Mediterranean-like ecosystem in South Africa. Here, the original ant fauna is pivotal in seed dispersal, as I saw in 1996 on a visit to these vibrantly colored heathlands with Cape Town myrmecologist Hamish Robertson. Beneath the blood red flower stalks of one of the resident proteas, Hamish showed me native *Anoplolepis custodiens* ants that were hauling the plants' seeds to their nest, where the workers would eat outgrowths on the seeds called elaiosomes. Afterward, the discarded seeds, like those of hundreds of other local plants, would have a chance to sprout in the ants' nutrient-rich trash heap. A short drive away, an encroaching Argentine ant supercolony had taken over a different area, wiping out the *Anoplolepis*. There, we found the ants eating the elaiosomes in the open but leaving the heavy seeds behind to dry out and die.[8]

The native plants of southern California face a similar threat. A small tree called the California bush poppy, which David pointed out on our visit to the besieged harvester ant colony, depends on the vanishing harvesters to carry off its hefty seeds.[9] This is only one of many native California species in decline in the areas encroached on by the Argentine ant. The result is that the flora and fauna of California and other affected regions are becoming increasingly uniform, in a tragic process known as biotic homogenization. In this battle, we are all the losers.

A DESTROYER ABROAD, BUT JUST ANOTHER BRUISER AT HOME

A year later, Melissa and I found ourselves shaking hands with David Holway again, this time in the stifling early morning heat of the Corrientes bus station in northern Argentina. We had just arrived from Buenos Aires, where the steaks were immense and I had managed a tango lesson without injuring either Melissa or the tiny but stern dance instructor. Still groggy from thirteen hours on a bus, we climbed into David's tin can rental car with his former student Ed LeBrun and drove another three hours to the field site. Relieved to finally stretch our legs, Melissa and I tromped with David and Ed through a thorny pastureland along a river, watching out for stray bulls. At our feet meandered thin columns of Argentine ants.

The discovery that California's Argentine ants join forces to create supercolonies had piqued David's curiosity as a postdoctoral student at UCSD. Was this ant, clearly an outrageously proficient fighter in California, as murderous in its native country? With this question in mind, in 1997 David and three colleagues had headed to the region Melissa and I were now visiting, where the species demonstrates its preference for moisture by living in river floodplains.

To the researchers' surprise, the Argentine ant is nondescript in its indigenous land, so innocuous and with the workers so sparsely distributed that the locals know it as the "sugar ant" (after one of its favorite foods) and hardly give it a glance. Moreover, workers transplanted between nearby nests often fight one another, suggesting the colonies are many but small. I confirmed this by putting some workers into a vial and carrying them to an adjacent field. There I dropped them on top of the local sugar ants, which went on the attack.

Even more unexpected, David and his colleagues discovered that the home-grown Argentine ants belong to a rich and sustainable community of ant species, among them relatives of California's besieged harvester ants. This isn't to say their relationships are harmonious. Just as in California, the Argentine ants wage territorial battles with other ant species that they then raid for their favorite snack, ant brood.[10] But in Argentina the resident ants put up a much better fight, so all the species persist.

Indeed, the Paraná River drainage in this part of Argentina is an exceptionally cutthroat environment for ants, forcing the indigenous species to hone their battle skills. The drainage is the original home of not only the sugar ant but also other species that are invading vast swaths of the world: *Pheidole obscurithorax, Pseudomyrmex gracilis, Paratrechina fulva, Solenopsis richteri, Solenopsis invicta,* and *Wasmannia auropunctata.*[11] As one friend put it, any ant colony exported from this region is like a World Cup–winning soccer team sent to suburban Ohio to compete in the junior-varsity soccer league.[12]

What has turned the sugar ant and these other species into unrivaled warriors? All of them harvest a broad range of foods through mass recruitment that brings great numbers of workers to a feeding site, creating intense competition. With many ant species, however, aggression between colonies declines after the combatants tussle, establish territorial borders, and then back off. Much as humans do in similar situations, the ants adjust to the other's presence in a kind of stalemate, even going so far as to establish a strip of land in limbo between disputed areas. Though the two sides continue occasionally to test each other, mortality drops—a behavior known as the dear enemy phenomenon, observed in ant species such as the weaver ant and in vertebrates from frogs to birds.[13]

Argentina's shifting floodplains allow for no such stalemate. Frequent rising waters repeatedly force colonies to high ground—even up trees—and into whatever temporary living space is available. Each time the floods drain away, the ants descend to their former homeland, where they must struggle to reestablish their territories from scratch. This training regime of agile movement and repeated and relentless marathon combat has turned the ants into expert globetrotters and efficient, cold-blooded killers. Shifting from place to place at the slightest opportunity, the Argentine ants and their floodplain adversaries move swiftly into any available living quarters and thus are able to hitchhike on ships bound for far corners of the world. Once a ship docks and the ants disembark, their aptitude for resettlement and conquest gives them control over the new landscape.

JUMPING AND BUDDING

A mastery of leapfrogging to far locations is called jump dispersal, and it is a highly developed skill among tramp species of ants. Before humans introduced oxcarts, boats, cars, and planes, the tramps had the potential to extend their ranges only when winged queens were blown by storms or colonies caught a ride on water-borne detritus. These stressful events rarely took the ants very far and had a low probability of success. The castaways often expired in transit.

A few days after our arrival, David took us to the murky Paraná. The river was so broad that the opposite bank was a barely visible green line, with masses of vegetation drifting on the currents and container ships passing in the middle distance. To the roar of howler monkeys in the trees, David led us on foot to the river bank along a route made treacherous by haphazard piles of massive driftwood and other dreck. I passed the toothy skull of a piranha being picked clean by Argentine ants. With the abundance of such resources at the water's edge, it's no wonder the ants often end up on flotsam. That must have been how they were swept away by the river for millennia, until the arrival of human forms of transportation accelerated successful jump dispersal for them a millionfold, and took them much greater distances as well. With a few workers and a queen or two, a splinter group of Argentine ants can survive in just about any carrier, whether a barrel of produce or the soil nourishing a small potted plant.

The Argentine ant's relationship to North Americans traces back to early trade out of Buenos Aires. The Plymouth Rock for these ants was likely a dock in New Orleans, the city where they were first recorded in 1891, in all probability arriving aboard a ship loaded with coffee. The first sightings of the species in California occurred circa 1907, in the San Francisco Bay area and around Los Angeles. The origin of these populations is not known, but since the Panama Canal had yet to open, it seems likely that they came from the southeastern United States rather than from Argentina, possibly stowing away in a railroad car.[14]

Slowly but surely, Argentine ants spread through southern California. Though the invaders never made off with our children, as some tabloid newspapers in the 1980s implied they

Argentine ants eating a piranha on the banks of the Paraná River in northern Argentina.

might, the ants have become a part of the landscape, and they continue to expand onto land as yet uninhabited by their species. Even the Very Large Colony is still growing. Where moist areas are continuous, a colony can diffuse outward as the ants migrate to previously untenanted sites, an incremental process of budding new nests that remain part of the same colony in this case. Such migrations occasionally take place across uninhabitable sites such as narrow roads, but the Lake Hodges Colony is split by two freeways, a more formidable obstacle. With other species, a queen on her mating flight could have traversed such a barrier to form a new colony on the other side, but the Argentine queen doesn't have that option. Either the Lake Hodges Colony is older than the freeways, or the colony was accidentally brought to the other side via human commerce—jump dispersal.

Jump dispersal has been essential for the Argentine ant's success at domination, as demonstrated within California. The spread of the species by budding nests is slow—up to 150 meters a year. If the Very Large Colony had expanded only by this method, it would have grown to just a tenth its current size. However, given the volume of cargo we haul not just across the oceans but between neighborhoods, modern-day humans must aid and abet ant stowaways in reaching thousands of new sites every day.

The Argentine workers are less crowded on their home turf, most likely because they contend with more numerous, and more dense, colonies of their own species as well as colonies of other equally aggressive ants. No doubt the competition forces colonies to divert labor and resources from foraging and colony growth, keeping the size and the density of the worker populations under control and allowing other species to settle and expand. Elsewhere in the world, less skilled fighters fail, and so the Argentine ant infests new territory in prodigious numbers.

When the ants first arrived in New Orleans and spread across the Southeast, their numbers were astronomical from the start.[15] In general, tramp species are expected to have few limits on their populations, thanks to what's known as ecological release: they are free not only from their competitors but also from the predators and parasites that bedevil any organism in its native range. Even though an Argentine supercolony's death toll is astonishing by human standards of warfare, mortality remains low overall because the vast majority of workers live far from borderland clashes. Once an Argentine ant colony lays claim to a plot of ground, the workers there may never experience conflict with their own or any other kind of ant. And so their populations shoot sky high.[16]

CONTROLLING THE LANDSCAPE

The success of the Argentine ant is facilitated by its fluid lifestyle. "The species is weird because nests are decentralized; they live anywhere," David Holway said. In this way, they resemble weaver ants with their interconnected roadways and leaf nests. But unlike weaver ants, Argentine ants invest little in infrastructure, which is one reason they easily expand their colonies without territorial limits. "They take whatever's around, usually near the ground surface," David explained. "If a nest under a stone becomes too hot, some workers and queens move under adjacent leaves. Some of them will move again as those spots dry out, or they'll shift locations to be near some aphids, all in the course of a day."

The ants bud new nests when moving not just beyond established borders but anywhere within the colony, and they do it all the time. The frequent shifting of domiciles within a territory is called seminomadism, a practice that seems to go hand in hand with living in temporary encampments. Such camps sound like an army ant bivouac, except, as David explained, instead of staying in one compact nest as army and marauder ants do, "Argentine ants are constantly nomadic, everywhere in a colony at once."

Supercolony sprawl and suburban sprawl are strongly linked by ants' and humans' thirst for water. If I could make the ants glow, irrigated human properties would phosphoresce to their edges, with additional illumination spreading along waterways and in moist natural habitats. These glowing patches would expand and contract as the workers came and went from nests, or created and abandoned nests based on temperature and water availability, a pattern tied to daily and seasonal cycles. But some luminosity would extend into the dark areas, as enclaves of the colony eke out a living up to 200 meters from obvious water sources.[17] It's in such corridors that the invaders do the most harm to California's scrub ecosystems.

The ants' constant redistribution of nests minimizes the time spent in foraging and commuting to meals, whatever the food is and no matter if it is clumped or scattered.[18] As a result,

Argentine ants have no commitment to a particular resource or site but control the useful part of the landscape absolutely. Their trails appear to emerge along lines of frequent passage, much as human paths form on beaten grass, though the ants are guided by scent rather than by the wear of footprints.[19] The densest columns develop between nests or from nests to productive food-harvesting sites, with the ants both reestablishing old routes and starting new ones swiftly and easily.

Fanning out from the routes, the workers behave a bit like army ants in acting through sheer force of numbers. Foragers advance over uncharted ground by laying exploratory trails, but they radiate out in a looser, more scattered way than army ants do at their raid fronts.[20] Physicist turned biologist Jean-Louis Deneubourg led the team that described this process, in which the whole group generates a trail behind it that leads back to a nest. The success of the operation depends on the workers departing a nest en masse while continually laying pheromones:

> The Argentine ants' exploratory behavior is exceptional in that they mark continually and explore collectively. Whereas other recruitment trails are constructed between two points (e.g., nest and food), their exploratory trails have no known destination, progressively advancing into the unknown. They rapidly lead new explorers to the frontier between the just explored and the about to be explored zones, avoiding situations where ants will end up exploring the same zone twice, and help returning explorers reach the nest directly. A wide corridor of the chemically unmarked area is thus systematically "swept" and marked in a minimum time with maximum economy.[21]

Each worker can wander at least half a meter from her neighbors, which suggests that, unlike an army ant, she is relatively free to explore on her own. Where their nestmates are scarce, the foragers take straight paths, which spreads them swiftly over new ground. As their numbers increase, the workers begin to take more irregular courses, such that each one's movements become limited to a smaller area, until the workers saturate the terrain.[22]

Argentine ants employ these exploratory patterns in order to muster a concentration of workers everywhere at once. They may not attain the aggregate densities and strength of raging army ants packed in a raid, or show the multipronged communication systems of the weaver ants, but the race is not to the swift, nor is the battle to the strong. The Argentine ants turn out to succeed despite a lack of many of the organizational skills we have come to expect from large societies in this book. The workers show a minimal division of labor, without polymorphism. They do not have assembly lines and teams, and they are not adept at moving food in a group (nor do they need to, since theft from competitors is so unlikely that they can eat the food where they find it). Yet they take to phenomenal extremes the rapid dominance military practices deployed by the marauder ant. Like a starfish that succeeds in prying open a clam through persistent application of pressure, these ordinary-looking imperialists wear down nasty rivals and prey many times their weight in wars of attrition staged over hours, days, weeks, and even years.[23]

17 the immortal society

In 1997 chemists Dangsheng Liang and Jules Silverman, working for Clorox, a company that makes baits for ants and cockroaches, were raising both kinds of insect in the laboratory. When their practical technician decided to feed their stock of Argentine ants a diet of the roaches on hand, what ensued was an example of scientific serendipity that parallels Jill Shanahand's discovery of Argentine ant warfare. At first, the ants happily ate their new food source. But then, as Dangsheng wrote me, "One day we noticed that instead of eating the roaches, the [ants] were trying to kill each other. Then we found out that the technician had switched the species of cockroaches he fed the ants, from German cockroaches to brownbanded cockroaches, which he had lots of that day." Within the hour the container was littered with dead ants.[1]

It turns out that Argentine ants will set upon any group of nestmates that has been in contact with a *Supella longipalpa,* or brownbanded cockroach, a pest introduced to the eastern United States from West Africa. The contaminated ants do not fight back because they still recognize their attackers as colonymates. But over the subsequent weeks, the surviving outcasts form their own group—essentially, a new society.

A society has been described as "a group of individuals" that is "organized in a cooperative manner."[2] But this description is incomplete: to create a stable, cooperative society, the members must also identify as a group, and to do so, they must see each other as similar and outsiders as different. To accomplish this, the members generate and recognize "labels"—shared signs of their identity, such as a common language or national flag for humans.[3] Hydrocarbon molecules on the body surface, detected as scents, are the labels ants use to form their societies, and individuals lacking the right ones may be ruthlessly killed. Dangsheng and Jules immediately understood that the exoskeleton of a brownbanded roach, perhaps by coincidence, has some critical component of the scent that Argentine ants use to cue in on one another. Contact with a roach transferred these hydrocarbons and caused the ants to be misidentified as belonging to an enemy colony.[4]

While confusing the scent of one colony with that of another can be a disaster, acquiring the right odor is like being given the key to a city: all is possible. In an orchard in Daintree, Australia, I tore apart a weaver ant nest to find an orange arachnid 5 millimeters long marked with clean white stripes. I recognized the species through the pain of the ant bites: *Cosmophasis bitaeniata,* a jumping spider that joins a weaver ant colony as if it were an ant itself. This identity theft is achieved when the spider takes on the colony's aroma by stealing and eating brood, after which it easily moves into the nest to seize more larvae from the nursing minor workers. In one sense, it's got it made. In another, though, this eight-legged interloper is now at great risk: because it has taken on the identity of that colony, it cannot travel to another weaver ant colony without being attacked as an invading ant.[5] Its plight is similar to that of an Amazon ant queen, who acquires the scent of the colony she enslaves by slaughtering and then licking the resident queen. Then she, in a way, is also enslaved by the ants she has conquered, unable to leave their nest for another.

In Queensland, Australia, the jumping spider *Cosmophasis bitaeniata* lives in weaver ant nests by taking on the same scent the ants use to identify their colony.

The ability to distinguish self from other (and friend from foe) has been a theme of evolution since the inception of life, beginning in earnest with the aggregation of cells into organisms and continuing with the grouping of organisms into societies (or "superorganisms," in societies with a sharply defined sense of self). Evolutionary turning points often require the components at one level of complexity (the bodies in a society, the cells in a body, even the parts of a cell) to come together to establish a group that takes on an identity of its own—a process of social bonding universal in nature from microbes on up.[6] When its identity is signaled clearly by all its constituents, an organism such as an ant or a society such as an ant colony is easily recognized, giving it a clear individuality.[7]

THE SCENT OF KINSHIP

Regardless of their usual hostility toward outsiders, colonies can be widely inclusive—sometimes without being tricked. Consider the carpenter ants that share space with acrobat ants in South American ant gardens, or the leafcutter ant and its fungus, which require each other absolutely; in these cases both species benefit from the association, and they (especially the latter pair) evolve together as integrated parts of one society.[8] Parasites, however, must trick the other species to join its society, to the detriment of the unwitting partner. Sometimes a society will trick an individual: in slavemaker colonies, a kidnapped worker accepts her captors as nestmates despite differences that seem obvious when we watch the ants through our microscopes. Actually, this acceptance is the least of it: to avoid fights within the nest, every slave and slavemaker must accept all the other slaves as well, no matter when or where they were captured. To further confuse matters, a slavemaker occasionally captures more than one kind of slave, yet there are no fights even with three or more species living in the same nest. None of them have a problem identifying the others as nestmates.

The slave example suggests ants can learn to recognize nestmates even when they come from a different environment or are genetically distinct. This must also be true for Argentine ants, whose colonies range over diverse environments and contain the progeny of multiple queens. Although all those queens arise within a single colony, their offspring exhibit some genetic differences.[9] Either the ants come to accept the varied odors present, or the odor cocktail of each colony member is diluted by food exchange or grooming between ants. That could be enough for the nestmates to achieve the same average scent—the only scent they need to function in their society.[10]

The changing assortment of offspring of the colony's many queens must create a changing scent profile, so Argentine workers must be able to refine their ability to distinguish friend and foe as they age. This is also seen in their adaptable response to enemies outside the colony. Familiarity with outsiders breeds contempt in Argentine ants, which do not exhibit the dear enemy phenomenon. In one experiment, workers were at first able to touch members of a foreign colony by sticking their antennae through a mesh barrier; afterward they attacked more ferociously than if they had met the enemy on the battlefield for the first time.[11]

SIZE AND IDENTITY

Ants don't break down their social identity into categories the way a person does. The same man can identify himself as an American, as a resident of Illinois and of Chicago, and as a fan of the Bears: one label doesn't invalidate another. But even the most discriminating ant can only distinguish other individuals by their caste: as queen, soldier, or small worker. Other associations between nestmates—a team carrying food or pinning down an enemy, for example—are temporary and impersonal. Nor is it likely for an ant to show an allegiance to a particular site or group of nestmates within a colony. In one study, Argentine ants were marked with a radioactive tracer and then allowed to disperse; within three days, the workers had spread outward to other nests at least 40 meters away.[12] That's a lot of "ant miles" they put between themselves.

Despite the ants' intermixing and flexibility in accommodating new odors, it is possible that the vast extent of a supercolony can mean differences in communication signals, leading to a breakdown in colonymate recognition. The mixture of scent signals used at one location might not apply elsewhere in the colony, for example. Local changes that affect the signals may arise by mutation, random shifts in gene frequency, or crossbreeding, as males fly in from other colonies to mate. The short-lived colonies of other species are unlikely to display genetic variability from place to place (residents can easily walk through the full territory of a colony, blending the population). But a supercolony has decades to accumulate genetic novelties, and because they spread slowly relative to the size of its territory, they will be limited to a particular area.[13] The spread is further inhibited by the fact that reproductive Argentine ants disperse little: queens stay in their birth colony and travel only on foot, while males are weak fliers and are usually killed if they land in another colony. These factors lead to inbreeding and result in regions of a supercolony becoming genetically distinct, as is the case within the Very Large Colony.[14]

Would the Very Large Colony fall apart if a genetic change altered the scent signals its members use to recognize each other? Not necessarily. Suppose that the gene affecting the colony's odor mutated in a queen. If this signal disrupted the colony identity, the workers in her birth nest would kill her, and the mutation would die with her. But if the mutation were subtle, she and her offspring would survive, and the colony would thereby incorporate the modest new aroma into its identity.

With discrete sites within a supercolony accumulating numbers of such small variations, we might expect ants from distant parts of the same supercolony to fight after being accidentally transported between suburbs in a rosebush or tossed together in a research tub.[15] Yet surprisingly, no such overt hostility has been recorded.[16] Researchers have brought together workers from hundreds of miles apart within the range of the Very Large Colony, from San Francisco and San Diego, and the ants accept their sisters-in-arms after taking at most a second to inspect them. Such perfect camaraderie amazes me. Argentine ants direct their aggression entirely toward outsiders, and none whatever toward their billions of colonymates.[17] Their smoothly run societies make ours, marred by meddling, sharp differences of opinion, cheating, selfishness, outright aggression, and occasional homicides, look positively dysfunctional. As frequent *New Yorker* contributor Clarence Day put it in 1920: "In a civilization of super-ants or bees there would have been no problem of the hungry unemployed, no poverty, no unstable government, no riots, no strikes for short hours, no derision of eugenics, no thieves, perhaps no crime at all."[18]

SOCIETIES WITHOUT END

The groups of Argentine ants that first arrived in the southeastern United States and, later, in California were each at most a few queens and workers from one colony. "It would be as if all of the people in the United States were descended from the Pilgrims who came here in 1620," says expert Neil Tsutsui.[19] Each supercolony therefore contains only a sampling of the genes that existed within the same colony at its home base in Argentina. This is known as the founder effect, which commonly occurs when a population is established in isolation from the rest of its species. It is widely thought that the founder effect explains how Argentine ants form supercolonies overseas: a pilgrim colony might be missing some of the genes that encode labels for colony identity or that are involved in the ants' ability to discriminate between labels and thereby distinguish colonymate from outsider. By simplifying the factors involved in colony identity, the loss of these genes might reduce the misidentifications that lead to civil unrest within a large population, allowing the invading colony to expand into a supercolony.[20]

This hypothesis is based on the premise that the complete allegiance of the overseas ants to their huge colonies is a result of genetic differences between supercolonies and their less impressive counterparts in Argentina. But do their much denser populations, vast territories, and capacity to wipe out other ant species indicate a change in the Argentine ants' behavioral abilities in other parts of the world?

I think not. While the colonies in Argentina are smaller than most supercolonies abroad, close inspection shows that they usually contain numerous queens and nests spread over hundreds of meters, a phenomenal area by any standard. Any of these colonies has the ability to grow to a mighty size, requiring only favorable conditions with no equally matched competitors.[21] The Argentine ant's flexible approach to food and shelter evidently trumps any limitations to its colonies caused by a dearth of genes. We saw that biological success emerges with little species diversity in weaver ants (of which there are only two kinds); Argentine ants go a step further and do surprisingly well with little genetic diversity.

How, then, do independent Argentine ant colonies with their own identity originate? In Argentina as in California, no airborne queen has been recorded. In the absence of mating flights, the intriguing possibility arises that there are no truly new colonies. When Argentine ants bud a nest, it remains part of the original society because all its workers and queens mix freely with residents of the nests from which they emigrated. The only way for another colony to appear at a location is for a fragment of a different colony, complete with queens and workers, to arrive at that spot by jump dispersal. Before people introduced more reliable forms of long-distance transportation, this was possible only by rafting on river debris to new locations, which yielded the intricate patchwork of colonies in Argentina.

All Argentine ants, both in Argentina and abroad, therefore must identify with a limited number of colonies that continue indefinitely and are largely inbred.[22] Each of California's four supercolonies, for example, originated from a different colony in Argentina, with its own social identity. Each is able to associate only with the populations it spins off and its mother colony, and not with the populations derived from any of the other supercolonies. The main reason the Very Large Colony is so very large is that it was first to arrive in California.

SOCIETIES AS SPECIES

It is reasonable, then, to think of California's four supercolonies as nothing less than the *very same societies* that invaded the state a century ago. Whereas most ant colonies go through a life cycle similar to that of an organism—being born when a queen rears her first brood and dying when the queen dies—Argentine ant societies are different. They have achieved a kind of immortality. Of course, both the queens and workers in them today are distant descendants of the original founders, much as the cells in our body are replaced many times in a lifetime.[23] But unlike the cells in a human being, the lines of descent within a supercolony constitute an ever-expanding body—a superorganism without discernible end.[24]

That's only the half of it. Like the protagonist of Gogol's story "The Nose," we don't expect our body parts to wander off. But a supercolony's ability to span space and time leads to quirks in individuality like nothing described before. Not only do Argentine workers move freely between interconnected nests, but because the pilgrim ants transported by nursery trucks in California or on floating debris in Argentina produce offspring that identify with the colony they came from, they spread their nationality. By leapfrogging about, each society re-creates itself in fragments. One part of the Lake Hodges Colony, for example, thrives 50 kilometers north of Escondido, around the town of Temecula, where it is as isolated as Alaska is from the Lower 48. New Zealand is occupied by a single supercolony, which—given the commerce between New Zealand and California, where the Very Large Colony controls the port cities of Richmond, Oakland, San Francisco, Long Beach, Los Angeles, and San Diego—is very likely an offshoot of the Very Large Colony. No one has checked.

Supercolonies confound our notions about societies, populations, and species like nothing else. An Argentine ant society is separated socially and reproductively from all other Argentine ants by an intolerance of outsiders. That differs from humans, whose cultures, though often violent toward one another, have a history of interbreeding.[25] Because there is almost no interbreeding between supercolonies, each effectively exists in isolation, as genetically isolated as lions are from tigers. In a very real sense, each Argentine ant society is its own species.[26]

Among Argentine ants, a new society (one with its own identity) may therefore be able to form only over the slow march of time. New species—"groups of interbreeding natural populations that are reproductively isolated from other such groups"—can originate as a result of the genetic differences that accumulate when a population becomes isolated from, and eventually unable to reproduce with, other populations of the original species.[27] Similarly, it should be possible for a supercolony to split in two over time as an isolated part of the society changes enough that the two groups would kill each other if they came back into contact.[28]

Back in Escondido, I extended a finger to touch the mêlée of workers on the borderlands between Lake Hodges and the Very Large Colony. I must have broken up a fight. Immediately the translucent ants scurried over my skin, their delicate bodies causing a barely perceptible tickling sensation before they fell harmlessly onto my feet. The Argentineans are right: what a bland beast this is! As a graduate student rummaging through Harvard's ant collection for novel morphology as a clue to fascinating behavior, I didn't give the feeble-looking Argentine ant workers a second glance. As the police would say, they have no distinguishing marks or features—or none that anyone but an ant nut would notice.

Yet these ants and human beings are the only animals capable of forming societies that can grow without bounds.[29] As a result, the Argentine ant, together with a few other invasive species and our own kind, has taken over vast tracts of the Earth. This achievement requires not individual strength but social coordination at a mass scale, made possible for the ants by their persistent group identity: a supercolony that truly lives up to the name superorganism.[30]

THE BATTLE OF THE SUPER ANTS

Only one thing appears to bring the Argentine ant to its tiny chitinous knees: an encounter with an invasive species more insidious than itself. This has already happened in the southeastern United States. The red imported fire ant, *Solenopsis invicta,* indigenous to the same floodplains of Argentina as the Argentine ant, entered North America in the 1930s via the port city of Mobile, Alabama.[31] Spreading across the same broad regions of the South controlled by the first wave of Argentine ants that had arrived in New Orleans four decades before, the fire ant originally organized its colonies as single nests each containing a single queen. But in the 1970s, a unicolonial form of the fire ant was recorded. Superior to both the single-nest form of their own species and the Argentine ant, these organizationally scaled-up fire ants have been taking the southern states by storm.

In Argentina, the Argentine ant and the red fire ant fight intensely, as do all the species in that region, but they seem equally matched. If anything, the Argentine ant, for unknown reasons, has the edge.[32] Melissa and I saw this for ourselves on our visit to the Paraná River drainage when David Holway dropped a dead grasshopper near some fire ants in the shade of an acacia tree. The fire ants swarmed the grasshopper, but a nearby stream of Argentine ants was diverted to the site as well. Melissa pointed to a fire ant that was waving her abdomen at an Argentine ant, a behavior called flagging. The flagging worker had extruded her stinger, which she slashed at the Argentine ants. Despite this deterrent, the Argentine ants increased in numbers, killed two of the fire ants, and in thirty minutes had taken control of the grasshopper.

An Argentine ant grabbing the leg of a fire ant in a fight to control the dead grasshopper they're standing on. The fire ant is exuding a drop of poison from her stinger, which she is about to slash across her attacker.

In the southeastern United States, however, the Argentine ant has a hard time because of occasional freezes, which don't affect the fire ant as adversely. As a result, the fire ant has beaten back the Argentine ant to pockets of resistance in places like Austin, Texas, and Athens, Georgia. The surviving Argentine colonies are smaller than those in California; indeed, they are often no bigger than the colonies in their homeland.[33] These diminutive supercolonies are also genetically distinct from one another, presumably reflecting a high frequency of stowaways entering this region from Argentina—no surprise, since the bulk of commerce from Argentina to the United States is to the southern states rather than to the West Coast.[34]

An editor once asked me to write an article on the red imported fire ant. Given their fiery stings, I sighed with relief when he cancelled the idea after I showed him a few preliminary photographs. Fire ants have little behavioral finesse. Each of my close-up images looked much like the next: dark orange workers, piled high and deep on one thing or another. But make no mistake, the fire ant is formidable. A few humans die each year from their toxins, most often because allergies cause the victim's throat to swell, inducing suffocation. The fire ant can easily overpower rival ants and even birds and some mammals. By gnawing through anything, edible or not, it is also destructive to crops, farm equipment, and electrical appliances such as air conditioners. As a result, the red imported fire ant poses a worse economic and ecological menace than the Argentine ant. It causes yearly losses in America's South amounting to $1 billion.[35]

In 1998, red fire ants were detected in a delivery of plants from a commercial nursery in California, triggering a massive government probe and the destruction of dozens of incipient populations. Now, all of Orange County and parts of nearby Los Angeles and Riverside Counties are under a quarantine enforced by the California Department of Food and Agriculture that regulates the shipment of soils, straw, and live plants—any of which could hide ant stowaways. But with millions of planes, trains, and automobiles entering the state each year, the assault of these ants seems inevitable. And then the trillion ants in the Very Large Colony will likely enter into statewide combat with the stinging red hordes in what will be the next phase of the conquest of California.

conclusion four ways of looking at an ant

We do not expect people to be deeply moved by what is not unusual. . . . If we had a keen vision and feeling of all ordinary human life, it would be like hearing the grass grow and the squirrel's heart beat, and we should die of that roar which lies on the other side of silence.

GEORGE ELIOT, *MIDDLEMARCH* (1874)

Ants fascinate me as individuals, and I have developed the patience to watch a single worker for an entire day. Yet to focus on the peculiarities of an individual ant is to miss the forest for the trees. Ants, in a sense, *are* their colonies. In recognition of this, I have explored, at various points in this book, three additional ways of looking at ants. These perspectives may be expressed as analogies: the ant colony is like a human society; the ant colony is like an organism; and the ant colony is like a mind. But before revisiting these, let's consider the single ant.

THE FIRST WAY: THE ANT AS AN INDIVIDUAL

On my belly in a field near my home in the village of Greenport, Long Island, I spy a worker of the Allegheny mound ant, *Formica exsectoides*. I approach carefully, anticipating from her dance-like movements what she might do next. The tilt of her head and the rigidity of her legs reveal her focus on the task before her (seeking prey, I decide). I recognize instantly when my presence becomes a distraction. She turns, tenses. Her antennae sweep in my direction, her mandibles gaping. I back off until she settles down. As I watch, by reflex I interpret the ant's actions in terms of her intentions, even her feelings, much as I would a dog's, or another human's.

When she first noticed me, had she felt afraid? Angry? Threatened? Murderous? Perhaps instead she was incapable of having feelings. Was she more like a machine, simply responding to stimuli in a predictable way?

It's easy for us to think of ants as robots, because we judge other creatures against the standard of what we see in ourselves. Anthropocentrism, the belief that humans are unique or central to the universe, has been challenged by scientists as far back as Copernicus.[1] And just as we make assumptions about other people based on their outward appearance—"The human body is the best picture of the human soul," writes Ludwig Wittgenstein—so we impute consciousness to other beings based on the expressiveness of their bodies, particularly their faces.[2] It's their segmented bodies and masklike faces that lead us to assume that ants do not have "human" qualities of character or intelligence.

But the astonishing truth is that the brains and central nervous systems of ants and human beings share closer evolutionary ties than was once believed.[3] In light of this, I disagree with one conclusion of the author who intrigued me with his superorganism ideas when I was a student.

Lewis Thomas writes in *The Lives of a Cell* that an ant "can't be imagined to have a mind at all, much less a thought."[4] I think it likely there is a mind in there, striving to understand the few things her genetic endowments allow her to. Is she intelligent? To my way of thinking, yes. We know a worker can evaluate the living space, ceiling height, entry dimensions, cleanliness, and illumination of a potential new home for her colony—a masterly feat, considering that she's a roving speck with no pen, paper, or calculator.

If ants possess intelligence, do they also possess personalities? Can we think of an individual ant as being somehow unique? It is true that an ant's caste or role in the colony limits the actions and choices that are available to her. But does it follow that, say, all minor workers of the marauder ant are interchangeable? Not necessarily. Other animals exhibit no greater variety of behaviors than do ants, even such vertebrates as the lions, tigers, and bears that we might think of as having personality. But personality is more subtle than what we can discern from simply counting and categorizing behaviors.[5] We pass above ants at airplane height, relative to the insect's size. Use a magnifier, become as intimate with the subject as Goodall was with Flo, Flint, and her other chimpanzees, and it's possible to notice much more.

At different times I have picked out, by quirks of movement and appearance, what I am confident is the same worker from a marauder ant swarm that I had observed an hour or a day before. Theory suggests that such distinct personas develop most readily in large ant colonies, where individual ants are less obliged to take on a range of responsibilities and have more opportunities to prefer a certain task and even perfect it through repetition, much like humans learning a trade suitable for city life. This process, combined with any hardwired caste differences, may have repercussions at the colony level, resulting in increased labor specialization that enhances the versatility of the colony; such specialization and versatility are expected to be general characteristics of large societies.[6] Still, I find individual ants are easier to distinguish when nests are tiny. Just as with students in a small classroom, I can quickly identify the slackers and the overachievers (the latter are known as colony elites, when *drudges* might be a better term).[7] So-called key individuals take on most of the labor, and in some situations serve as a catalyst, stimulating others to join in. They may be the first to notice that a job needs to be done, just as the same person may always wash the dishes piling in the sink before their spouse gets around to them. Remove an elite, and productivity plummets. Sound familiar? The same thing happens in any office, factory, ball team, or family.

THE SECOND WAY: THE ANT COLONY AS A SOCIETY

Wherever we notice parallels between ant colonies and our own societies, we should remember that the ant societies came first. Ants formed coordinated labor forces of expert homemakers and superb soldiers millions of years before we came on the scene. The leafcutters invented agriculture eons before we did. The army ants have long outdone Attila the Hun. No wonder there has been a tendency since King Solomon not only to empathize with ants, but also to view them as diminutive versions of ourselves.[8] In Ovid's *Metamorphoses*, Zeus transforms an army of ants into a horde of human warriors. The poet describes these warriors, the Myrmidons, as

> True to their origin. You have seen their bodies,
> And they still have their customary talents,
> Industry, thrift, endurance; they are eager
> For gain, and never easily relinquish
> What they have won.[9]

Just as humans lend their ears to friends, relations, and countrymen, ants are responsive, largely by means of chemical signals, first and foremost to nestmates. They, like us, are the descendants of successful cooperators, and their pursuits are largely social. The commonalities between ants and people are striking. Both alter nature to build nurseries, fortresses, stockyards, and highways, while nurturing friends and livestock and obliterating enemies and vermin. Both ants and humans express tribal bonds and basic needs through ancient, elaborate codes. Both create universes of their own devising through the scale of their domination of the environment. As inveterate organizers, ants and people face similar problems in obtaining and distributing resources, allocating labor and effort, preserving civil unity, and defending communities against outside forces. But compared to humans, ants perform these tasks with a single-minded savagery, and they use anatomical and behavioral tools unique to their size and insect ancestry. Moreover, while human traditions pass from one generation to the next largely by social mechanisms, ants encode their colony's social systems primarily in their genes.

The variation in size and scale of ant populations matches that of people, from the handful of individuals in a readily movable band to several tens of millions in a vast city. It turns out to be possible to look at an *Acanthognathos* trapjaw ant colony of a few ants nesting in a twig using the paradigms that anthropologists apply to hunter-gatherers, and to examine megalopolises such as those of weaver ants the way a sociologist would study a human city-state.[10] Mature ant societies exhibit many of the same interrelated trends observed in both increasingly complex and increasingly populous human societies: a faster tempo of life and correspondingly higher information flow; more complex and nuanced communications; greater regulation and control of the environment; declining individual self-reliance and more specialization; a growing tendency for populations to subdivide into teams and form assembly lines and other labor crews; greater surpluses of energy, food, and labor; amplified risk-taking and the emergence of large-scale warfare; and the inception of social mechanisms unknown and unnecessary in small communities, such as elaborate infrastructure, efficient mass transit, and even features of a market economy, such as the collection and distribution of goods for consumers based on popularity and need.

Beyond the similarities, I have tried in this book to point out ways in which ant societies perform better than ours. The ant's self-sacrifice can be a little frightening; we have seen how readily the workers of some ant societies put themselves at risk and even condemn themselves to death when it serves the interests of the colony. Also, the lack of centralized control and the redundancy of operations in ant communities allow for fast responses to local situations and social

An *Acanthognathus* trapjaw ant colony lodged within a single twig in Costa Rica. The workers, like those of most ant species with small colonies, are slow, methodical, and capable of working independently.

Ants along the trail of a large *Crematogaster* nest in Ghana. In ant species with large mature colonies, workers tend to move fast, constantly gather information from nestmates, and rely more on joint action.

continuity even when individuals make errors or die.[11] These features also make it more difficult for parasites, predators, and competitors to bring down an ant society (in contrast, human terrorists can find easy targets in key buildings and leaders). Human hunter-gatherers and some of the earliest farming communities appear to have had a similar egalitarian social structure, without hereditary commanders. That's because members of these societies were unlikely to accumulate resources and wealth, and therefore power, so that leadership, when it emerged at all, was weak and fluid, and easily trounced by the collective will of the group.[12] As in some ant societies, there could even have been several leaders at a time, but they led by example, never by decree.

After five millennia in which despots have ruled civilizations around the world, the rise of modern democracies with systems of checks and balances represents, in a sense, a return to the ant style of governance, yet we remain dependent on hierarchies of political power. Some argue that the Internet and cell phones have enabled people everywhere to reexert collective influence over their societies, however. Without the bottlenecks that hamper bureaucracies, networks of people can handle masses of data and act on them more efficiently, like the community of scientists who weed through all the published ideas on a topic to find and follow up on the best few.[13] "Smart mobs," communicating, for example, with text messages, have disseminated ideas and combated fraud with almost antlike speed, even among people who don't know each other well.[14] Such "weak ties"—wide-ranging connections that take us beyond the tight-knit groups we interact with regularly—are likely of special importance in organizing both ants and people.[15]

Has this collective mode of organization improved the quality of life for the industrious ant? Measured by longevity, it certainly has for the queens: successful ones can live for many years, especially those belonging to species in which a single queen rears a large colony. For ant workers, however, existence appears to be "poore, nasty, brutish, and short," to borrow a phrase used by seventeenth-century English philosopher Thomas Hobbes—a matter of weeks to a couple of years. Individuals who engage in lower-risk behaviors, such as the replete workers who store food in their bodies, tend to live the longest. And yet, despite the cannon-fodder expendability of some members of large colonies, their life spans are still an improvement over those of other insects the size of an ant.[16]

THE THIRD WAY: THE ANT COLONY AS AN ORGANISM

Many ancient peoples likened their settlements to the anatomy of the human body. The Greeks reformulated this view as the *body politic,* a likeness between the body and the state, and from these comparisons the idea of the superorganism was born.[17] While the bedlam of modern human societies can make the idea of a body politic seem strained, an ant colony does often seem to act as an individual, once you get to know it. I have detected differences in temperament between marauder ant societies, with one nest appearing more aggressive or hardworking than another. Researchers have trained whole colonies of the British ant *Leptothorax albipennis* to be more proficient at migrating to new nests: the workers learn collectively, after repeated practice, perhaps in part by improving their individual performances and in part by interacting more effectively with their fellows.[18]

Some parallels to organisms are easy to see: ant colonies can be like human bodies, composed of nonreproducing workers that sustain the whole (equivalent to the somatic cells that compose the organs of the human body, such as the lungs and the heart) and a permanent reproductive queen that produces the next generation (like the human body's ovaries and testes).[19]

Still, people who see the resemblance between ant and human societies often find the similarities between a colony and an organism less apparent. Certainly a colony is a kind of individual, in the same way that a university is one entity even when it occupies many buildings. But most of us think of an organism as an integrated being with a body of a specific size and shape. A colony, which may seem nothing more than a scattered assemblage of ants, lacks this feature—but then so do organisms such as mats of fungi or ivy, which grows in a rambling manner and, as it turns out, reproduces flexibly—budding flowers here and there something like an Argentine ant colony with its ongoing production of new queens.[20]

It's easiest to grasp the likeness of an ant colony to a simple organism, such as a freshwater *Volvox,* which contains up to fifty thousand cells arranged in a sphere that can reach the size of a small ant. Some of the cells are big and capable of reproduction, but the majority of them are tiny and sterile. These sterile cells, like worker ants, collaborate to transport nutrients and work dynamically as a team, much as marauder ants do around prey, to move the sphere toward or away from light. Smaller and simpler still, because it has no differentiated sexual cells—or any other clear labor specialization—is the species *Eudorina elegans,* another swimming organism that is able to perform most of the same coordinated activities as *Volvox,* and even, like an ant colony that develops as the workers gradually emerge, goes through a simple embryonic transformation.[21]

Volvox and *Eudorina* are composed of just one or two cell types, but the human body is made up of more than two hundred varieties of cells. Complexity—usually measured by this kind of division of labor—generally increases as size increases, whether the organisms in question are individuals composed of cells or societies made up of ants. Most small colonies, like the *Acanthognathus* nests I collected in Costa Rica, have a single worker type, but a marauder ant colony, reaching a much greater size, contains a number of worker castes, including categories differentiated by both size and age. Similarly, while people in a nomadic hunter-gatherer society are essentially nonspecialists, today even a midsized town has dozens of job descriptions, and Manhattan has hundreds. The more jobs there are, the more the members of a society begin to function like tissues in a living organism, by being assembled into social networks and work groups.

Why is this so? For a small organism or group, specialization is typically unnecessary, and it might even be dangerous: with excessive division of labor, a few deaths could wipe out all the specialists, leaving jobs undone. On the other hand, a large organism or group has to have specialists for the same reasons it usually requires more intricate methods of communication and transportation: processing and distributing resources is logistically complex for a larger populace spread over a wider and more varied space. Even so, the number of specialties is always smaller than the number of chores to be done. Whether it be in colonies, cities, or organisms, creating an expert or team is complicated and expensive. Every new function must be coordinated with the others, which, in bigger, more complex bodies and groups, can require a lot of retooling.[22] Argentine ants, for one, get away with very little worker specialization despite their prodigious colony sizes.

THE FOURTH WAY: THE ANT COLONY AS A MIND

A superorganism is able to gather and use information. Like a computer, which uses segments of code to handle chunks of data, and brains, which use neurons, the colony assigns information processing to subunits, the workers. In each case the subunits are simple and redundant, which allows the whole to function even with sloppiness and local failure. The ability to process information, however, is not the same as consciousness. Neither computers nor ant colonies need consciousness to make smart choices. We have seen, for example, that individually ignorant workers are able, as a group, to select the closest or richest source of food, without any individual knowing a choice was made. In a way, the group as a whole could be said to be thinking. Cognition, of course, is hard to assess, even in big-brained vertebrates.[23] Still, it seems likely that an ant colony is more like a human mind than may at first be evident. Brains consist of neurons that, like ants, interact without direction from a central authority; thoughts emerge from these interactions in what consciousness expert Marvin Minsky describes as "a society of mind."[24]

But while the neuron occupies a fixed position and is capable only of simple responses—it functions like an on/off switch in a machine—each individual ant processes a lot of information, communicates with coworkers using an assortment of signals, performs labor, may specialize, and moves around. Does mobility give a collection of ants an advantage over the neurons in a brain? We have seen that engineers have had success with swarm-bots, groups of simple robots that self-organize like ants do to solve complex problems, such as recruiting to resources.[25] But for processing data, such mobility can be a drawback. The all-but-hardwired communication channels between neurons in the brain allow simple messages to convey complex meanings. A worker ant, if we consider her as a subunit of the collective mind, has to convey more generic information to be understood by the ever-changing workers around her.

Even accounting for their body size, ant workers have small brains when compared to mammals.[26] Still, a large nest has no shortage of processing power. The nerve cells of an army ant colony, distributed among a million or more bodies, easily outnumber those in the human cerebral cortex. However, while the superorganism may deploy a kind of swarm intelligence, with workers responding quickly to conditions at a local level, the flow of information through a whole system of roaming bodies can be slow and imprecise.[27] It's no wonder ant colonies have never been able to invent calculus or write a symphony.

Of course, humans can function extraordinarily well both individually and collectively, so our species can produce both Beethoven and the San Francisco Symphony. There can be elements of the "emergent brain" in the synergy between musicians playing a sonata. When we brainstorm with others, we are engaging in the same kind of activity that ants do when they collectively decide to focus on the closest or richest food source, and in some cases a group reaches a viable solution to a problem that no individual would have dreamed up.[28]

UNITY AND DESTINY

Despite their refined ability to work together, ants do not always live in harmony. There can be discord, most commonly over reproductive rights.[29] Typically the largest mature colonies show the least obvious internal friction and the most violence toward outsiders; consider the Argentine ant supercolonies, with their disciplined yet expendable armies of billions. The situation is reversed in small societies. A hundred or so species, most of them belonging to the ponerine group, have a fluid division of labor, even in sex, with multiple queens that are not clearly distinguishable from workers, or workers that can act as queens.[30] These societies don't exactly have traitors, but they do experience domestic strife. The *Diacamma* ants, for example, have no distinct queen. A mated worker gnaws off tiny growths on the backs of her nestmates, a mutilation that demotes these "marked for life" individuals into non-egg-laying foragers, whose ovaries shrink. Fights ensue following the death of this gamergate "queen" until a new one emerges.[31]

In some ponerine species, those individuals most physiologically ready to be queen can take over the queen's role when she dies; with this comes the danger of being mistaken for a potential competitor to the queen while she is still alive, and being harried or killed. The colonies of such species are virtual police states, in which ants root out nestmates with the potential to become egg layers.[32] This type of persecution is rare when workers and queens are so different in their morphologies that the queens can monopolize reproduction. Marauder ant workers, for example, lack ovaries altogether and therefore have no prospects for procreation. Differentiation of this kind accelerates the continued evolution of differences between workers and queens, resulting in adaptations that streamline efficiency within the workforce. In some species this has allowed for colony growth into the many thousands and beyond, as has been the case for the central characters of this book. But even species with a distinct queen caste aren't immune from conflict: when a colony has multiple queens, they may fight each other or (as we saw in the Argentine ant) be culled by workers.

Comparing ant colonies with human societies, organisms, and minds may give us insights into the question of conflict among ants. It turns out that resolving discord is a feature of biology at every level. Our own bodies are sites of strife, much of it imperceptible to us. "The unity of the organism is an approximation," write evolutionary biologists Austin Burt and Robert Trivers.

> The genes in an organism sometimes "disagree" over what should happen. That is, they appear to have opposing effects. In animals, for example, some genes may want (or act as if they want) a male to produce lots of healthy sperm, but other genes in the same male want half the sperm to be defective. Some genes in a female want her to nourish all her embryos; others want her to abort half of them. Some genes in a fetus want it to grow quickly, others slowly, and yet others at an intermediate level. Some genes want it to become a male, others a female.[33]

Often conflict can be a useful tool. Neurobiologists find that even our thoughts emerge from a cacophony of competing mental elements.[34] The vigilance of ponerine workers against upstarts, for example, resembles the way humans have wielded power through political oversight.

Citizens in a democracy may vehemently express opinions over a controversial issue yet reach a collective decision by casting votes; as we've seen, worker ants use a voting system called quorum sensing to reach a decision about where to nest.

Nevertheless, whereas ponerines like *Diacamma* can be abundant and successful, the nestmates of most ant species lead less contentious lives. We do not yet know if equanimity is essential for ants to develop large-scale societies. Instead, it may be that a worker is so unlikely to profit from conflict in a large society that social discord all but disappears, bred out over time from the choices that individuals can make. After all, when the worker is just one among thousands, what are the chances that she will take over the queen's role? It could also be that subversive behaviors exist in large colonies but are harder for human observers to recognize. For example, workers of some species, though never mated, can surreptitiously lay unfertilized eggs, which develop into male ants.[35]

Dissension among its ranks and simple organization notwithstanding, even a ponerine colony can be viewed as a superorganism. We've seen that while organisms may look like harmonious beings, conflict can be part of any healthy body. And while most familiar living things are complex, there exist simple organisms without division of labor or sophisticated communications; judging from *Eudorina elegans,* whose cells live and die as a single generation, even a clear separation of reproductive duties is not absolutely required. What all organisms do possess in common with all ant colonies, however, is that the parts are tied absolutely to the whole: no ant, not even a ponerine worker persecuted by her nestmates, has the option to get up and leave. It's the unbreakable binding force of their shared group identity that makes the colonies of all ant species superorganisms.[36]

That said, the marauder ant, certain army ants, and the Argentine ant (and perhaps some other invasive species) represent clear pinnacles of superorganism biology, showing the most parallels to biological organisms. These species lack the weaver ants' versatile social exchanges and the leafcutter ants' intricate organizational skills—but then so do the cells of such simple organisms as *Eudorina elegans* and *Volvox*. What they exhibit strongly is an integration in which the individual ant, as the basic subunit of the superorganism, exhibits a minimal degree of autonomy. She is incapable of learning much on her own, and never wanders more than an inch or two from her sisters. Yet the coordinated feats of the whole colony are remarkable. Despite the fact that army ants don't build permanent nests, but rather rest en masse, often exposed to the elements, the collective body of interlinked workers is as well regulated and homeostatic as the body of a warm-blooded mammal. In one species, the metabolism and spacing of workers keep a colony's temperature to within a degree or so of 83.5 degrees Fahrenheit.[37]

Army ant colonies also have a very low rate of reproduction, investing heavily in one large offspring at a time; this ensures that colonies are as well formed from the start as a newborn mammal.[38] Army ants even manage to forgo the infrastructure that keeps most large societies rooted in place and wander the environment with an agility unusual for such a massive social group. As a result of their cohesion, these ants in particular come closest to attaining what the Belgian poet Maurice Maeterlinck described as a "masked power, sovereignly wise." Of the honeybee, a species that has achieved a similar level of coordination, Maeterlinck asked in 1901, "What is this 'spirit of the hive'—where does it reside?"

> It comes to pass with bees as with most of the things in this world; we remark some few of their habits; we say they do this, they work in such and such fashion, their queens are born thus, their workers are virgin, they swarm at a certain time. And then we imagine we know them, and ask nothing more. . . . Their life seems very simple to us, and bounded, like every life, by the instinctive cares of reproduction and nourishment. But let the eye draw near, and endeavour to see; and at once the least phenomenon of all becomes overpoweringly complex; we are confronted by the enigma of intellect, of destiny, will, aim, means, causes; the incomprehensible organization of the most insignificant act of life.[39]

Maeterlinck's enigmas arise by means both simpler and more universal than he could have imagined. Recent investigations across the sciences and humanities are in fact proving how commonalities among colonies, cities, organisms, and minds run deep, with principles and constraints operating in a similar manner whether we look at a cell, a brain, a body, or a superorganism.[40] At each level the system exhibits a personal identity and a separateness from outsiders. Each system requires a means of distributing energy, nutrients, and information, and of removing waste.

What most excites me are the unexpected insights that emerge from comparisons between these different levels of organization. In the study of ants, our insights thus far have been limited by the visual prominence of individual workers: imagine trying to grasp the totality of a person and being overwhelmed by the sight of neurons and blood cells. I expect the superorganism metaphor will come to permeate and enrich the biological, social, and information sciences. For this to happen, it will be necessary to first understand the basic functioning of an ant colony in the same way a physician understands a human body: its metabolism and mass, its anatomy and internal integration, its growth and development, its ability to reproduce, its responsiveness to stimuli, its physiological stability and self-repair mechanisms, its capacity to distinguish self from other, and its ability to move and explore, achieve goals, glean nutrients, communicate with others, and adapt to a changing world.

A few months after seeing the Argentine ant battlefield with David Holway in San Diego, Melissa and I decided to pursue a more exotic adventure. In January 2008 we joined a research team on Easter Island led by John Loret, director of the Science Museum of Long Island. John, who at eighty still has the muscular physique of Popeye, first traveled to Easter Island in the 1950s with Thor Heyerdahl, famous for his journeys on the raft *Kon-Tiki*. He was going back now to explore caves and, with my assistance, to look for invasive ants.

Remote islands seldom harbor native ant species because ants are not skilled at crossing oceans without human help. When Ed Wilson studied the ants of Easter Island in 1973, the only species on record had been brought by commerce, and the Argentine ant was not among those found.[41] Thirty-five years later, Argentine ants had swept the island, transforming it into one wide ant hill. Melissa and I collected workers from half a dozen localities and deposited them among their sisters elsewhere, from the sleepy village of Hanga Roa to a stone wall near Mahatua to the giant moai heads of Akahanga. Everywhere the ants mixed blissfully. The entire island proved to be a single supercolony—the product, we assumed, of one introduction by ship from mainland Chile.

Melissa had another reason for traveling with me to Easter Island. With the help of a former governor of the island, the archeologist Sergio Rapu, we had arranged to be married at the edge of the Rano Kao volcano. It was an ancient ceremony, one that had not been conducted on the island for several decades. Rapa Nui tribesmen in loincloths brought us to the precipice, stripped us naked, then clothed and painted us in beaten bark, feathers, and shells. As the winds threatened to tear off these scant garments, we exchanged marital rocks selected from the volcano rim. We took our vows in a ceremony of beating drums and shouted chants that hadn't changed in centuries. At least one thing had changed, however. I looked down at one point and saw Argentine ants racing across my toes.

Experiencing untouched nature is all but impossible now. From the depths of the oceans to the farthest reaches of the atmosphere, there is no corner of the planet that humans have not explored, no place that has not been altered by our presence. Neither is there any corner of the globe that the ant cannot invade. Consider Biosphere 2, a $200 million, eight-story structure erected in the Sonoran desert of Arizona to demonstrate the power of technology over nature. Sealed off from the surrounding environment in 1991, Biosphere 2 was intended to be a closed ecological system, from which people would learn how to be self-sufficient in outer space. But a colony of *Paratrechina longicornis,* an invasive species from the Old World known as the crazy ant for its mad zigzag dashes, somehow found a way inside the otherwise impregnable glass-and-steel bubble.

Crazy ants hijacked this attempt to create an ecological utopia. By the time the project closed down three years later, they were everywhere. According to a contemporary press report, "Swarms of them crawled over everything in sight: thick foliage, damp pathways littered with dead leaves, and even a bearded ecologist in the humid rain forest."[42] To greater effect than intended, the project builders had created a microcosm of the Earth, complete with the human-induced traumas and foibles our planet faces—including the ants that hitch rides with us wherever we go.

Like it or not, ants and humans are in this together.

acknowledgments and a note on content

My fervor for ant watching has increased over years spent in rainforests, savannas, and deserts. In its scientific manifestation this fervor is myrmecology, the study of ants. In 1990, Bert Hölldobler and Edward O. Wilson published the encyclopedic *The Ants,* a technical review of the discipline that won the Pulitzer Prize. Their popular treatment of the subject, *Journey to the Ants,* published the following year, is highly recommended, as is *Ants of North America,* a field guide by Brian Fisher and Stefan Cover. (Hölldobler and Wilson have since addressed more specialized matters of social insect evolution in *The Superorganism,* published in 2008.) The availability of these works has given me the latitude to be pointedly eclectic in my coverage. I take the reader into the field to meet a few extraordinary ants, and I share my passionate interest in them, particularly their foraging and defense behavior, while ignoring many other topics, such as ant guests and reproductive ecology. I try to give a sense of how field scientists think about their subjects, while making no attempt to be comprehensive. My notes and citations concentrate on selected literature published since Hölldobler and Wilson's magnum opus.

Writing this book involved an unholy amount of correspondence concerning not just ants but also wolves' inability to recruit assistance to the kill, human walking speeds in cities, the domestication of sheep, problem solving in slime molds, and the history of trade routes between Buenos Aires and the United States. I thank the following people for their patience with my questions, and especially those whose names are in italics, who read whole chapters or more (and I beg the forgiveness of any whose names were lost when a computer crash in Australia's Daintree Rainforest erased four months of effort): Kirsti Abbott, Ehab Abouheif, Kamariah Abu Salim, John Acorn, Eldridge Adams, Yasmine Akky, Lenoir Alain, *John Alcock,* Leeanne Alonso, Gary Alpert, Ronald Amundson, Kellar Autumn, Leticia Aviles, *Stefanie Berghoff,* Samuel Beshers, Luis Bettencourt, Johan Billen, Nico Blüthgen, Chris Boehm, Eric Bonabeau, John Bonner, Paul Bosu, Andrew Bourke, Sean Brady, *Michael Breed,* Charles Brewer-Carias, Rodney Brooks, Brian V. Brown, Donald E. Brown, Jeff Brown, Mark J. F. Brown, *Stephen Brush, Stephen Buchmann,* Gordon Burghardt, Austin Burt, Alfred Buschinger, Michael Caplan, James Carey, Robert Carneiro, *Deby Cassill,* Joel Chadabe, Mark Chappell, Eric Charnov, Jonathan Cole, Ross Cole, Ray Coppinger, Dora L. Costa, *Jim Costa, Stefan Cover,* John W. Crawford, Ross Crozier, Cameron Currie, Tomer Czaczkes, Wojciech Czechowski, Diane Davidson, Lloyd Davis, Stéphane De Greef, Alain Dejean, Terezinha Della Lucia, Phil DeVries, Mark Deyrup, Tony Dixon, David Donoso, Marco Dorigo, Anna Dornhaus, Knut Drescher, *Robert Dudley,* Mark Elgar, Mark Elliott, David Emmett, Terry Erwin, Xavier Espadaler, Alejandro Farji-Brener, Donald Feener Jr., Faerthen Felix, Javier Fernández-Busquets, *Brian Fisher, Charles Fleming,* Henri Joseph Folse, Kevin Foster, Harold Fowler, André Francoeur, Megan Frederickson, Tsukasa Fukushi, Dorian Fuller, David Furth, Raghavendra Gadagkar, Paul Gepts, Rosemary Gillespie, Ronald Glasser, Paul Goldberg, Jeremy Goldbogen, Jane Goodall, William Gotwald Jr., Gunnar Grah, Donato Grasso, *Nick Griffin,* Wulfila Gronenberg, Alexandra Grutter, David Haig, James F.

Hancock, James Hare, Adam Hart, Brian Hayden, *Bernd Heinrich,* Robert Heinsohn, Jürgen Heinze, Dirk Helbing, Joan Herbers, Kristina Hillesland, Jane Hirshfield, Owen Holland, *David Holway, Kathy Horton,* Jerome Howard, Sarah Hrdy, David Hu, Krista Ingram, Benoît Jahyny, Rudolf Jander, Robert Jeanne, Raphael Jeanson, Christine Johnson, Clara Jones, Pedro Jover, Ken Kamler, Michael Kaspari, *Kevin Kelly,* Raymond Kelly, Robert Kelly, Mimi Koehl, Ron Kube, Jennie Kuzdzal-Fick, John La Polla, Steven LeBlanc, Ed LeBrun, Sally Leys, Dangsheng Liang, John Lighton, *John Longino,* Lloyd Loope, Margaret Lowman, Tom Loynachan, Don Mabry, *Richard Machalek,* Jonathan Majer, George Markin, Michael Martin, Norman Mason, Keiichi Masuko, Terry McGlynn, Stuart McKamey, Brian McNab, Raymond Mendez, Amy Mertl, Alexander Mikheyev, Peter Molnar, Nicholas Money, Corrie Moreau, John C. Moser, *Ulrich Mueller,* Dennis H. "Paddy" Murphy, *Peter Nonacs,* Sean O'Donnell, Joachim Offenberg, Paulo Oliveira, Jaak Panksepp, John Allen Paulos, Roger Payne, Christian Peeters, Renkang Peng, *Dale Peterson,* Michael Poulsen, Scott Powell, Mary Power, *Stephen Pratt,* Frederick Prete, Mary Price, David Queller, Neel Kamal Rastogi, Alan Rayner, Jim Reichman, the late and great *Carl Rettenmeyer, Howard Rheingold,* Steve Rissing, *Simon Robson,* Flavio Roces, Michael Rosenberg, Nathan Rosenstein, Stephen Roxburgh, Kari Ryder Wilkie, Bob Sacha, Victor Sadras, Guillermo Sanchez, Nathan Sanders, Joe Sapp, *Leslie Saul,* Paul Schmid-Hempel, Thomas Schoener, *Caspar Schöning,* Ruud Schoonderwoerd, Timothy Schowalter, *Ted Schultz,* Thomas Seeley, Jose Serrano, Cosma Shalizi, Paul Sherman, Peter Simmons, *Monica Smith,* Myron Smith, Cristian Solari, Frederick Spiegel, Dan Stahler, Sky Stephens, Nigel Stork, *Robin Stuart, Andrew Suarez, Frank Sulloway,* Adam Summers, Kelly Swing, Robert Taylor, Guy Theraulaz, Melissa Thomas, Barbara Thorne, Adam Tofilski, *Howard Topoff, James Traniello,* Walter Tschinkel, Neil Tsutsui, *Jeffery Scott Turner,* Stephen Vander Wall, Paul Van Mele, *Steven Vogel,* David Wardle, Christopher Waters, James Waters, Steven Watts, Bruce Webber, Rüdiger Wehner, Anja Weidenmüller, Phyllis Weintraub, Michael Weiser, John Wenzel, Ronald M. Weseloh, *James Wetterer,* Diana Wheeler, *Alex Wild,* Paul Williams, Bastow Wilson, David Sloan Wilson, Gina Wimp, Neville Winchester, Jon Wraith, *Richard Wrangham,* Peter Howard Wrege, Andrew Yang, Stephen Yanoviak, Elsa Youngsteadt, *Douglas Yu,* and Melinda Zeder. My adventure companion, the lady of my life, Melissa Wells, took pleasure in my text no matter how often she read it and offered all the encouragement I could wish for.

The National Geographic Society contributed greatly to this book in both grants from the Expeditions Council and assignments for the magazine. I'm especially thankful to Rebecca Martin and Chris Johns for their generous support.

A special thanks to University of California Press for the meticulous care and attention to craftsmanship that saw the book through to completion. My extraordinary UCP team included my sponsoring editor, Blake Edgar, who snapped up my book idea enthusiastically when he had the chance; three vigilant copyeditors, Anne Canright, Chalon Emmons, and Juliana Froggatt; the text designer, Jody Hanson; and the jacket designer, Lia Tjandra. Production editor Dore Brown in particular went the extra mile for me every day, asking incisive questions about each line of text. I thank my agent, Katinka Matson of Brockman, Inc., for connecting me with these fine people.

notes

Introduction

Epigraph adapted from Rolf Jacobsen, "Country Roads," translated by Robert Bly, with change to feminine by permission of Robert Bly, in Robert W. Bly, ed., *News of the Universe: Poems of Twofold Consciousness* (San Francisco: Sierra Club Books, 1995).

1. Ant population in the ground and litter is estimated at 8 million per hectare in EJ Fittkau, H Klinge 1973, On biomass and trophic structure of the central Amazonian rain forest ecosystem, *Biotropica* 5: 2–14. Canopy ant population is put at 16 million per hectare by Terry Erwin (personal communication). In total, there are more than 6 trillion ants in a square mile.
2. This isn't to imply that ants evolve from one of these kinds of societies to the next. Nor do human societies necessarily take this path; for example, even hunter-gatherer bands exist today.

A Brief Primer on Ants

1. In ants, part of the abdomen is united with the thorax, so among ant specialists these body parts are more correctly referred to as the gaster and the trunk (or mesosoma). The waist is formally known as the petiole; it may also have a second segment, called the postpetiole.
2. I have also seen bulldog ants turn to watch me go by before sprinting after me and leaping onto my legs—an undesirable situation given their swordlike stingers. See MW Moffett 2007, Bulldog ants: Lone huntress, *National Geographic* 211: 140–149.
3. As Deby Cassill puts it, "Our hands have segmented digits (fingers) and a one-segmented palm. Ants have single segmented digits (spines) and multiple segmented palms (tarsi)" (personal communication). See also D Cassill, A Greco, R Silwal, X Wang 2007, Opposable spines facilitate fine and gross object manipulation in fire ants, *Naturwissenschaften* 94: 326–332.
4. For other examples of "eusocial" animals and a discussion of this term and others, see James T. Costa, *The Other Social Insects* (Cambridge, MA: Harvard University Press, 2006); and Nigel C. Bennett and Cris G. Faulkes, *African Mole-Rats: Ecology and Eusociality* (Cambridge: Cambridge University Press, 2000). For views on the original conditions of ants and termites, see BL Thorne, JFA Traniello 2003, Comparative social biology of basal taxa of ants and termites, *Annu. Rev. Entomol.* 48: 283–306.
5. Benoit Jahyny, personal communication; B Jahyny, S Lacau, JHC Delabie, D Fresneau, Le genre *Thaumatomyrmex*, cryptique et prédateur spécialiste de *Diplopoda Penicillata*, in *Sistemática, biogeografía y conservación de las hormigas cazadoras de Colombia*, ed. E Jiménez, F Fernández, T Milena Arias, FH Lozano-Zambrano (Bogotá: Instituto de Investigación de Recursos Biológicos Alexander von Humboldt, 2007), pp. 329–346.
6. One ant species has dispensed with males entirely; see AG Himler, EJ Caldera, BC Baer, HF Marín, UG Mueller, No sex in fungus-farming ants or their crops, *Proc. R. Soc. Lond. Ser. B* 276: 2611–2616. Is there a take-home message for feminists from the ant sisterhoods? Probably not. The careers of ants, male or female, queen or worker, are largely immutable. Channeled into an occupation and career path from the onset of adulthood, an ant has few options in life—which brings to mind the sign posted at the entrance of the ant colony in T. H. White's *The Once and Future King:* "Everything not forbidden is compulsory."
7. EO Wilson 2005, Kin selection as the key to altruism: Its rise and fall, *Soc. Res.* 72: 159–166; EO Wilson, B Hölldobler 2005, Eusociality: Origin and consequences, *Proc. Natl. Acad. Sci.* 102: 13367–13371. For other views, see P Nonacs, KM Kapheim 2007, Social heterosis and the maintenance of genetic diversity, *J. Evol. Biol* 20: 2253–2265; and KR Foster, T Wenseleers, FLW Ratnieks 2006, Kin selection is the key to altruism, *Trends Ecol. Evol.* 21: 57–60.
8. In rare cases genetic differences are critical in caste determination; see KE Anderson, TA Linksvayer, CR Smith 2008, The causes and consequences of genetic caste determination in ants, *Myrmecol. News* 11: 119–132.
9. Different age classes are referred to as "temporal castes," though recent work suggests young workers focus on nursing chores not because they are specialists but rather because they are developmentally immature and tend to stay on the brood piles where they are born, while older workers are broadly competent at tasks both inside and outside the nest; see ML Muscedere, TA Willey, JFA Traniello 2009, Age and task efficiency in the ant *Pheidole dentata:* Young minor workers are not specialist nurses, *Anim. Behav.* 77: 911–918; and MA Seid, JFA Traniello 2006, Age-related repertoire expansion and division of labor in *Pheidole dentata:* A new perspective on temporal polyethism and behavioral plasticity in ants, *Behav. Ecol. Sociobiol.* 60: 631–644.
10. Edward O. Wilson, *Success and Dominance in Ecosystems: The Case of the Social Insects* (Oldendorf/Luhe, Germany: Ecology Institute, 1990).

11. CS Moreau, CD Bell, R Vila, SB Archibald, NE Pierce 2006, Phylogeny of the ants: Diversification in the age of angiosperms, *Science* 312: 101–104; and EO Wilson, B Hölldobler 2005, The rise of the ants: A phylogenetic and ecological explanation, *Proc. Natl. Acad. Sci.* 102: 7411–7414.

1. Strength in Numbers

1. C. T. Bingham, *Hymenoptera*, vol. 2, *Ants and Cuckoo-Wasps*, in *The Fauna of British India, Including Ceylon and Burma*, ed. W. T. Blanford (London: Taylor & Francis, 1903), p. 161.
2. It's curious that the largest marauder ants don't take a role in transporting food; see FD Duncan 1995, A reason for division of labor in ant foraging, *Naturwissenschaften* 82: 293–296. Similar arguments have been made for why small leafcutter ants ride on foliage carried by their large sisters, though in that case the hitchhikers have other functions (see p. 183).
3. In this book, the term *army ant* is used to indicate membership in the ant subfamilies Dorylinae, Ecitoninae, and Aenictinae, issues of whether they share a common ancestor aside. For general reviews, see DJC Kronauer 2008, Recent advances in army ant biology, *Myrmecol. News* 12: 51–65; and WH Gotwald Jr., *Army Ants: The Biology of Social Predation* (Ithaca, N.Y.: Cornell University Press, 1995). For views on their evolution, see SG Brady 2003, Evolution of the army ant syndrome: The origin and long-term evolutionary stasis of a complex of behavioral and reproductive adaptations, *Proc. Natl. Acad. Sci.* 100: 6575–6579; and SG Brady, TR Schultz, BL Fisher, PS Ward 2006, Evaluating alternative hypotheses for the early evolution and diversification of ants, *Proc. Natl. Acad. Sci.* 103: 18172–18177.
4. Herbert Spencer, *The Principles of Sociology* (New York: Appleton, 1876), 1: 447–600; and WM Wheeler 1911, The ant-colony as an organism, *J. Morphol.* 22: 307–325. The mid-nineteenth-century German beekeeper Johannes Mehring was perhaps the first to compare colonies (of honeybees) with whole animal bodies; see J Mehring, *Das neue Einwesensystem als Grundlage zur Bienenzucht, oder Wie der rationelle Imker den höchsten Ertrag von seinen Bienen erzielt. Auf Selbsterfahrungen gegründet* (Frankenthal: Albeck, 1869).
5. See, e.g., Rudolf Virchow, *Cellular Pathology*, 2nd English ed. (New York: Robert De Witt, 1860), pp. 12–13: "The composition of the major organism, the so-called individual, must be likened to a kind of social arrangement or society, in which a number of separate existences are dependent upon one another, in such a way, however, that each element possesses its own peculiar activity and carries out its own task by its own powers."
6. Lewis Thomas, *The Lives of a Cell* (New York: Viking Press, 1974), p. 12.
7. The identity of the signal remains unknown; see EJH Robinson, DE Jackson, M Holcombe, FLW Ratnieks 2005, Insect communication: "No entry" signal in ant foraging, *Nature* 438: 442.
8. CJ Kleineidam, W Rössler, B Hölldobler, F Roces 2007, Perceptual differences in trail-following leafcutting ants relative to body size, *J. Insect Physiol.* 53: 1233–1241.
9. The telecommunications industry may one day take advantage of this system of mass communication with a program to, in effect, release digital ants into the telecom network that will leave digital pheromones to reinforce a path where travel is easy. The electronic pheromones will decay over time, so the current best route will always be marked most strongly, limiting traffic tie-ups. See E Bonabeau, C Meyer 2001, Swarm intelligence: A whole new way to think about business, *Harvard Business Review* 79: 106–114.
10. Unfortunately, most scientists use the word *forager* to designate any worker outside the nest. It takes care to ascertain whether a worker is searching for food or carrying out such activities as looking for enemies, constructing a trail, or cleaning. Sometimes it's a matter of probabilities: while workers focus their efforts in areas that are likely to yield food, or even certain kinds of food, they can take on other functions as the need arises, as if a switch has been flipped in their head. For example, in northern Argentina, David Holway, Edward LeBrun, and I observed *Ectatomma* workers after their nest entrance was inundated: the first workers out of the nest ignored prey insects for several minutes while they cleared the entrance. I have observed Amazon ants switch from "move away from the nest" behavior, in which they ignore food, to a "foraging phase," in which they search for an ant nest to raid the brood (see chapter 13). Deborah Gordon describes how the first ants to leave a harvester nest each day assessed the local conditions but didn't return with seeds; the foragers emerged only later (Deborah Gordon, *Ants at Work: How an Insect Society Is Organized* [New York: Free Press, 1999], p. 34).
11. *Webster's Unabridged* defines "army ant" as "any species of ant that goes out in search of food in companies" (*Webster's Unabridged International Dictionary*, 2d ed. [New York, 1939]). Foraging "in companies" is the unique and archetypal trait of the three army ant subfamilies, though there are ancillary characteristics widely associated with an army ant "syndrome," such as group predation (the catching of prey in a group), group retrieval (used to describe both group transport—see chapter 5—and the retrieval of food along a common path), nomadism, queen morphology, and mode of colony foundation (see chapter 4), as originally discussed by EO Wilson 1958, The beginnings of nomadic and group-predatory behavior in ponerine ants, *Evolution* 12 (1958): 24–36.

12. *Group foraging* and *group hunting* are the most common terms, though *group* too easily conjures the discrete packs (i.e., with circumscribed memberships) of recruited workers characteristic of some "raiding" ant species that shouldn't be confused with army ants. Certain ants display intermediate strategies between solitary and group foraging, such as the Argentine ant (see chapter 16).
13. KG Facurel, AA Giarettal 2009, Semi-terrestrial tadpoles as vertebrate prey of trap-jaw ants (*Odontomachus*, Formicidae), *Herpetol. Notes* 2: 63–66.
14. Several categories of forager (scout) and recruit have been described, such as recruits that find food other than that to which they were recruited, but most of the distinctions seem limited in value. See, e.g., JC Biesmeijer, H de Vries 2001, Exploration and exploitation of food sources by social insect colonies, *Behav. Ecol. Sociobiol.* 49: 89–99.
15. Admittedly, distinguishing these situations is more easily said than done. Rather than choose their routes at random, for example, solitary foragers may favor sites where they found food before; see, e.g., JFA Traniello, V Fourcassié, TP Graham 1991, Search behavior and foraging ecology of the ant *Formica schaufussi:* Colony-level and individual patterns, *Ethol. Ecol. Evol.* 3: 35–47. As a result, multiple workers may come to one place through individual choice, rather than as a result of a coordinated action. It's also unlikely that foragers take completely independent courses, entirely ignoring any nestmates encountered in their travels; see, e.g., DM Gordon 1995, The expandable network of ant exploration, *Anim. Behav.* 50: 995–1007. To my mind, as long as workers don't strongly constrain or guide each other throughout the food search, for all intents and purposes they are acting solitarily.
16. More accurately, no worker travels without guidance from other workers for more than a minute fraction of the span explored by all the participants during the course of a raid.

2. The Perfect Swarm

1. M Moffett 1984, Swarm raiding in a myrmicine ant, *Naturwissenschaften* 71: 588–590.
2. Many details of this chapter are discussed in MW Moffett 1988, Foraging dynamics in the group-hunting ant, *Pheidologeton diversus*, *J. Insect Behav.* 1: 309–331.
3. Chapter 8 will describe the subterranean army ant *Dorylus laevigatus,* studied since my work on the marauder ant; this ant has raid speeds and travels distances similar to those of the marauder ant.
4. A swarm and column raider like *Pheidologeton diversus, P. silenus* could also be called a "marauder ant," but for clarity in this book I restrict this term to *diversus. Pheidologeton silenus* is in some ways more convergent with army ants: colonies lack stable trails, often abandon half-eaten bonanzas at the end of a raid, and are decisively carnivorous, mostly carving prey into pieces that are then carried by two or three workers. Even with its faster armies, *silenus* takes less and less diverse food than *diversus,* with a preference for poky insect larvae. See MW Moffett 1988, Foraging behavior in the Malayan swarm-raiding ant *Pheidologeton silenus, Ann. Entomol. Soc. Am.* 81: 356–361.
5. MW Moffett 1986, Behavior of the group-predatory ant *Proatta butteli, Insectes Soc.* 33: 444–457.
6. Other ant species catch prey in a group this way, for example *Myrmicaria opaciventris,* which has workers that move near their trails in such numbers that they jointly seize prey ten times their length, a tactic similar to that of *Proatta,* but perhaps with less "sitting" or "waiting"; see A Dejean, B Schatz, M Kenne 1999, How a group foraging myrmicine ant overwhelms large prey items, *Sociobiology* 34: 407–418. It is unclear in these species whether the ants should be described as foraging as a "group," in the sense that they may be constrained or guided in some way by their nestmates. That could be the case if workers clump together by actively orienting toward one another or a feature in the environment, as suggested for another species by HC Morais 1994, Coordinated group ambush: A new predatory behavior in *Azteca* ants, *Insectes Soc.* 41: 339–342. Alternatively, the ants could ignore each other but, being inactive, end up close enough together to jointly catch prey.
7. A Dejean, C Evraerts 1997, Predatory behavior in the genus *Leptogenys:* A comparative study, *J. Insect Behav.* 10: 177–191.
8. The widely spread out searches of the solitary foragers in most large ant colonies tend to result in a steadier intake of food; see D Naug, J Wenzel 2006, Constraints on foraging success due to resource ecology limit colony productivity in social insects, *Behav. Ecol. Sociobiol.* 60: 62–68.
9. Forget that all the workers in a raid constantly come and go; the effect would be the same if they all stayed within the swarm.
10. Sun Tzu, *The Art of War,* in *Roots of Strategy: The Five Greatest Military Classics of All Time,* ed. Thomas Raphael Phillips (Mechanicsburg, Pa.: Stackpole Books, 1985), pp. 21–63.
11. These swarm raiders are essentially all African and American, leaving the marauder ant to occupy the swarm-raider niche in Asia.
12. In a sense, what this means is that not only don't individuals serve as scouts, but a group of workers can't jointly serve as a scout for other raids, either. A tactic with some characteristics of mass foraging occurs in species with a trunk trail that rotates gradually at a speed that depends on food availability.

See S Goss, J-L Deneubourg 1989, The self-organising clock pattern of *Messor pergandei*, *Insectes Soc.* 36: 339–346; and RA Bernstein 1975, Foraging strategies of ants in response to variable food density, *Ecology* 56: 213–219.

13. See, e.g., H Topoff, J Mirenda, R Droual, S Herrick 1980, Behavioural ecology of mass recruitment in the army ant *Neivamyrmex nigrescens*, *Anim. Behav.* 28: 779–789. It could also be that recruitment and exploratory signals are the same, but the pheromone is deposited at a higher concentration and thus is more attractive. Workers in some solitary-foraging species can lay exploratory trails to investigate novel terrain on their own; see, e.g., EO Wilson 1962, Chemical communication among workers of the fire ant *Solenopsis saevissim*, 1: The organization of mass-foraging, *Anim. Behav.* 10: 135–147. See also chapter 16.

14. In army ants, the existence of exploratory trails is inferred from the posture of the foragers at the front: each presses her body against the ground in a manner that suggests she is releasing a pheromone on the walking surface; see J Billen, B Gobin 1996, Trail following in army ants, *Neth. J. Zool.* 46: 272–280. The only solid proof of exploratory trails exists in certain Malayan *Leptogenys*, currently the best-studied examples of convergence with army ant raids. *Leptogenys distinguenda*, for example, has swarm raids a few yards wide, and pioneers reaching unmarked ground lay secretions different from recruitment signals to food. See V Witte, U Maschwitz 2002, Coordination of raiding and emigration in the ponerine army ant *Leptogenys distinguenda*: A signal analysis, *J. Insect Behav.* 15: 195–217. I predict much more will be learned about foraging strategies from other species of *Leptogenys*, which show a striking diversity of behaviors up to and including trunk trails and true mass foraging.

15. EC Yip, KS Powers, L Avilés 2008, Cooperative capture of large prey solves scaling challenge faced by spider societies, *Proc. Natl. Acad. Sci. USA* 105: 11818–11822; and DE Jackson 2007, Social spiders, *Curr. Biol.* 17: R650–R652.

16. JC Bednarz 1988, Cooperative hunting Harris' hawks *(Parabuteo unicinctus)*, *Science* 239: 1525–1527.

17. D Kaiser 2003, Coupling cell movement to multicellular development in myxobacteria, *Nat. Rev. Microbiol.* 1: 45–54.

18. John T. Bonner, *Why Size Matters: From Bacteria to Blue Whales* (Princeton: Princeton University Press, 2006).

19. Edward O. Wilson, *Sociobiology: The New Synthesis* (Cambridge, MA: Harvard University Press, 1975), p. 53.

20. John T. Bonner, *The Social Amoebae* (Princeton: Princeton University Press, 2008); and JJ Kuzdzal-Fick, KR Foster, DC Queller, JE Strassmann 2007, Exploiting new terrain: An advantage to sociality in the slime mold *Dictyostelium discoideum*, *Behav. Ecol.* 18: 433–437.

21. T Nakagaki, H Yamada, Á Tóth 2000, Intelligence: Maze-solving by an amoeboid organism, *Nature* 407: 470.

22. J Reinhard, M Kaib 2001, Trail communication during foraging and recruitment in the subterranean termite *Reticulitermes santonensis*, *J. Insect Behav.* 14: 157–171; and J Reinhard, H Hertel, M Kaib 1997, Systematic search for food in the subterranean termite *Reticulitermes santonensis*, *Insectes Soc.* 44: 147–158.

23. Terrence D. Fitzgerald, *The Tent Caterpillars* (Ithaca, NY: Cornell University Press, 1995).

24. AN Radford, AR Ridley 2006, Recruitment calling: A novel form of extended parental care in an altricial species, *Curr. Biol.* 16: 1700–1704.

25. TM Judd, PW Sherman 1996, Naked mole-rats recruit colony mates to food sources, *Anim. Behav.* 52: 957–969.

26. B Heinrich, T Bugnyar 2007, Just how smart are ravens? *Sci. Am.* 296: 64–71.

27. For humans, see Keith F. Otterman, *How War Began* (College Station: Texas A&M University, 2004). The two activities commonly share metaphors. See, e.g., David Livingstone Smith, *The Most Dangerous Animal: Human Nature and the Origins of War* (New York: St. Martin's Press, 2007); and Bradley A. Thayer, *Darwin and International Relations: On the Evolutionary Origins of War and Ethnic Conflict* (Lexington: University Press of Kentucky, 2004).

28. The density of army ants tends to be greatest for those species raiding largely underground, presumably because space is cramped there, though they often continue to be concentrated when exposed.

29. William M. Wheeler, *Ants: Their Structure, Development, and Behavior* (New York: Columbia University Press, 1910), p. 246.

30. JH Fewell 2003, Social insect networks, *Science* 301: 1867–1870.

31. As expressed in this statement: "Men do not fight for a cause but because they do not want to let their comrades down" (Samuel L.A. Marshall, *Men against Fire: The Problem of Battle Command* [Washington, D.C.: William Morrow, 1947], pp. 42–43).

32. The only exception I have seen documented is the ant *Paraponera*, whose workers reportedly identify each other as individuals and also identify one another's trails; see MD Breed, JM Harrison 1987, Individually discriminable recruitment trails in a ponerine ant, *Insectes Soc.* 34: 222–226. Another researcher claims that workers identify individuals at times; see Zhanna Reznikova, *Animal Intelligence: From Individual to Social Cognition* (Cambridge: Cambridge University Press, 2007). This behavior is known for the queens of one ant; see S Dreier, JS van Zweden, P D'Ettorre 2007, Long-term memory of individual identity in ant queens, *Biol. Lett.* 3: 459–462. Some ants form dominance hierarchies, but they do so by recognizing not individuals per se but each other's reproductive or competitive status; see, e.g., J

Heinze 2008, Hierarchy length in orphaned colonies of the ant *Temnothorax nylanderi*, *Naturwissenschaften* 95: 757–760.
33. The slow growth of army ant colonies, for example, may reflect their high worker death rate; see S Powell 2004, Polymorphism and ecology in the New World army ant genus *Eciton*, Ph.D. thesis, University of Bristol, England, 2004.
34. M Hammond 1980, A famous "exemplum" of Spartan toughness, *Classical J.* 75: 97–109.

3. Division of Labor

1. MW Moffett 1986, Marauders of the jungle floor, *National Geographic* 170: 272–286.
2. Formally, only the minor workers, being a discrete size group, should be called a "caste" in this species; the medias, the majors, and possibly the giants, being distinguished only by their relative frequency in a size continuum, are "subcastes." Marauder ant division of labor is described in MW Moffett 1987, Division of labor and diet in the extremely polymorphic ant *Pheidologeton diversus*, *Natl. Geogr. Res.* 3: 282–304.
3. AL Mertl, JFA Traniello 2009, Behavioral evolution in the major worker subcaste of twig-nesting *Pheidole*: Does morphological specialization influence task plasticity? *Behav. Ecol. Sociobiol.* 63: 1411–1426; and EO Wilson 1984, The relation between caste ratios and division of labor in the ant genus *Pheidole*. *Behav. Ecol. Sociobiol.* 16: 89–98.
4. See, e.g., DE Wheeler, HP Nijhout 1984, Soldier determination in *Pheidole bicarinata:* Inhibition by adult soldiers, *J. Insect Physiol.* 30: 127–135. Caste sizes and frequencies can vary with locality, suggesting evolution to suit local needs: L Passera, E Roncin, B Kaufmann, L Keller 1996, Increased soldier production in ant colonies exposed to intraspecific competition, *Nature* 379: 630–631.
5. The individuals of a size that is scarce, such as those falling at the midpoint between the peak sizes for media and major workers of *Pheidologeton diversus*, might carry out the same tasks as workers of the more common sizes but accomplish those tasks relatively poorly, or they may be of a size relatively less often needed. Alternatively, they could execute different tasks than other workers—tasks that happen to require relatively few individuals.
6. Quoted in Drew G. Faust, *This Republic of Suffering* (New York: Alfred Knopf, 2008), p. 59.
7. Nathan Rosenstein, personal communication. For more examples relating to humans, see Robert L. O'Connell, *Ride of the Second Horseman: The Birth and Death of War* (New York: Oxford University Press, 1997).
8. U Maschwitz, M Hahn, P Schönegge 1979, Paralysis of prey in ponerine ants, *Naturwissenschaften* 66: 213–214.
9. MW Moffett 1989, Trap-jaw ants, *National Geographic* 175: 394–400. Related species have a broader diet; see W Gronenberg, CRF Brandao, BH Dietz, S Just 1998, Trap-jaws revisited: The mandible mechanism of the ant *Acanthognathus*, *Physiol. Entomol.* 23: 227–240.
10. SN Patek, JE Baio, BL Fisher, AV Suarez 2006, Multifunctionality and mechanical origins: Ballistic jaw propulsion in trap-jaw ants, *Proc. Natl. Acad. Sci.* 103: 12787–12792.
11. MW Moffett 1986, Trap-jaw predation and other observations on two species of *Myrmoteras*, *Insectes Soc.* 33: 85–99.
12. George F. Oster and Edward O. Wilson, *Caste and Ecology in the Social Insects* (Princeton: Princeton University Press, 1978), pp. 281–286.
13. This is true for both cold- and warm-blooded animals; see, e.g., AA Heusner 1985, Body size and energy metabolism, *Annu. Rev. Nutr.* 5: 267–293. James Waters (pers. comm.) found a similar decrease in metabolic rate with colony size in *Pogonomyrmex californicus* (expressed per capita). My "energy to spare" hypothesis is supported, for example, by the greater metabolic scope available to large animals, i.e., the factor by which basal rate can rise to maximum activity (Steven Vogel, pers. comm.).
14. R Jeanson, JH Fewell, R Gorelick, SM Bertram 2007, Emergence of increased division of labor as a function of group size, *Behav. Ecol. Sociobiol.* 62: 289–298.
15. See, e.g., JM Herbers 1981, Reliability theory and foraging by ants, *J. Theor. Biol.* 89: 175–189.
16. T Pham 1924, Sur le régime alimentaire d'une espèce de fourmi Indochinoise (*Pheidologeton diversus* Ierdon), *Ann. Sci. Nat. Zool.* 10: 131–135.
17. ML Roonwal 1975, Plant-pest status of root-eating ant, *Dorylus orientalis*, with notes on taxonomy, distribution, and habits, *J. Bombay Nat. Hist. Soc.* 72: 305–313.
18. Granivory has in turn been hypothesized to have evolved from carnivory, in areas of limited protein sources; see JH Brown, OJ Reichman, DW Davidson 1979, Granivory in desert ecosystems, *Annu. Rev. Ecol. Syst.* 10: 201–227.
19. On this military tactic, see Basil H. Liddell Hart, *Strategy*, 2d ed. (London: Faber & Faber, 1967).
20. Judged by the marauder ant's cousin, *affinis*; see SM Berghoff, U Maschwitz, KE Linsenmair 2003, Influence of the hypogaeic army ant *Dorylus (Dichthadia) laevigatus* on tropical arthropod communities, *Oecologia* 135: 149–157.
21. Harlan K. Ullman and James P. Wade, *Shock and Awe: Achieving Rapid Dominance* (Washington, D.C.: Center for Advanced Concepts and Technology, 1996), p. 25.
22. NR Franks, LW Partridge 1993, Lanchester battles and the evolution of combat in ants, *Anim. Behav.* 45: 197–199.

23. Another extraordinary exception is the jumping spider *Portia*, which catches a diversity of spider prey; see DP Harland, RR Jackson, *Portia* perceptions: The umwelt of an araneophagic jumping spider, in *Complex Worlds from Simpler Nervous Systems*, ed. Frederick R. Prete (Cambridge, MA: MIT Press, 2004), pp. 5–40. For further information on ecological specialization, see DP Vázquez, RD Stevens 2004, The latitudinal gradient in niche breadth: Concepts and evidence, *Am. Nat.* 164: E1–E19; and DJ Futuyma, G Moreno 1988, The evolution of ecological specialization, *Annu. Rev. Ecol. Syst.* 19: 207–233.
24. Similar experiments have since been done on a New World army ant—moving prey around in the litter and providing dead insects, for example—showing that the ants can subdivide their swarm raids to create a multipronged search pattern that is more efficient in dealing with food patches; see NR Franks, N Gomez, S Goss, J-L Deneubourg 1991, The blind leading the blind in army ant raid patterns: Testing a model of self-organization, *J. Insect Behav.* 4: 583–607.
25. Marauder ant raids can begin by recruitment overrun in response to food found on or near a trail by a single worker (chapter 2). Other ants use solitary foragers to investigate areas in this way, including excess recruited ants and ants lost from trails. One army ant in the southwestern United States exhibits similar behavior: after a raid begins to plunder an ant colony for its brood, further column raids spread outward from there in all directions, and as a result the ants often find more nest entrances nearby; see HR Topoff, J Mirenda, R Droual, S Herrick 1980, Behavioural ecology of mass recruitment in the army ant *Neivamyrmex nigrescens*, *Anim. Behav.* 28: 779–789.
26. S Garnier, J Gautrais, G Theraulaz 2007, The biological principles of swarm intelligence, *Swarm Intell.* 1: 3–31; and Marco Dorigo and Thomas Stützle, *Ant Colony Optimization* (Cambridge, MA: MIT Press, 2004).
27. Charles Darwin, *Descent of Man* (London: John Murray, 1872), 1: 140.
28. Lewis Thomas, *Lives of the Cell* (New York: Viking Press, 1974), p. 12; for ant colonies as brains, see W Gronenberg 2008, Structure and function of ant brains: Strength in numbers, *Myrmecol. News* 11: 25–36.
29. MW Moffett 1987, Ants that go with the flow: A new method of orientation by mass communication, *Naturwissenschaften* 74: 551–553.
30. DE Jackson, M Holcombe, FLW Ratnieks 2004, Trail geometry gives polarity to ant foraging networks, *Nature* 432: 907–909.
31. Among trail-laying ants, the density of traffic seldom overwhelms; rather, the push-and-shove causes the workers to add backup routes until the roads are commodious enough to accommodate the traffic; see, e.g., A Dussutour, V Fourcassié, D Helbing, J-L Deneubourg 2004, Optimal traffic organization in ants under crowded conditions, *Nature* 248: 70–73.
32. Theodore Schneirla, the father of all army ant research, thought that army ants could find their way home in more or less this manner, though he proposed that odor may also influence their directional choices at trail intersections; see Theodore C. Schneirla, *Army Ants: A Study in Social Organization* (San Francisco: W. H. Freeman, 1971).
33. A more complicated theory of "lane formation" is developed in ID Couzin, NR Franks 2003, Self-organized lane formation and optimized traffic flow in army ants, *Proc. R. Soc. Lond. Ser. B* 270: 139–146.
34. A John, A Schadschneider, D Chowdhury, K Nishinari 2008, Characteristics of ant-inspired traffic flow, *Swarm Intell.* 2: 25–41; A Dussutour, J-L Deneubourg, V Fourcassié 2005, Temporal organization of bi-directional traffic in the ant *Lasius niger*, *J. Exp. Biol.* 208: 2903–2912; and D Helbing, P Molnár, IJ Farkas, K Bolay 2001, Self organizing pedestrian movement, *Envir. Plann. B* 28: 361–383.

4. Infrastructure

This chapter derives mostly from MW Moffett, Nesting, emigrations, and colony foundation in two group-hunting myrmicine ants, in *Advances in Myrmecology*, ed. James C. Trager and George C. Wheeler (New York: EJ Brill, 1989), pp. 355–370; and MW Moffett 1987, Division of labor and diet in the polymorphic species *Pheidologeton diversus*, *Nat. Geogr. Res.* 3: 282–304.
1. H Samaniego, ME Moses 2008, Cities as organisms: Allometric scaling of urban road networks, *J. Trans. Land Use* 1: 21–39; and John Tyler Bonner, *The Evolution of Complexity* (Princeton: Princeton University Press, 1988). Unlike vertebrate circulatory systems, human highway systems tend to be decentralized, as are trails for species such as the weaver ant (see chapter 9).
2. The lower cost to ants of sticking to clear paths rather than going over rough ground is addressed in JH Fewell 1988, Energetic and time costs of foraging in harvester ants, *Pogonomyrmex occidentalis*, *Behav. Ecol. Sociobiol.* 22: 401–408.
3. See, e.g., AS Aleksiev, B Longdon, MJ Christmas, AB Sendova-Franks, NR Franks 2007, Individual choice of building material for nest construction by worker ants and the collective outcome for their colony, *Anim. Behav.* 74: 559–566.
4. R Beckers, OE Holland, J-L Deneubourg 1994, From local actions to global tasks: Stigmergy and collective robotics, in *Artificial Life IV: Proceedings of the Fourth Workshop on the Synthesis and Simulation of Living Systems*, ed. Rodney A. Brooks and Pattie Maes (Cambridge, MA: MIT Press, 1994), pp. 181–189.

For further examples of stigmergy, see G Theraulaz, J Gautrais, S Camazine, J-L Deneubourg 2003, The formation of spatial patterns in social insects: From simple behaviours to complex structures, *Philos. Trans. R. Soc. Lond. Ser. A*, 361: 1263–1282; and Thomas D. Seeley, *The Wisdom of the Hive* (Cambridge, MA: Harvard University Press, 1995).

5. MA Elloitt 2007, Stigmergic collaboration, Ph.D. thesis, University of Melbourne, Australia.

6. Smaller majors also perform the shoving task, but with more limited effectiveness. Such is the irony of being proficient at a job in a society: rather than growing in number as a result, as do urban pigeons adept at building nests on every window ledge, an efficiently run colony gets by with fewer extreme specialists—and possibly, when their duties are rare enough, perilously few; see EO Wilson 1968, The ergonomics of caste in the social insects, *Am. Nat.* 102: 41–66.

7. DL Cassill, K Vo, B Becker 2008, Young fire ant workers feign death and survive aggressive neighbors, *Naturwissenschaften* 95: 617–624.

8. For defensive behaviors generally, see A Buschinger, U Maschwitz 1984, Defensive behavior and defensive mechanisms in ants, in *Defensive Mechanisms in Social Insects*, ed. Henry R. Hermann (New York: Praeger, 1984), pp. 95–150; and MHJ Möglich, GD Alpert 1979, Stone dropping by *Conomyrma bicolor:* A new technique of interference competition, *Behav. Ecol. Sociobiol.* 6: 105–113.

9. CJ Lumsden, B Hölldobler 1983, Ritualized combat and intercolony communication in ants, *J. Theor. Biol.* 100: 81–98.

10. NR Franks, LW Partridge 1993, Lanchester battles and the evolution of combat in ants, *Anim. Behav.* 45: 197–199; and SD Porter, CD Jorgensen 1981, Foragers of the harvester ant, *Pogonomyrmex owyheei:* A disposable caste? *Behav. Ecol. Sociobiol.* 9: 247–256. A sad fact of human military history is that putting the most expendable soldiers in the front lines often pays off. Although the modern military does what it can to protect its soldiers (conducting saturation bombing before sending them in, for example), it is no accident that at the time of this writing, the last war in which a U.S. general died in combat was the Vietnam War—and it was caused by a helicopter crash in dense fog, not enemy fire.

11. Similar shifts back and forth between nests also occur in army ants; see WH Gotwald Jr. 1978, Emigration behavior of the East African driver ant *Dorylus (Anomma) molestus*, *J.N.Y. Entomol. Soc.* 86: 290.

12. RS Savage 1847, On the habits of the "drivers" or visiting ants of West Africa, *Trans. R. Entomol. Soc. Lond.* 5: 1–15; the quotation appears on p. 4.

13. See, e.g., C. Schöning, WM Njagi, NR Franks 2005, Temporal and spatial patterns in the emigrations of the army ant *Dorylus (Anomma) molestus* in the montane forest of Mt. Kenya, *Ecol. Entomol.* 30: 532–540; and EO Wilson 1958, The beginnings of nomadic and group-predatory behavior in the ponerine ants, *Evolution* 12: 24–31.

14. H Topoff, J Mirenda 1980, Army ants on the move: Relation between food supply and emigration frequency, *Science* 207: 1099–1100. Certain *Leptogenys* ants that, like *diversus,* are now known to raid like army ants are thought to have become nomadic prior to evolving into mass foragers, but apparently for a different reason: species of this genus move due to the frequent disturbances they experience from nesting in the leaf litter; see V Witte, U Maschwitz 2002, Coordination of raiding and emigration in the ponerine army ant *Leptogenys distinguenda:* A signal analysis, *J. Insect Behav.* 15: 195–217.

15. See, e.g., JM Leroux 1982, *Ecologie des populations de dorylines Anomma nigricans dans la région de Lamto (Côte d'Ivoire),* Publications du Laboratoire de Zoologie, no. 22, Ecole Normale Supérieure, Paris. One New World army ant has even been recorded encamped at one site for at least eight months; see HG Fowler 1979, Notes on *Labidus praedator* in Paraguay, *J. Nat. Hist.* 13: 3–10.

16. J Smallwood 1982, Nest relocations in ants, *Insectes Soc.* 29: 138–147. Even some large, entrenched colonies can migrate often; see, e.g., HG Fowler 1981, On the emigration of leaf-cutting ant colonies, *Biotropica* 13: 316.

17. V Witte and U Maschwitz 2008, Mushroom harvesting ants in the tropical rain forest, *Naturwissenschaften* 95: 1049–1054.

18. In a variant of fission called budding, far less than half the colony departs with a newly crowned queen to start their own colony elsewhere, leaving the original queen with her nest. Terry McGlynn thinks that budding and fission may be widespread among ants with small nests in rainforest leaf litter, where competition can be too intense for queens to start colonies alone. See TP McGlynn 2006, Ants on the move: Resource limitation of a litter-nesting ant community in Costa Rica, *Biotropica* 38: 419–427. Over much of pre-agricultural human history, hunter-gatherer groups arose in much the same way. Bands split as they grew beyond a few dozen individuals and either began experiencing discord or had trouble searching widely for meals as a foraging group; without infrastructure and stockpiles, such divisions were easy. These groups also fused more readily than do ant colonies. See FW Marlowe 2005, Hunter-gatherers and human evolution, *Evol. Anthropol.* 14: 54–67.

19. Though specialized egg-carrying by young queens suggests that they are able to start colonies from scratch, a possibility is that marauder ants can also re-adopt mated queens from their own or other colonies; these queens might then produce colonies by fission or budding (see note 18).

5. Group Transport

1. Much of the information on group transport in this and the next section is from MW Moffett 1992, Group transport and other behavior in *Daceton armigatum* ants in Venezuela, *Nat. Geogr. Res.* 8: 220–231; and MW Moffett 1988, Cooperative food transport by an Asiatic ant, *Nat. Geogr. Res.* 4: 386–394.
2. I am grateful to Jane Goodall, George Schaller, John Eisenberg, and Frans de Waal for these examples.
3. WL Brown Jr. 1960, Contributions toward a reclassification of the Formicidae, III: Tribe Amblyoponini, *Bull. Mus. Comp. Zool.* 122: 145–230.
4. That's a problem with wood, too; as a result, beavers often drag logs alone, and termites eat wood on the spot or slice it into bits that they then carry solo. African harvester termites, however, slice grass into appropriately sized segments and, on rare occasions, move them as a group.
5. VS Banschbach, A Brunelle, KM Bartlett, JY Grivetti, RL Yeamans 2006, Tool use by the forest ant *Aphaenogaster rudis:* Ecology and task allocation, *Insectes Soc.* 53: 463–471.
6. B Hölldobler 1981, Foraging and spatiotemporal territories in the honey ant *Myrmecocystus mimicus, Behav. Ecol. Sociobiol.* 9: 301–314.
7. Galileo Galilei, "Dialogues Concerning Two New Sciences" (1638), in *On the Shoulders of Giants,* ed. Stephen Hawking (Philadelphia: Running Press, 2002), p. 498.
8. John H. Sudd, *An Introduction to the Behaviour of Ants* (London: Edward Arnold, 1967).
9. Owen Holland, personal communication; J Halloy et al. 2007, Social integration of robots into groups of cockroaches to control self-organized choices, *Science* 318: 1155–1158; and F Mondada, LM Gambardella, D Floreano, S Nolfi, J-L Deneubourg, M Dorigo 2005, The cooperation of swarm-bots, *IEEE Robot. Automat. Mag.* 12: 21–28.
10. Even ants that perform rudimentary group transport probably balance their loads at least crudely; see NR Franks 1986, Teams in social insects: Group retrieval of prey by army ants *(Eciton burchelli), Behav. Ecol. Sociobiol.* 18: 425–429.
11. In most sports, players rotate in and out during a game, so participation isn't constant; the same holds true for workers entering and leaving a raid. The idea of noninterchangeability as a basis for recognizing "teams" was proposed by George F. Oster and Edward O. Wilson in *Caste and Ecology in the Social Insects* (Princeton: Princeton University Press, 1978).
12. Originally this model was described for army ant transport teams; e.g., C Anderson, NR Franks 2003, Teamwork in animals, robots, and humans, *Adv. Study Behav.* 33: 1–48; and NR Franks, AB Sendova-Franks, C Anderson 2001, Division of labour within teams of New World and Old World army ants, *Anim. Behav.* 62: 635–642. Their descriptions of teams don't resemble our everyday notions of teams, either the teams people form in which tasks are not done at the same time, as in baseball, or teams in which each player does the same thing, as in bowling.
13. The ant's name has been changed, having been described as *Formica schaufussi* in the original research; see SK Robson, JFA Traniello, Key individuals and the organisation of labor in ants, in *Information Processing in Social Insects,* ed. Claire Detrain, Jean-Louis Deneubourg, and Jacques M. Pasteels (New York: Springer, 1999); and SK Robson, JFA Traniello 1998, Resource assessment, recruitment behavior, and organization of cooperative prey retrieval in the ant *Formica schaufussi, J. Insect Behav.* 11: 1–22.
14. This army ant behavior is described in the references in note 12.
15. Adam Smith, *Wealth of Nations* (London: Strahan & Cadell, 1776); see also Emile Durkheim, *The Division of Labor in Society* (New York: Free Press, 1964).
16. Members of socially complex ant species tend to be simpler than members of simpler societies; e.g., K Jaffe, MJ Hebling-Beraldo 1993, Oxygen consumption and the evolution of order: Negentropy criteria applied to the evolution of ants, *Experientia* 49: 587–592.
17. J Bequaert 1922, Predaceous enemies of ants, *Bull. Am. Mus. Nat. Hist.* 45: 271–331.

6. Big Game Hunters

1. Driver ants are classified as the surface-active, swarm-raiding army ant species in *Anomma,* a subgenus within the genus *Dorylus.*
2. Ogden Nash, *Ogden Nash's Zoo* (New York: Stewart, Tabori & Chang, 1986), p. 41.
3. Y Möbius, C Boesch, K Koops, T Matsucawa, T Humle 2008, Cultural differences in army ant predation by West African chimpanzees? A comparative study of microecological variables, *Anim. Behav.* 76: 37–45; C Schöning, T Humle, Y Möbius, WC McGrew 2008, The nature of culture: Technological variation in chimpanzee predation on army ants revisited, *J. Hum. Evol.* 20: 48–59.
4. WH Gotwald Jr. 1984, Death on the march: Army ants in action, *Rotunda* 17: 37–41.
5. EO Wilson, NI Durlach, LM Roth 1958, Chemical releasers of necrophoric behavior in ants, *Psyche* 65: 108–114.
6. The segregation of tasks based on size, as occurs in more labor-intensive road-construction activities in the marauder ant, is less evident in driver ants: Caspar Schöning, personal communication; C Schöning, W Kinuthia, NR Franks 2005, Evolution of allometries in the worker caste of *Dorylus* army ants, *Oikos* 110: 231–240.
7. Marauders bring in little food during dry spells; however, they cut back foraging accordingly, with

few and weaker raids. Driver ant raids are more richly rewarded after rains, when insect abundance skyrockets. But as I later saw for myself on a trip to rainy Ghana, even then escapees are common. In the fifty or so raids I eventually observed of seven driver ant species in Nigeria and Ghana, less than one in a hundred ants, and often more like one in a thousand, returned with food in her jaws, compared to 10 to 40 percent of marauder ants. It may be that some populations are more efficient; according to one report (A Raignier, J van Boven, Étude taxonomique, biologique et biométrique des *Dorylus* du sous-genre *Anomma*, Annales du Musée Royal du Congo Belge, n.s. 4, *Sciences Zoologiques*, vol. 2 [1955]: 1–359), from 7 to 22 percent of *Dorylus wilverthi* workers carry back food, though Caspar Schöning tells me this estimate seems high.

8. EO Wilson 1958, The beginnings of nomadic and group-predatory behavior in the ponerine ants, *Evolution* 12: 24–36.

9. This runs contrary to expectations that larger colonies show a steadier food intake and most likely occurs, I believe, because of the nonindependence of raiding ants during mass foraging, which makes an army ant colony the equivalent of a single large animal; see JW Wenzel, J Pickering 1991, Cooperative foraging, productivity, and the central limit theorem, *Proc. Natl. Acad. Sci.* 88: 36–38.

10. This idea is pursued in WH Gotwald Jr. 1974, Predatory behavior and food preferences of driver ants in selected African habitats. *Ann. Entomol. Soc. Am.* 67: 877–886.

11. These are not members of the replete caste, just well-fed ordinary workers.

12. Overlapping ideas are discussed for leafcutter ants in J Röschard, F Roces 2003, Cutters, carriers, and transport chains: Distance-dependent foraging strategies in the grass-cutting ant *Atta vollenweideri*, *Insectes Soc.* 50: 237–244; JJ Howard 2001, Costs of trail construction and maintenance in the leaf-cutting ant *Atta columbica*, *Behav. Ecol. Sociobiol.* 49: 348–356; and F Roces, JA Núñez 1993, Information about food quality influences load-size selection in recruited leaf-cutting ants, *Anim. Behav.* 45: 135–143.

13. LMA Bettencourt, J Lobo, D Helbing, C Kühnert, GB West 2007, Growth, innovation, scaling, and the pace of life in cities, *Proc. Natl. Acad. Sci.* 104: 7301–7306.

14. Deborah M. Gordon, *Ants at Work: How an Insect Society Is Organized* (New York: Free Press, 1999). To study another instance of apparent "counting," researchers chopped an ant's legs short or lengthened them with straw to demonstrate that distance traveled was measured by an internal odometer; see M Wittlinger, R Wehner, H Wolf 2006, The ant odometer: Stepping on stilts and stumps, *Science* 312: 1965–1967.

15. Scott Camazine, Jean-Louis Deneubourg, Nigel R. Franks, James Sneyd, Guy Theraulaz, and Eric Bonabeau, eds., *Self-Organization in Biological Systems* (Princeton: Princeton University Press, 2001).

16. William F. Joyce, Nitin Nohria, and Bruce Roberson, *What Really Works: The 4+2 Formula for Sustained Business Success* (New York: HarperCollins, 2003); James Surowiecki, *The Wisdom of Crowds: Why the Many Are Smarter than the Few and How Collective Wisdom Shapes Business, Economies, Societies, and Nations* (New York: Doubleday, 2003).

17. The idea of homeostasis and its application to both bodies and societies was originally developed in Walter B. Cannon, *The Wisdom of the Body*, 2d ed. (New York: W. W. Norton, 1939).

18. See, e.g., JW Wenzel, J Pickering 1991, Cooperative foraging, productivity, and the central limit theorem, *Proc. Natl. Acad. Sci.* 88: 36–38; R Rosengren, W Fortelius, K Lindström, A Luther 1987, Phenology and causation of nest heating and thermoregulation in red wood ants of the *Formica rufa* group studied in coniferous forest habitats in southern Finland, *Ann. Zool. Fenn.* 24: 147–155; and Edward O. Wilson, *The Insect Societies* (Cambridge, MA: Harvard University Press, 1971).

19. Admittedly, these raids include a variety of army ants, which may select different ants as prey. See S O'Donnell, J Lattke, S Powell, M Kaspari 2007, Army ants in four forests: Geographic variation in raid rates and species composition, *J. Anim. Ecol.* 76: 580–589; and M Kaspari, S O'Donnell 2003, High rates of army ant raids in the Neotropics and implications for ant colony and community structure, *Evol. Ecol. Res.* 5: 933–939.

20. NR Franks, CR Fletcher 1983, Spatial patterns in army ant foraging and migration: *Eciton burchelli* on Barro Colorado Island, Panama, *Behav. Ecol. Sociobiol.* 12: 261–270. Because raids are only a few meters wide, the overlap avoided by foraging in this pattern is very slight, involving only the first few of the many meters traveled from the nest.

21. For army ants, "a long time" is any period longer than the normal length of a raid.

22. Such patches can be exploited even by species otherwise known for harvesting patches of dispersed items; see, e.g., F Lopez, JM Serrano, FJ Acosta 1992, Intense reactions of recruitment facing unusual stimuli in *Messor barbarus*, *Dtsch. Entomol. Z.* 39: 135–142.

23. JFA Traniello 1989, Foraging strategies of ants, *Annu. Rev. Entomol.* 34: 191–210.

24. This defensive function of trunk trails has been described for other ant species, including ones in which foragers travel solitarily from the trails; see B Hölldobler 1976, Recruitment behavior, home range orientation and territoriality in harvester ants, *Pogonomyrmex*, *Behav. Ecol. Sociobiol.* 1: 3–44.

25. Whether the ants were drawn in this direction because of the success of the previous day's raid is an open question. This was one of the four cases I documented in Nigeria in which routes were partially reused after one to three days of neglect. Twice workers reestablished a twisty trail on rocky ground, including sections that had been subterranean.
26. Carl Rettenmeyer told me he saw trail-following of prey by *Eciton hamatum* many times, but the only published evidence that this is the column raider's strategy is presented in RL Torgerson, RD Akre 1970, Interspecific responses to trail and alarm pheromones by New World army ants, *J. Kans. Entomol. Soc.* 43: 395–404.

7. Clash of the Titans

1. Caspar Schöning tells me, however, that springtails show up in driver ant prey samples in Kenya.
2. JT Longino 2005, Complex nesting behavior by two neotropical species of the ant genus *Stenamma*, *Biotropica* 37: 670–675.
3. Driver ants would prove to be among the most common chimpanzee prey; see C Schöning, D Ellis, A Fowler, V Sommer 2007, Army ant prey availability and consumption by chimpanzees *(Pan troglodytes vellerosus)* at Gashaka (Nigeria), *J. Zool.* 271: 125–133.
4. C Schöning, MW Moffett 2007, Driver ants invading a termite nest: Why do the most catholic predators of all seldom take this abundant prey? *Biotropica* 39: 663–667. Among the New World army ants, the largest recorded attacks may be the subterranean plunderings of giant leafcutter ant nests; see S Powell, E Clark 2004, Combat between large derived societies: A subterranean army ant established as a predator of mature leaf-cutting ant colonies, *Insectes Soc.* 51: 342–351.
5. C Seignobos, JP Deguine, HP Aberlenc 1996, Les Mofu et leurs insectes, *J. d'Agri. Tradition. Bot. Appl.* 33: 125–187.
6. D Inward, G Beccaloni, P Eggleton 2007, Death of an order: A comprehensive molecular phylogenetic study confirms that termites are eusocial cockroaches, *Biol. Lett.* 3: 331–335.
7. There was an indication that the ants might process termites for short-term storage rather than eating them at once. Termite reproductives were taken to the old nest with their limbs hanging loosely. By the time they were transported to the new nest they resembled streamlined lozenges, each leg and wing having been trimmed. Not that this helped: after the migration, the ants still had more food than they could handle, and the termites piled in the new nest were already starting to stink.
8. The only seeds army ants take are those with elaiosomes, digestible outgrowths similar in composition to insect prey, which plants use to lure ants to carry seeds off without eating the embryo within; see L Hughes, M Westoby, E Jurado 1994, Convergence of elaiosomes and insect prey: Evidence from ant foraging behavior and fatty acid composition, *Funct. Ecol.* 8: 358–365.
9. The narrow defense function of *Eciton* soldiers is affirmed by H Topoff, K Lawson, P Richards 1973, Trail following in two species of the army ant genus *Eciton:* Comparison between major and intermediate-sized workers, *Ann. Entomol. Soc. Am.* 66: 109–111.
10. Probably the migrations of some army ants are more commonly triggered by problems at the nest (e.g., inserting a stick) than by a dearth of food. Colonies choose a nest site over the course of a few hours at most, necessarily based on limited information. Expropriation by a returning animal, floods, and unstable support branches or cavity walls are likely.
11. For the original report, see A Dejean, PJ Solano, J Ayroles, B Corbara, J Orivel 2005, Insect behaviour: Arboreal ants build traps to capture prey, *Nature* 434: 973.
12. My research was done on the ant shrub *Cordia nodosa,* which has arcades that appear identical to those of *Hirtella physophora,* the host plant to *Allomerus decemarticulatus* in French Guiana, where the study by Dejean et al. took place (see note 11). For my experiments I used Orthoptera 6 to 15 mm long—all smaller than the grasshopper shown in the photograph in the original study. To determine how well *Allomerus* used this tactic to catch prey, I forced the issue by holding forty grasshoppers in forceps against a covered trail for fifteen seconds—longer than any healthy insect would stay put while being stung and bitten. All escaped the ants easily thereafter, except a small one, which was killed and eaten. I thank Frederick Prete for advice on the limitations of orthopteran vision and locomotion.

8. Notes from Underground

1. Aldous Huxley, *Brave New World Revisited* (New York: Harper, 1958), p. 23.
2. A Raignier, J van Boven 1955, Etude taxonomique, biologique et biométrique des *Dorylus* du sous-genre *Anomma, Annales du Musée Royal du Congo Belge,* n.s. 4, *Sciences Zoologiques* 2: 1–359.
3. For Old World army ants, see C Schöning, WM Njagi, NR Franks 2005, Temporal and spatial patterns in the emigrations of the army ant *Dorylus (Anomma) molestus* in the montane forest of Mt. Kenya, *Ecol. Entomol.* 30: 532–540. Fights have been documented for Asian subterranean species; see SM Berghoff, J Gadau, T Winter, KE Linsenmair, and U Maschwitz 2003, Sociobiology of hypogaeic army ants: Characterization of two sympatric *Dorylus* species on Borneo and their colony conflicts, *Insectes Soc.* 50: 139–147. Also, African *Typhlopone* army ants

are focused predators of driver ants; see JM Leroux 1982, Ecologie des populations de dorylines *Anomma nigricans* dans la région de Lamto (Côte d'Ivoire), *Publications du Laboratoire de Zoologie*, no. 22, Ecole Normale Supérieure, Paris.
4. MK Peters, S Likare, M Kraemer 2008, Effects of habitat fragmentation and degradation on flocks of African ant-following birds, *Ecol. Appl.* 18: 847–858. In tropical America, the birds snatch enough food to reduce the ants' raid productivity; see PH Wrege, M Wikelski, JT Mandel, T Rassweiler, ID Couzin 2005, Antbirds parasitize foraging army ants, *Ecology* 86: 555–559.
5. C Schöning, W Njagi, W Kinuthia 2008, Prey spectra of two swarm-raiding army ant species in East Africa, *J. Zool.* 274: 85–93. For a possible prey specialist army ant, see WH Gotwald Jr. 1978, Trophic ecology and adaptation in tropical Old World ants of the subfamily Dorylinae, *Biotropica* 10: 161–169.
6. For an excellent experiment on dietary specialization, see A Dejean, B Schatz, J Orivel, G Beugnon, JP Lachaud, B Corbara 1999, Feeding preferences in African ponerine ants: A cafeteria experiment, *Sociobiology* 34: 555–568.
7. B Schatz, J-P Suzzoni, B Corbara, A Dejean 2001, Selection and capture of prey in the African ponerine ant *Plectroctena minor*, *Acta Oecol.* 22: 55–60.
8. Leafcutters are the ultimate example of learned preferences in ants; they tolerate only the single strain of fungus in their own nest and not the strains of their neighbors (see chapter 15).
9. A Dejean, R Fénéron 1999, Predatory behaviour in the ponerine ant, *Centromyrmex bequaerti:* A case of termitolesty, *Behav. Processes* 47: 125–133.
10. Because most ants at an army ant raid front are small, size matching during foraging per se may not be significant. Size matching and differences in size preference can occur both within and between colonies of the same and different species; see, e.g., M Kaspari 1996, Worker size and seed size selection by harvester ants in a neotropical forest, *Oecologia* 105: 397–404; and JFA Traniello, SN Beshers 1991, Polymorphism and size-pairing in the harvester ant *Pogonomyrmex badius:* A test of the ecological release hypothesis, *Insectes Soc.* 38: 121–127. The efficiency of polymorphism can be addressed using a "pseudomutant technique" of removing selected size classes. This method has shown that in leafcutter ants, the young colonies contain the minimal size range of workers to perform efficiently; although their division of labor is flexible, a colony with the "optimal" worker size-frequency distribution may still be most efficient; see EO Wilson 1985, The sociogenesis of insect colonies, *Science* 228: 1489–1495.
11. C Schöning, W Kinuthia, NR Franks 2005, Evolution of allometries in the worker caste of *Dorylus* army ants, *Oikos* 110: 231–240; M Kaspari, MD Weiser 1999, The size-grain hypothesis and interspecific scaling in ants, *Funct. Ecol.* 13: 530–538.
12. SM Berghoff, A Weissflog, KE Linsenmair, R Hashim, U Maschwitz 2002, Foraging of a hypogaeic army ant: A long neglected majority, *Insectes Soc.* 49: 133–141; SM Berghoff, A Weissflog, KE Linsenmair, M Mohamed, U Maschwitz 2002, Nesting habits and colony composition of the hypogaeic army ant *Dorylus (Dichthadia) laevigatus*, *Insectes Soc.* 49: 380–387.
13. A Weissflog, E. Sternheim, WHO Dorow, S Berghoff, U Maschwitz 2000, How to study subterranean army ants: A novel method for locating and monitoring field populations of the South East Asian army ant *Dorylus (Dichthadia) laevigatus* with observations on their ecology, *Insectes Soc.* 47: 317–324.
14. Henry Walter Bates, *Naturalist on the River Amazon* (London: Bradbury & Evans, 1864). As we shall see in chapter 14, this is an example of task partitioning; see C Anderson, FLW Ratnieks 2000, Task partitioning in insect societies: Novel situations, *Insectes Soc.* 47: 198–199.
15. TD Seeley, PK Visscher 2003, Choosing a home: How the scouts in a honeybee swarm perceive the completion of their group decision making, *Behav. Ecol. Sociobiol.* 54: 511–520.
16. NR Franks, SC Pratt, EB Mallon, NF Britton, DJT Sumpter 2002, Information flow, opinion polling, and collective intelligence in house-hunting social insects, *Philos. Trans. R. Soc. Lond. Ser. B* 357: 1567–1584; PK Visscher 2007, Group decision making in nest-site selection among social insects, *Annu. Rev. Entomol.* 52: 255–275; and Bert Hölldobler and Edward O. Wilson, *The Superorganism* (New York: W. W. Norton, 2008).
17. J Hickson, SD Yamada, J Berger, J Alverdy, J O'Keefe, B Bassler, C Rinker-Schaeffer 2009, Societal interactions in ovarian cancer metastasis: A quorum-sensing hypothesis, *Clin. Exp. Metastasis*, 26: 67–76.
18. A likely absence of the army ant characteristics of mass foraging and prey carriage in the Nigerian sub may represent an evolutionary loss of those features, given that it is an unidentified species belonging to *Dorylus* subgenus *Dorylus*, a group not considered "primitive" (that is, basal to other army ants); see DJC Kronauer, C Schöning, LB Vilhelmsen, JJ Boomsma 2007, A molecular phylogeny of *Dorylus* army ants provides evidence for multiple evolutionary transitions in foraging niche, *BMC Evol. Biol.* 7: 56–66.
19. This "looping" behavior was first described for an army ant, but it is widespread among recruiting ants; see H Topoff, J Mirenda, R Droual, S Herrick 1980, Behavioural ecology of mass recruitment in the army ant *Neivamyrmex nigrescens*, *Anim. Behav.* 28: 779–789.
20. R Chadab, CW Rettenmeyer 1975, Mass recruitment by army ants, *Science* 188: 1124–1125.
21. The species was either *affinis* itself or a close relative. Its subterranean tendencies are documented

by M Masayuki, CE Heng, AH Ahmad 2005, Marauder ant *(Pheidologeton affinis)* predation of green turtle *(Chelonia mydas)* nest in Chagar Hutang, Redang Island and measures to protect the nests, *Proc. 2nd Int. Symp. SEASTAR2000 Asian Bio-logging Sci.*, 55–62.

22. This is another example of recruitment overrun, which occurs in both mass-foraging and solitary-foraging species. In a similar chain reaction I've seen, *Solenopsis invicta* fire ants will depart from the food to which they've been recruited, fanning out as searching individuals to recruit to other food, and sometimes enough are present to catch large prey along the way.

23. Perhaps if I'd been less distracted by *affinis*, I would also have noticed in the same forest a particular *Leptogenys* that later was described as having raids convergent with army ants; see, e.g., U Maschwitz, S Steghaus-Kovac, R Gaube, H Hänel 1989, A South East Asian ponerine ant of the genus *Leptogenys* with army ant life habits, *Behav. Ecol. Sociobiol.* 24: 305–316.

24. The detailed developmental unfolding of trunk trails is largely unknown. To transition to a raid process, trail production must become disengaged from the discovery of food, as would be the case if foragers departing from them were to employ exploratory trails in their solitary searches for food (see chapter 16). The closest to such an "intermediate" strategy is described for a species well on its way to foraging like an army ant: see FD Duncan, RM Crewe 1994, Group hunting in a ponerine ant, *Leptogenys nitida*, *Oecologia* 97: 118–123.

9. Canopy Empires

1. Generally, I won't distinguish the two species, as their ecology is similar.

2. WH Gotwald Jr. 1972, *Oecophylla longinoda*, an ant predator of *Anomma* driver ants, *Psyche* 79: 348–356.

3. The driver ants' response was a ratcheted-up version of the runaway reaction that occurs when the big-headed *Pheidole dentata* ant comes under attack from fire ants and explosively abandons its nest, with hundreds fleeing in all directions; see EO Wilson 1976, The organization of colony defense in the ant *Pheidole dentata*, *Behav. Ecol. Sociobiol.* 1: 63–81.

4. More weaver ants are recruited to snatch workers from *Dorylus* columns than are required to catch the driver ants for food, suggesting the thefts may be defensive; see B Hölldobler 1979, Territories of the African weaver ant *(Oecophylla longinoda)*, *Z. Tierpsychol.* 51: 201–213.

5. Sloppiness can be good, and it can even be built into the system: the advantage of a honeybee making errors about the direction to which she was recruited is comparable to a navy ship that "should fire salvoes with a considerable scatter, in the hope that at least one shell will hit a hostile ship and slow it down"; see JBS Haldane, H Spurway 1954, A statistical analysis of communication in "Apis mellifera" and a comparison with communication in other animals, *Insectes Soc.* 1: 247–283. See also RP Fletcher, C Cannings, PG Blackwell 1995, Modelling foraging behaviour of ant colonies, in *Advances in Artificial Life*, ed. Federico Moran, Alvaro Moreno, Juan J. Merelo, and Pablo Chacón (Berlin: Springer Verlag, 1995), pp. 772–783; and J-L Deneubourg, JM Pasteels, JC Verhaeghe 1983, Probabilistic behaviour in ants: A strategy of errors? *J Theor. Biol.* 105: 259–271.

6. The living chains superficially resemble those made by army ants to form their nests, except in army ants the workers link toe to toe rather than jaw to waist, and army ant chains are passive—they are not used to pull objects together. Other ants can use their own bodies to make bridges, ladders, flanges, curtains, walls, plugs, rafts, and other constructions. See C Anderson, G Theraulaz, J-L Deneubourg 2002, Self-assemblages in insect societies, *Insectes Soc.* 49: 99–110; and C Anderson, DW McShea 2001, Intermediate-level parts in insect societies: Adaptive structures that ants build away from the nest, *Insectes Soc.* 48: 291–301.

7. B Hölldobler, EO Wilson 1983, The evolution of communal nest-weaving in ants, *Am. Sci.* 71: 490–499.

8. B Hölldobler, CJ Lumsden 1980, Territorial strategies in ants, *Science* 210: 732–739.

9. RK Peng, K Christian, K Gibb 1998, Locating queen ant nests in the green ant, *Oecophylla smaragdina*, *Insectes Soc.* 45: 477–480.

10. In Australia, mature colonies can also have multiple queens; see RK Peng, K Christian, K Gibb 1998, How many queens are there in mature colonies of the green ant, *Oecophylla smaragdina*?, *Aust. J. Entomol.* 37: 249–253.

11. ML Smith 2007, Territories, corridors, and networks: A biological model for the premodern state, *Complexity* 12: 28–35.

12. G Beugnon, A Dejean 1992, Adaptive properties of the chemical trail system of the African weaver ant *Oecophylla longinoda*, *Insectes Soc.* 39: 341–346; and A Dejean, G Beugnon 1991, Persistent intercolonial trunkroute-marking in the African weaver ant *Oecophylla longinoda:* Tom Thumb's versus Ariadne's orienting strategies, *Ethology* 88: 89–98.

13. B Hölldobler, EO Wilson 1978, The multiple recruitment systems of the African weaver ant *Oecophylla longinoda*, *Behav. Ecol. Sociobiol.* 3: 19–60.

14. For other examples of ritualized behaviors in ants, see Bert Hölldobler and Edward O. Wilson, *The Superorganism* (New York: W.W. Norton, 2008).

15. Bert Hölldobler and Edward O. Wilson, *Journey to the Ants* (Cambridge, MA: Harvard University Press, 1994), p. 43.

16. D Leston 1971, Ants, capsids and swollen-shoot in Ghana: Interactions and the implications for pest control, *Proc. 3rd Int. Cocoa Res. Conf. Accra* (Ghana) (1969), pp. 205–221.
17. Defense of their termite resources is bolstered by preemptive attacks on competing colonies located closest to these prey: CJ Lumsden, B Hölldobler 1983, Ritualized combat and intercolony communication in ants, *J. Theor. Biol.* 100: 81–98.
18. Some human pastoralists can be viewed, like honeypot ants, as defending their resources (i.e., animal herds) rather than territorial lands per se; Robert L. Kelly and Michael Rosenberg, personal communication; M Rosenberg 1998, Cheating at musical chairs: Territoriality and sedentism in an evolutionary context, *Curr. Anthropol.* 39: 653–682.
19. Robert L. O'Connell, *Ride of the Second Horseman: The Birth and Death of War* (New York: Oxford University Press, 1995), p. 49.
20. B Hölldobler 1983, Territorial behavior in the green tree ant *(Oecophylla smaragdina)*, *Biotropica* 15: 241–250.
21. L Lefebvre, SM Reader, D Sol 2004, Brains, innovations, and evolution in birds and primates, *Brain Behav. Evol.* 63: 233–246.
22. RL Carneiro 2000, The transition from quantity to quality: A neglected causal mechanism in accounting for social evolution, *Proc. Nat. Acad. Sci.* 97: 12926–12931. These novel mechanisms can in turn create further surpluses and increases in population; see Elman R. Service, *Origins of the State and Civilization* (New York: W.W. Norton, 1975).
23. For ants generally, see, e.g., M Beekman, DJT Sumpter, FLW Ratnieks 2001, Phase transition between disordered and ordered foraging in pharaoh's ants, *Proc. Nat. Acad. Sci.* 98: 9703–9706; and R Beckers, S Goss, J-L Deneubourg, JM Pasteels 1989, Colony size, communication, and ant foraging strategy, *Psyche* 96: 239–256.
24. This estimate is based on a small colony with twelve nests bringing in 45,000 prey per year; see A Dejean 1991, Adaptation d'*Oecophylla longinoda* aux variations spatio-temporelles de la densité de proies, *Entomophaga* 36: 29–54.
25. A Dejean, Prey capture strategy of the African weaver ant, in *Applied Myrmecology: A World Perspective,* ed. RK Vander Meer, K Jaffe, A Cedeno (Boulder, CO: Westwood Press, 1990), pp. 472–481.
26. Dale Peterson and Richard W. Wrangham, *Demonic Males: Apes and the Origins of Human Violence* (New York: Houghton Mifflin, 1996), pp. 5–21.
27. W Federle, W Baumgartner, B Hölldobler 2004, Biomechanics of ant adhesive pads: Frictional forces are rate- and temperature-dependent, *J. Exp. Biol.* 207: 67–74.
28. J Wojtusiak, E Godzinska, A Dejean 1995, Capture and retrieval of very large prey by workers of the African weaver ant, *Oecopylla longinoda, Trop. Zool.* 8: 309–318.
29. N Rastogi 2000, Prey concealment and spatio-temporal patrolling behaviour of the Indian tree ant *Oecophylla smaragdina, Insectes Soc.* 47: 92–93.
30. DL Cassill, J Butler, SB Vinson, DE Wheeler 2005, Cooperation during prey digestion between workers and larvae in the ant *Pheidole spadonia, Insectes Soc.* 52: 339–343.
31. A number of related ants show similar behavior; see C Saux, BL Fisher, GS Spicer 2004, Dracula ant phylogeny as inferred by nuclear 28S rDNA sequences and implications for ant systematics, *Mol. Phylogen. Evol.* 33: 457–468.
32. K Masuko 1989, Larval hemolymph feeding in the ant *Leptanilla japonica* by use of a specialized duct organ, the "larval hemolymph tap," *Behav. Ecol. Sociobiol.* 24: 127–132.
33. The fastest and most voluminous liquid feeders tend to be nondominant ants (see chapter 10) that grab meals and run; see DW Davidson, SC Cook, RR Snelling 2004, Liquid-feeding performances of ants: Ecological and evolutionary implications, *Oecologia* 139: 255–266; and DW Davidson 1997, The role of resource imbalances in the evolutionary ecology of tropical arboreal ants, *Biol. J. Linn. Soc.* 61: 153–181.
34. B Hölldobler 1985, Liquid food transmission and antennation signals in ponerine ants, *Isr. J. Entomol.* 19: 89–99.
35. See, e.g., Walter R. Tschinkel, *The Fire Ants* (Cambridge, MA: Harvard University Press, 2006), pp. 332–333.
36. MW Moffett 1985, An Indian ant's novel method for obtaining water, *Natl. Geogr. Res.* 1: 146–149.
37. EO Wilson, RW Taylor 1964, A fossil ant colony: New evidence of social antiquity, *Psyche* 71: 93–103.
38. LU Gadrinab, M Belin 1981, Biology of the green spots in the leaves of some dipterocarps, *Malay. For.* 44: 253–266.
39. Weaver ants are known to catch and eat the pollinators of one tree species, so the value of their protective services may be mixed (though some flowers may secrete repellants to get around problems with the ants); see K Tsuji, A Hasyim, H Nakamura, K Nakamura 2004, Asian weaver ants, *Oecophylla smaragdina,* and their repelling of pollinators, *Ecol. Res.* 19: 669–673; and J Ghazoul 2001, Can floral repellents pre-empt potential ant-plant conflicts?, *Ecol. Lett.* 4: 295–299.
40. GM Wimp, TG Whitham, Host plants mediate ant-aphid mutualisms and their effects on community structure and diversity, in *Ecological Communities: Plant Mediation in Indirect Interaction Webs,* ed. T Ohgushi, TP Craig, PW Price (Cambridge: Cambridge University Press, 2007), pp. 683–738.
41. Many taxonomists now call these insects Sternorrhyncha. Before Homoptera excrete sap, they remove a proportion of its amino-nitrogens and convert many of its simple sugars into

polysaccharides; such changes are considered minor, however, and the ants that feed on honeydew are therefore widely treated as "primary consumers."

42. See, e.g., JAH Benzie 1985, Selective positioning of arboreal tents by weaver ants *Oecophylla smaragdina:* A possible co-evolutionary development with Maha-dan trees, *Syzygium cumini, Aust. Entomol. Mag.* 12: 17–19. At colony borders, pavilions can serve simultaneously as barrack nests, housing both warriors and Homoptera.

43. JR Malcolm, Insect biomass in Amazonian forest fragments, in *Canopy Arthropods,* ed. NE Stork, J Adis, RK Didham (London: Chapman & Hall, 1997), pp. 510–533.

44. Deby Cassill, personal communication; D Cassill 2003, Rules of supply and demand regulate recruitment to food in an ant society, *Behav. Ecol. Sociobiol.* 54: 441–450; and Thomas D. Seeley, *The Wisdom of the Hive* (Cambridge, Mass.: Harvard University Press, 1995).

45. AA Sorensen, TM Busch, SB Vinson 1985, Control of food influx by temporal subcastes in the fire ant, *Solenopsis invicta, Behav. Ecol. Sociobiol.* 17: 191–198.

46. The body takes what you give it: only nutrients such as iron and calcium are subject to shifts in absorption based on need (though communication between the gut and brain can help regulate appetite); see OB Chaudhri, V Salem, KG Murphy, SR Bloom 2008, Gastrointestinal satiety signals, *Annu. Rev. Physiol.* 70: 239–256.

47. Marauder ants and army ants kill Homoptera rather than collect honeydew, though for examples to the contrary, see William H. Gotwald Jr., *Army Ants: The Biology of Social Predation* (Ithaca, NY: Cornell University Press, 1995). Because nomadic ants are unlikely to return to the same spot, eating the honeydew yields these ants a lower payoff than predation.

10. Fortified Forests

1. Some of my discussion of canopy biology in this chapter is adapted from MW Moffett 2001, Nature and limits of canopy biology, *Selbyana* 22: 155–179; and Mark W. Moffett, *The High Frontier: Exploring the Tropical Rainforest Canopy* (Cambridge, MA: Harvard University Press, 1994).

2. DW Yu 1994, The structural role of epiphytes in ant gardens, *Biotropica* 26: 222–226.

3. At least one of the epiphytes has seeds with a scent attractive to the ants; see E Youngsteadt, S Nojima, C Häberlein, S Schulz, C Schal 2008, Seed odor mediates an obligate ant-plant mutualism in Amazonian rainforests, *Proc. Natl. Acad. Sci.* 105: 4571–4575.

4. A Vantaux, A Dejean, A Dor, J Orivel 2007, Parasitism *versus* mutualism in the ant-garden parabiosis between *Camponotus femoratus* and *Crematogaster levior, Insectes Soc.* 54: 95–99; and DW Davidson 1988, Ecological studies of neotropical ant gardens, *Ecology* 69: 1138–1152.

5. The most abundant ant species, measured in numbers of individuals, are among the most successful organisms; see JE Tobin, Ecology and diversity of tropical forest canopy ants, in *Forest Canopies,* ed. Margaret D. Lowman and Nalini M. Nadkarni (St. Louis, MO: Academic Press, 1995), pp. 129–147. It would be interesting to study the abundance of ant species in terms of numbers of colonies, as it perhaps should be measured.

6. AY Harada, J Adis, The ant fauna of tree canopies in central Amazonia: A first assessment, in *Canopy Arthropods,* ed. Nigel E. Stork, Joachim Adis, and Raphael K. Didham (London: Chapman & Hall, 1998), pp. 382–400.

7. JE Tobin, Ants as primary consumers: Diet and abundance in the Formicidae, in *Nourishment and Evolution in Insect Societies,* ed. James H. Hunt and Christine A. Nalepa (Boulder, CO: Westwood Press, 1994), pp. 129–147.

8. The data on the relative abundance and biomass of ants in the ground are scarce and may be skewed by the numbers of minute mites and springtails in the soil; but see EJ Fittkau, H Klinge 1973, On biomass and trophic structure of the central Amazonian rain forest ecosystem, *Biotropica* 5: 2–14. Many ground-dwelling species depend on nectaries or Homoptera residing on shrubs, herbs, and the roots of larger plants.

9. This shift in diet occurred several thousand years prior to agriculture (see chapter 15). Of course, both weaver ant colonies and humans in similarly entrenched settlements catch large prey when they can; see, e.g., B Hayden 1990, Nimrods, piscators, pluckers, and planters: The emergence of food production, *J. Anthropol. Archaeol.* 9: 31–69.

10. A few other species may on occasion come close to engaging in warfare. Honeybees attack weak colonies to steal honey (Tom Seeley, personal communication). Researchers have observed piles of dead nasute termites, suggesting recent battles involving mostly the worker caste, and hundreds of workers will respond en masse to kill even a single conspecific (James Traniello, personal communication; SC Levings, ES Adams 1984, Intra- and interspecific territoriality in *Nasutitermes* in a Panamanian mangrove forest, *J. Anim. Ecol.* 53: 705–714; BL Thorne 1982, Termite-termite interactions: Workers as an agnostic caste, *Psyche* 89: 133–150). However, workers in some termite species that live in huge colonies mix freely without aggression (Scott Turner, personal communication). Male chimpanzees are known to conduct predatory raids to kill neighboring chimpanzees, usually one at a time, which if repeated can devastate adjacent communities

7. PD Haemig 1997, Effects of birds on the intensity of ant rain: A terrestrial form of invertebrate drift, *Anim. Beh.* 54: 89–97; and B Hölldobler 1965, Springende Ameisen, *Mitt. Schweiz. entomol. Ges.* 30: 80–81.
8. SP Yanoviak, R Dudley, M Kaspari 2005, Directed aerial descent in canopy ants, *Nature* 433: 624–626.
9. For details, see R Dudley, G Byrnes, SP Yanoviak, B Borrell, R Brown, J McGuire 2007, Gliding and the functional origins of flight: Biomechanical novelty or necessity? *Annu. Rev. Ecol. Syst.* 38: 179–201.
10. Eberhard Horn, Gravity, in *Comprehensive Insect Physiology, Biochemistry, and Pharmacology*, vol. 6, ed. Gerald A. Kerkut and Lawrence I. Gilbert (Oxford: Pergamon, 1985), pp. 557–576; and Hubert Markl 1974, Perception of gravity and of angular acceleration in invertebrates, in *Handbook of Sensory Physiology*, vol. 6/1, ed. Hans H. Kornhuber (Heidelberg: Springer Verlag, 1974), pp. 17–74. Even ants dwelling in a relatively flat desert environment can be trained to climb specific distances, though they tend to overshoot the correct height; see G Grah, R Wehner, B Ronacher 2007, Desert ants do not acquire and use a three-dimensional global vector, *Front. Zool.* 4: 12; and G Grah, R Wehner, B Ronacher 2005, Path integration in a three-dimensional maze: Ground distance estimation keeps desert ants *Cataglyphis fortis* on course, *J Exp. Biol.* 208: 4005–4011.
11. Francis Hallé, Roelof A. A. Oldeman, and P. Barry Tomlinson, *Tropical Trees and Forests: An Architectural Analysis* (New York: Springer-Verlag, 1978).
12. Crestline orientation is untested in windblown conditions when the gravity vector would be difficult for ants to measure; see JH Klotz, BL Reid 1992, The use of spatial cues for structural guideline orientation in *Tapinoma sessile* and *Camponotus pennsylvanicus*, *J. Insect Behav.* 5: 71–82; R. Jander 1990, Arboreal search in ants: Search on branches, *J. Insect Behav.* 3: 515–527; and JH Klotz, SL Cole, HR Kuhns 1985, Crest-line orientation in *Camponotus pennsylvanicus*, *Insectes Soc.* 32: 305–312. One temperate ant shows a more random leaf and branch search pattern; see RM Weseloh 2001, Patterns of foraging of the forest ant *Formica neogagates* on tree branches, *Biol. Control* 20: 16–22. Ants use similar rules to navigate low-growing plants; see KN Ganeshaiah, T Veena 1988, Plant design and non-random foraging by ants on *Croton bonplandianum*, *Anim. Behav.* 36: 1683–1690.
13. CJ Kleineidam, M Ruchty, ZA Casero-Montes, F Roces 2007, Thermal radiation as a learned orientation cue in leaf-cutting ants *(Atta vollenweideri)*, *J. Insect Phys.* 53: 478–487.
14. R Jander, U Jander 1998, The light and magnetic compass of the weaver ant, *Oecophylla smaragdina*, *Ethology* 104: 743–758.
15. B Hölldobler 1980, Canopy orientation: A new kind of orientation in ants, *Science* 210: 86–88.
16. B Hölldobler 1983, Territorial behavior in the green tree ant *(Oecophylla smaragdina)*, *Biotropica* 15: 241–250.
17. MD Breed, JM Harrison 1987, Individually discriminable recruitment trails in a ponerine ant, *Insectes Soc.* 34: 222–226.
18. AP Baader 1996, The significance of visual landmarks for navigation of the giant tropical ant, *Paraponera clavata*, *Insectes Soc.* 43: 435–450. Navigating ants can identify the same object from different angles; see SPD Judd, TS Collett 1998, Multiple stored views and landmark guidance in ants, *Nature* 392: 710–714.
19. JF Harrison, JH Fewell, TM Stiller, MD Breed 1988, Effects of experience on use of orientation cues in the giant tropical ant, *Anim. Behav.* 37: 869–871.
20. A Lioni, C Sauwens, G Theraulaz, J-L Deneubourg 2001, Chain formation in *Oecophylla longinoda*, *J. Insect Behav.* 14: 679–696.
21. U Maschwitz, J Moog 2000, Communal peeing: A new mode of flood control in ants, *Naturwissenschaften* 87: 563–565.
22. J Moog, T Drude, U Maschwitz, D Agosti 1997, Flood control by ants: Water-bailing behaviour in the Southeast Asian plant-ant genus *Cladomyrma*, *Naturwissenschaften* 84: 242–245; and RW Klein, U Maschwitz, D Kovac 1993, Flood control by ants: A South-East Asian bamboo-dwelling *Tetraponera* bails water from its internode nests, *Insectes Soc.* 40: 115–118.
23. Carton can also be waterproof because the surface tension of water transforms its porous surface into a complete shield, though buffeting by rain and wind can cause breakage. Thus many carton-nesting species are restricted to the understory, with ant-garden ants (chapter 10) shielding their nest beneath garden foliage.
24. N Rastogi 2004, Behavioral strategy of returning foragers of the arboreal ant *Oecophylla smaragdina* during the monsoon, *J. Bombay Nat. Hist. Soc.* 101: 388–391.
25. Deby Cassill, personal communication, and WL Morrill 1974, Dispersal of red imported fire ants by water, *Fla. Entomol.* 57: 38–42.
26. Joachim Adis, Herbert O.R. Schubart 1984, Ecological research on arthropods in central Amazonian forest ecosystems with recommendations for study procedures, in *Trends in Ecological Research for the 1980s*, ed. June H. Cooley and Frank B. Golley (New York: Plenum Press, 1984), pp. 111–144.
27. Michael Goulding, *Amazon: The Flooded Forest* (New York: Sterling, 1990), pp. 26–27.
28. J Adis 1982, Eco-entomological observations from the Amazon, III: How do leafcutting ants of inundation forests survive flooding? *Acta Amazon.* 12: 839–840.
29. Information on both carpenter ants is from MB DuBois, R Jander 1985, Leg coordination and

swimming in an ant, *Camponotus americanus, Physiol. Entomol.* 10: 267–270.
30. Simon Robson, personal communication, and MG Nielsen, Nesting biology of the mangrove mud-nesting ant *Polyrhachis sokolova* in northern Australia, *Insectes Soc.* 44 (1997): 15–21.
31. CM Clarke, RL Kitching 1995, Swimming ants and pitcher plants: A unique ant-plant interaction from Borneo, *J. Trop. Ecol.* 11: 589–602.
32. HF Bohn, W Federle 2004, Insect aquaplaning: *Nepenthes* pitcher plants capture prey with the peristome, a fully wettable water-lubricated anisotropic surface, *Proc. Natl. Acad. Sci.* 101: 14138–14143.
33. MW Moffett 1989, Notes on the behavior of the dimorphic ant *Oligomyrmex overbecki, Psyche* 93: 107–116.
34. First proposed for animals in B Rensch 1956, Increase in learning with increase in brain size, *Am. Nat.* 90: 81–95; for ants, see BL Cole 1985, Size and behavior in ants: Constraints on complexity, *Proc. Natl. Acad. Sci.* 82: 8548–8551. I doubt that idea applies to most polymorphic species, in which majors tend to be limited in their repertoires. For other intriguing ideas about the relation between colony size, worker size, and productivity, see M Kaspari 2005, Global energy gradients and size in colonial organisms: Worker mass and worker number in ant colonies, *Proc. Natl. Acad. Sci.* 102: 5079–5083.
35. A baleen whale's trawling may most closely resemble foraging by army ants, except army ants raid together to catch prey larger than they could individually, whereas whales trawl to catch prey smaller than normal for an animal of their size. Trawling behavior can be very effective for large individuals, and both army ant colonies and whales have evolved to become larger over time. Ironically, the terrestrial vertebrate with the closest equivalent to this behavior may be the anteater; in general, the effectiveness of trawling for a land vertebrate is reduced by the intake of detritus (Brian McNab, personal communication).
36. At a smaller scale, superabundant mites consume plant pests too minute for ants and live in microscopic protective retreats provided by the plants; see GQ Romero, WW Benson 2005, Biotic interactions of mites, plants, and leaf domatia, *Curr. Opin. Plant Biol.* 8: 436–440; and DE Walter, Hidden in plain sight: Mites in the canopy, in *Forest Canopies,* 2d ed., ed. Margaret D. Lowman and H. Bruce Rinker (St. Louis, Mo.: Academic Press, 2004), pp. 224–241.
37. Technically, workers operate in "series parallel"; see George F. Oster and Edward O. Wilson, *Caste and Ecology in the Social Insects* (Princeton: Princeton University Press, 1978), pp. 12–14.
38. These systems of control are probably necessary to manage unions of self-interested, often distantly related humans who mostly don't know each other; see Stephen K. Sanderson, ed., *Social Transformations: A General Theory of Historical Development,* expanded ed. (Oxford: Blackwell, 1995).
39. Raphael D. Sagarin and Terence Taylor, eds., *Natural Security: A Darwinian Approach to a Dangerous World* (Berkeley: University of California Press, 2008).
40. One World Bank official seeking solutions to global problems has proposed that hierarchical governments be replaced by networked governance; see J-F Rischard, Global issues networks: Desperate times deserve innovative measures, *Washington Quarterly* 26 (2002): 17–33.
41. A few ants (most of them with small colonies) survive the queen's death by having the workers take over reproduction; see the conclusion to this book and Bert Hölldobler and Edward O. Wilson, *The Superorganism* (New York: W.W. Norton, 2008), pp. 356–426.
42. A Buschinger, U Maschwitz, Defensive behavior and defensive mechanisms in ants, in *Defensive Mechanisms in Social Insects,* ed. Henry R. Hermann (New York: Praeger Scientific, 1984), pp. 95–150; the quote appears on p. 124.

12. Slaves of Sagehen Creek

1. I focus here on *breviceps,* but there is excellent work on the other *Polyergus* species.
2. Donato Grasso, personal communication. Another fascinating possibility is that the *Formica* bring up some brood to make them accessible to raiding parties in a tactic that diverts the *Polyergus* from more critical parts of the nest, in a kind of "lizard loses its tail" sacrifice.
3. Apparently colony identity odors can be detected not just at contact but also over distances of a few millimeters; see AS Brandstaetter, A Endler, CJ Kleineidam 2008, Nestmate recognition in ants is possible without tactile interaction, *Naturwissenschaften* 95: 601–608.
4. Propaganda substances have so far been studied only in the European Amazon ant; see R. Visicchio, A Mori, DA Grasso, C Castracani, F Le Moli 2001. Glandular sources of recruitment, trail, and propaganda semiochemicals in the slave-making ant *Polyergus rufescens, Ethol. Ecol. Evol.* 13: 361–372.
5. Those *Formica* species that have a shorter history of conflict with the Amazons tend to fight back more, and as a consequence they suffer higher mortality from raids. The species raided can vary from place to place, even from colony to colony; see, e.g., JM Bono, R Blatrix, MF Antolin, JM Herbers 2007, Pirate ants *(Polyergus breviceps)* and sympatric hosts *(Formica occulta* and *Formica* sp. cf. *argentea):* Host specificity and coevolutionary dynamics, *Biol. J. Linn. Soc.* 91: 565–572.

6. JM Herbers, S Foitzik 2002, The ecology of slavemaking ants and their hosts in north temperate forests, *Ecology* 83: 148–163; and RJ Stuart 1988, Collective cues as a basis for nestmate recognition in polygynous leptothoracine ants, *Proc. Natl. Acad. Sci.* 85: 4572–4575.
7. TO Richardson, PA Sleeman, JM McNamara, AL Houston, NR Franks 2007, Teaching with evaluation in ants, *Curr. Biol.* 17: 1520–1526; and E. Leadbeater, N Raine, L Chittka 2006, Social learning: Ants and the meaning of teaching, *Curr. Biol.* 16: R323–R325.
8. RJ Stuart 1986, An early record of tandem running in leptothoracine ants: Gottfrid Adlerz 1896, *Psyche* 93: 103–106; and TM Alloway 1979, Raiding behaviour of two species of slave-making ants, *Harpagoxenus americanus* and *Leptothorax duloticus*, *Anim. Behav.* 27: 202–210.
9. J Beibl, RJ Stuart, J Heinze, S Foitzik 2005, Six origins of slavery in formicoxenine ants, *Insectes Soc.* 52: 291–297.
10. MW Moffett 1989, Life in a nutshell, *National Geographic* 175: 783–796.
11. A Lenoir, P D'Ettorre, C Errard 2001, Chemical ecology and social parasitism in ants, *Annu. Rev. Entomol.* 46: 573–599. Slavemaker colonies act more like predators than parasites, which feed on a single host; see M Brandt, S Foitzik, B Fischer-Blass, J Heinze 2005, The coevolutionary dynamics of obligate ant social parasite systems: Between prudence and antagonism, *Biol. Rev. Camb. Philos. Soc.* 80: 251–267. We'll see in the next chapter that the Amazons collect some of the colony's food, in that the slaves eat some of the raided brood.
12. John Lubbock, *Ants, Bees, and Wasps* (New York: Appleton, 1883). Thief ants belong to *Solenopsis* subgenus *Diplorhoptrum*.
13. Pierre Huber, *Recherches sur les moeurs des fourmis indigènes* (Geneva: J. J. Paschoud, 1810), pp. 219–224.
14. Charles Darwin, *Origin of Species* (London: John Murray, 1859), pp. 219–224.
15. Milton Meltzer, *Slavery: A World History* (Chicago: Da Capo Press, 1993); and Orlando Patterson, *Slavery and Social Death: A Comparative Study* (Cambridge, MA: Harvard University Press, 1982).
16. A Achenbach, S Foitzik 2009, First evidence for slave rebellion: Enslaved ant workers systematically kill the brood of their social parasite *Protomognathus americanus*, *Evolution* 63: 1068–1075.
17. W Czechowski 2006, Route of *Formica polyctena* as a factor promoting emancipation of *Formica fusca* slaves from colonies of *Polyergus rufescens*, *Pol. J. Ecol.* 54: 159–162.
18. Some fleeting tussles between slaves and slavemakers have also been recorded, most prevalently in species in the early stages of the evolution of slavemaking; see Bert Hölldobler and Edward O. Wilson, *The Ants* (Cambridge, MA: Harvard University Press, 1990), p. 463; and EO Wilson 1975, *Leptothorax duloticus* and the beginnings of slavery in ants, *Evolution* 29: 108–119.
19. Karl Marx's idea of false consciousness was given its name by Friedrich Engels; see György Lukács, *History and Class Consciousness: Studies in Marxist Dialectics* (Cambridge, MA: MIT Press, 1971).
20. Even the adult workers may shift over from a losing colony; see GB Pollock, SW Rissing 1989, Intraspecific brood raiding, territoriality, and slavery in ants, *Am. Nat.* 133: 61–70.
21. For similar reasons, the myth of Tarzan, raised by apes, is more plausible than that of Romulus and Remus, the legendary wolf-raised founders of Rome. A stunning exception for the ants is a parasitic species that never rears its own brood but rather sneaks it into the nurseries of a distantly related species; see U Maschwitz, C Go, E Kaufmann, A Buschinger 2004, A unique strategy of host colony exploitation in a parasitic ant: Workers of *Polyrhachis lama* rear their brood in neighbouring host nests, *Naturwissenschaften* 91: 40–43. Should employing a related species for labor be called slavery? In *Planet of the Apes*, at least, we accept that apes have turned humans into "slaves" without the word sounding a false note.
22. M Stoneking 2003, Widespread prehistoric human cannibalism: Easier to swallow, *Trends Ecol. Evol.* 18: 489–490.
23. Thanks to Sarah Hrdy for examples. One author proposes that kidnapping by primates is "ancestral" to slavery among humans, a view I have trouble believing: "Surrounded by strangers I thought were my friends," *Ethology* 113: 499–510.
24. These birds do not imprint on their parents and therefore will follow whatever authority figure gives the right signal; see RG Heinsohn 1991, Kidnapping and reciprocity in cooperatively breeding white-winged choughs, *Anim. Behav.* 41: 1097–1100.
25. R Blatrix, JM Herbers 2003, Coevolution between slave-making ants and their hosts: Host specificity and geographic variation, *Mol. Ecol.* 12: 2809–2816.
26. JM Herbers 2006, The loaded language of science, *Chron. High. Educ.* 52: B5. A technical alternative is the term *dulosis*, derived from the Greek word for slave.
27. H Topoff, Slave-making queens, *Scientific American* 281: 84–90.
28. Chris Boehm, *Hierarchy in the Forest* (Cambridge, MA: Harvard University Press, 1999); and Robert L. O'Connell, *Ride the Second Horseman: The Birth and Death of War* (New York: Oxford University Press, 1995).
29. Bert Hölldobler and Edward O. Wilson, *Journey to the Ants* (Cambridge, MA: Harvard University Press, 1994), p. 9.
30. Frans de Waal, *Good Natured: The Origins of Right and Wrong in Humans and Other Animals* (Cambridge, MA: Harvard University Press, 1997).
31. This section summarizes Topoff, Slave-making queens, cited in n. 27.

32. DA Grasso, A Mori, F Le Moli, J Billen 2005, Morpho-functional comparison of the Dufour gland in the female castes of the Amazon ant *Polyergus rufescens*, *Zoomorphology* 124: 149–153.
33. Topoff, Slave-making queens, p. 87, cited in n. 27.
34. It is conceivable that the queen plays a primary role in determining the odor used to signal colony identity in these ants, as has been shown with the related carpenter ants; see NF Carlin, B Hölldobler 1983, Nestmate and kin recognition in interspecific mixed colonies of ants, *Science* 222: 1027–1029. See also H Topoff, E Zimmerli 1993, Colony takeover by a socially parasitic ant, *Polyergus breviceps:* The role of chemicals obtained during host-queen killing, *Anim. Behav.* 46: 479–486.
35. A Buschinger 1986, Evolution of social parasitism in ants, *Trends Ecol. Evol.* 1: 155–160.

13. Abduction in the Afternoon

1. DA Grasso, A Mori, P D'Ettorre, F Le Moli 1994, Intraspecific raids and territoriality in *Polyergus rufescens*, *Ethol. Ecol. Evol.* 3: 81–87; and H Topoff, B LaMon, L Goodloe, M Goldstein 1984, Social and orientation behavior of *Polyergus breviceps* during slave-making raids, *Behav. Ecol. Sociobiol.* 15: 273–279.
2. William H. McNeill, *The Pursuit of Power: Technology, Armed Force, and Society since A.D. 1000* (Chicago: University of Chicago Press, 1982), p. 131.
3. Dan Stahler, personal communication; Scott Creel and Nancy Marushka Creel, *The African Wild Dog: Behavior, Ecology, and Conservation* (Princeton: Princeton University Press, 2002), p. 76.
4. H Topoff, D Bodoni, P Sherman, L Goodloe 1987, The role of scouting in slave raids by *Polyergus breviceps*, *Psyche* 94: 261–270.
5. See, e.g., J Dobrzańska, J Dobrzański 1989, Controversies on the subject of slave-raids in Amazon ants (genus *Polyergus*), *Acta Neurobiol. Exp.* 49: 367–379.
6. See, e.g., E Janssen, B Hölldobler, HJ Bestmann 1999, A trail pheromone component of the African stink ant, *Pachycondyla (Paltothyreus) tarsata*, *Chemoecology* 9: 9–11.
7. Free-living *Formica* ants separate their foraging behavior into similar phases of linear travel and irregular searching, which suggests this strategy has ancient roots; see JFA Traniello, V Fourcassié, TP Graham 1991, Search behavior and foraging ecology of the ant *Formica schaufussi:* Colony-level and individual patterns, *Ethol. Ecol. Evol.* 3: 35–47.
8. In fact, the chemical trail of a successful raid lasts long enough for the Amazons to reuse it during the next day or two, if a raid is very productive. It is unknown whether a scout is required for these subsequent raids.
9. EO Wilson 1975, Slavery in ants, *Sci. Am.* 232: 32–36. According to one interpretation, the Amazon ants continue this tradition of territorial warfare: if Amazons identified their *Formica* slaves as nestmates, they would raid *Formica* colonies as if they were attacking their own species: H Topoff 1990, The evolution of slave-making behavior in the parasitic ant genus *Polyergus*, *Ethol. Ecol. Evol.* 2: 284–287.
10. Part of the reason for the absence of slavery among army ants is that few of their species attack other army ants, which means that successfully rearing slaves, which are typically close relatives (see chapter 12), from prey booty is unlikely.
11. Their proclivity for territorial battle may explain how it is that enslaved workers of these species sometimes join their masters on slave raids.
12. For a general discussion of tactical deception, see RW Byrne, N Corp 2004, Neocortex size predicts deception rate in primates, *Proc. R. Soc. Lond. B* 271: 1693–1699. In another example of tactical deception, honeypot ants that find termite prey will instigate tournaments near competing nests to distract their rivals; see B Hölldobler 1981, Foraging and spatiotemporal territories in the honey ant *Myrmecocystus mimicus*, *Behav. Ecol. Sociobiol.* 9: 301–314. Bert Hölldobler informs me that genetic studies show slaves taken after such tournaments are common. Honeypot ants were first introduced in chapter 4 in a discussion of how the combatants determine if they are outnumbered.
13. E Cool-Kwait, H Topoff 1984, Raid organization and behavioral development in the slave-making ant *Polyergus lucidus*, *Insectes Soc.* 31: 361–374.
14. H Topoff 1985, Effect of overfeeding on raiding behavior in the western slave-making ant *Polyergus breviceps*, *Natl. Geogr. Res.* 1: 437–441.
15. RJ Stuart, MW Moffett 1994, Recruitment communication and pheromone trails in the neotropical ants *Leptothorax (Nesomyrmex) spininodis* and *L. (N.) echinatinodis*, *Experientia* 50: 850–852.
16. This hypothesis as well as the importance of a seasonal brood cycle are reviewed by Bert Hölldobler and Edward O. Wilson, *The Ants* (Cambridge, MA: Harvard University Press, 1990), p. 447.
17. As we saw for the capacity of ants to delay eating in order to group-transport food to a nest (chapter 5), such forbearance is rare among animals generally; see JR Stevens, DW Stephens 2008, Patience, *Curr. Biol.* 18: R11–R12.
18. See Stephen B. Vander Wall, *Food Hoarding in Animals* (Chicago: University of Chicago Press, 1990), p. 64. This advantage for slavery in temperate ants could exist even though ants generally do better than solitary insects in hunkering down in their nests to wait out hard times.
19. M Kaspari, L Alonso, S O'Donnell 2000, Three energy variables predict ant abundance at a geographical scale, *Proc. R. Soc. Lond. B* 267: 485–489.

14. A Fungus Farmer's Life

1. In this book I do not distinguish leafcutter species unless there are known differences between them. For a general review, see Rainer Wirth, Hubert Herz, Ronald J. Ryel, Wolfram Beyschlag, and Bert Hölldobler, *Herbivory of Leaf-Cutting Ants: A Case Study on* Atta colombica *in the Tropical Rainforest of Panama* (Berlin: Springer-Verlag, 2003).
2. Leafcutter workers of all sizes respond to disturbances in the nest, but the largest, the "soldiers," are particularly effective at repelling vertebrates.
3. SEF Evison, FLW Ratnieks 2007, New role for majors in *Atta* leafcutter ants, *Ecol. Entomol.* 32: 451–454. Occasionally the majors (soldiers) help clear trails as well.
4. MW Moffett 1986, Observations of *Lophomyrmex* ants from Kalimantan, Java, and Malaysia, *Malay. Nat. J.* 39: 207–211.
5. F Roces, JRB Lighten 1995, Larger bites of leaf-cutting ants, *Nature* 373: 392–393.
6. AJ Edwards, JD Fawke, JG McClements, SA Smith, P Wyeth 1993, Correlation of zinc distribution and enhanced hardness in the mandibular cuticle of the leaf-cutting ant *Atta sexdens rubropilosa, Cell Biol. Int.* 17: 697–698.
7. See, e.g., JM van Breda, DJ Stradling 1994, Mechanisms affecting load size determination in *Atta cephalotes, Insectes Soc.* 41: 423–434.
8. James K. Wetterer, unpublished manuscript.
9. There is some evidence that leafcutter ants are unable to cooperate in carrying any object, no matter what its shape; see JJ Howard 2001, Costs of trail construction and maintenance in the leaf-cutting ant *Atta columbica, Behav. Ecol. Sociobiol.* 49: 348–356.
10. Still, larger ants have relatively shorter legs and cut fragments small for their size; see JK Wetterer 1995, Forager polymorphism and foraging ecology in the leaf-cutting ant, *Atta colombica, Psyche* 102: 131–145; and JK Wetterer 1991, Allometry and the geometry of leaf-cutting in *Atta cephalotes, Behav. Ecol. Sociobiol.* 29: 347–351.
11. M Burd, JJ Howard 2005, Global optimization from suboptimal parts: Foraging *sensu lato* by leaf-cutting ants, *Behav. Ecol. Sociobiol.* 59: 234–242.
12. OT Lewis, MM Martin, TJ Czaczkes 2008, Effects of trail gradient on leaf tissue transport and load size selection in leaf-cutter ants, *Behav. Ecol.* 19: 805–809.
13. RJ Quinlan, JM Cherrett 1979, The role of fungus in the diet of the leaf-cutting ant *Atta cephalotes, Ecol. Entomol.* 4: 151–160.
14. Michael M. Martin, *Invertebrate-Microbial Interactions: Ingested Fungal Enzymes in Arthropod Biology* (Ithaca, N.Y.: Comstock Publishing Associates, 1987), pp. 107–124.
15. PF Dowd 1992, Insect fungal symbionts: A promising source of detoxifying enzymes, *J. Ind. Microbiol. Biotech.* 9: 149–161.
16. AB Abril, EH Bucher 2004, Nutritional sources of the fungus cultured by leaf-cutting ants, *Appl. Soil Ecol.* 26: 243–247.
17. All termites have microbes in their gut to digest cellulose in wood and dried leaves, but Africa's fungus-growing termites take this process a step further. Their colonies are basically cows within cows. Workers eat the plant matter, then use their feces to create gardens that look surprisingly like the ones leafcutters construct from foliage. The fungi the termite workers eat pass through their digestive systems intact and do not themselves serve as food. Instead, when the fungi are combined with the feces in the gardens, they degrade the feces into a form the termites can eat and fully digest. There are also two families of ambrosia beetle that feed their larvae on hyphae they rear in dead or dying wood, which makes them lumber pests.
18. For descriptions of large nests, see, e.g., AA Moreira, LC Forti, APP Andrade, MAC Boaretto, JFS Lopes 2004, Nest architecture of *Atta laevigata, Stud. Neotrop. Fauna Environ.* 39: 109–116; and JCM Jonkman 1980, The external and internal structure and growth of nests of the leafcutting ant *Atta vollenweideri*, Part II, *Z Angew. Entomol.* 89: 217–246.
19. C Kleineidam, R Ernst, F Roces 2001, Wind-induced ventilation of the giant nests of the leaf-cutting ant *Atta vollenweideri, Naturwissenschaften* 88: 301–305.
20. H Herz, W Beyschlag, B Hölldobler 2007, Herbivory rate of leaf-cutting ants in a tropical moist forest in Panama at the population and ecosystem scales, *Biotropica* 39: 482–488.
21. M Bass, JM Cherrett 1995, Fungal hyphae as a source of nutrients for the leaf-cutting ant *Atta sexdens, Physiol. Entomol.* 20: 1–6.
22. JC Moser 2006, Complete excavation and mapping of a Texas leafcutting ant nest, *Ann. Entomol. Soc. Am.* 99: 891–897.
23. Some *Acromyrmex* with colonies of a few thousand have minors and majors that differ little in size, but there is still a division of labor: majors cut and carry leaves and minors tend the gardens.
24. George F. Oster and Edward O. Wilson, *Caste and Ecology in the Social Insects* (Princeton: Princeton University Press, 1978), p. 22.
25. William H. Davidow and Michael S. Malone, *The Virtual Corporation: Structuring and Revitalizing the Corporation for the 21st Century* (New York: Harper Collins, 1992), p. 167.
26. There is no full-time gardener worker caste. Description based on EO Wilson 1980, Caste and division of labor in leaf-cutting ants, I: The overall pattern in *A. sexdens, Behav. Ecol. Sociobiol.* 7: 143–156.
27. Such microbes remain in the gardens at low levels and are so little known that some may prove to be vital to the ants; see RJ Quinlan, JM Cherrett 1978, Studies

on the role of the infrabuccal pocket of the leaf-cutting ant *Acromyrmex octospinosus, Insectes Soc.* 25: 237–245.

28. M Bass, JM Cherrett 1996, Leaf-cutting ants prune their fungus to increase and direct its productivity, *Funct. Ecol.* 10: 55–61.

29. M Bass, JM Cherrett 1996, Fungus garden structure in the leaf-cutting ant *Atta sexdens, Symbiosis* 21: 9–24.

30. Data on the size of the trail systems and the costs of trail construction and mantenance are from JJ Howard 2001, cited in n. 9.

31. Bert Hölldobler, personal communication; and M Autori 1947, Combate a formiga saúva, *Biológico* 13: 196–199.

32. AG Farji-Brener, C Sierra 1998, The role of trunk trails in the scouting activity of the leaf-cutting ant *Atta cephalotes, Ecoscience* 5: 271–274.

33. These ideas about what has been described earlier as "recruitment overrun" are based on leaf baits left on the ground near trunk trails. See JD Sheperd 1982, Trunk trails and the searching strategy of a leaf-cutter ant, *Atta colombica, Behav. Ecol. Sociobiol.* 11: 77–84.

34. A Trewavas 2005, Green plants as intelligent organisms, *Trends Plant Sci.* 10: 413–419; and F López, JM Serrano, FJ Acosta 1994, Parallels between the foraging strategies of ants and plants, *Trends Ecol. Evol.* 9: 150–153.

35. See, e.g., B Hölldobler 1976, Recruitment behavior, home range orientation and territoriality in harvester ants, *Pogonomyrmex, Behav. Ecol. Sociobiol.* 1: 3–44. Some trails are marked with colony-specific territorial odors; see JFA Traniello, Recruitment communication, in *Encyclopedia of Insects*, 2nd ed., ed. Vincent H. Resh and Ring T. Cardé (New York: Academic Press, 2009), pp. 980–987.

36. JC Crick, JP Grime 1987, Morphological plasticity and mineral nutrient capture in two herbaceous species of contrasted ecology, *New Phytol.* 107: 403–414.

37. SEF Evison, AG Hart, DE Jackson 2008, Minor workers have a major role in the maintenance of leafcutter ant pheromone trails, *Anim. Behav.* 75: 963–969.

38. AG Farji-Brener, G Barrantes, O Laverde, K Fierro-Calderón, F Bascopé, A López 2007, Fallen branches as part of leaf-cutting ant trails: Their role in resource discovery and leaf transport rates in *Atta cephalotes, Biotropica* 39: 211–215.

39. LL Rockwood, SP Hubble 1987, Host-plant selection, diet diversity, and optimal foraging in a tropical leafcutting ant, *Oecologia* 74: 55–61.

40. EO Wilson 1984, Clockwork lives of the Amazonian leafcutter army, *Smithsonian* 15: 92–101.

41. M Burd, D Archer, N Aranwela, DJ Stradling 2002, Traffic dynamics of the leaf-cutting ant, *Atta cephalotes, Am. Nat.* 159: 283–293.

42. A Dussutour, S Beshers, J-L Deneubourg, V Fourcassié 2007, Crowding increases foraging efficiency in the leaf-cutting ant *Atta colombica, Insectes Soc.* 54: 158–165.

43. Delia Goetz and Sylvanus G. Morley, *Popol Vuh: The Sacred Book of the Ancient Quiché Maya,* from the Spanish translation by Adrián Recinos (Norman: University of Oklahoma Press, 1950), p. 83.

44. C Anderson, JLV Jadin 2001, The adaptive benefit of leaf transfer in *Atta colombica, Insectes Soc.* 48: 404–405.

45. HL Vasconcelos, JM Cherrett 1996, The effect of wilting on the selection of leaves by the leaf-cutting ant *Atta laevigata, Entomol. Exp. Appl.* 78: 215–220.

46. AG Hart, FLW Ratnieks 2001, Leaf caching in the leafcutting ant *Atta colombica:* Organizational shift, task partitioning and making the best of a bad job, *Anim. Behav.* 62: 227–234.

47. SP Hubbell, LK Johnson, E Stanislav, B Wilson, H Fowler 1980, Foraging by bucket-brigade in leaf-cutter ants, *Biotropica* 12: 210–213.

48. SG Rudolph, C Loudon 1986, Load size selection by foraging leaf-cutter ants *(Atta cephalotes), Ecol. Entomol.* 11: 401–410.

49. JJ Bartholdi III, LA Bunimovich, DD Eisenstein 1999, Dynamics of two- and three-worker "bucket brigade" production lines, *Oper. Res.* 47: 488–491.

50. AG Hart, FLW Ratnieks 2001, cited in n. 46.

51. SP Hubbell, LK Johnson, E Stanislav, B Wilson, H Fowler 1980, cited in n. 47.

52. Thomas Belt, *The Naturalist in Nicaragua* (London: John Murray, 1874), p. 76. The migration had been forced upon the ants by carbolic acid poured into their nest to stop them from invading a garden. What intrigues is that they managed to partition the labor of transporting their gardens in this way under such an unnatural circumstance.

53. JJ Howard, ML Henneman, G Cronin, JA Fox, G Hormiga 1996, Conditioning of scouts and recruits during foraging by a leaf-cutting ant, *Atta colombica, Anim. Behav.* 52: 229–306. Others have shown that unfamiliar food can be preferred; see, e.g., JM Cherrett, Chemical aspects of plant attack by leaf-cutting ants, in *Phytochemical Ecology*, ed. JB Harbourne (New York: Academic Press, 1972), pp. 13–24.

54. F Roces, B Hölldobler 1994, Leaf density and a trade-off between load-size selection and recruitment behavior in the ant *Atta cephalotes, Oecologia* 97: 1–8.

55. Bees can also minor in other flowers; see, e.g., B Heinrich 1979, "Majoring" and "minoring" by foraging bumblebees, *Bombus vagans:* An experimental analysis, *Ecology* 60: 245–55.

56. PD Coley, TM Aide 1989, Red coloration of tropical young leaves: A possible antifungal defence? *J. Trop. Ecol.* 5: 293–300.

57. JJ Howard, Resource quality and cost in the foraging of leaf-cutter ants, in *Ant-Plant Interactions*, ed. Camilla R. Huxley and David F. Cutler (New York: Scientific, 1991), pp. 42–50.

58. CM Nichols-Orians, JC Schultz 1990, Interactions among leaf toughness, chemistry, and harvesting by attine ants, *Ecol. Entomol.* 15: 311–320.
59. The loss of toxins during domestication is often mentioned but needs documentation for human crops (Dorian Fuller, personal communication).
60. K Jaffe, Leaf-cutting ants, in *Encyclopedia of Entomology*, 2nd edition, ed. John L. Capinera (New York: Springer-Verlag, 2008), pp. 2151–2160; and HL Vasconcelos 1999, Levels of leaf herbivory in Amazonian trees from different stages in forest regeneration, *Acta Amazon.* 29: 615–623.
61. Andreas Schaller, ed., *Induced Plant Resistance to Herbivory* (New York: Springer-Verlag, 2008).
62. JJ Howard 1990, Infidelity of leafcutting ants to host plants: Resource heterogeneity or defense induction? *Oecologia* 82: 394–401.
63. JM Cherrett, Resource conservation by the leaf-cutting ant *Atta cephalotes* in tropical rain forest, in *Tropical Rain Forest: Ecology and Management*, ed. SL Sutton, TC Whitmore (Oxford: Blackwell, 1983), pp. 253–263. In chapter 6, we saw that army ants similarly "prune" populations of their prey.
64. See Wirth, Herz, Ryel, Beyschlag, and Hölldobler, *Herbivory of Leaf-Cutting Ants*, cited in n. 1; and B Haines, Impact of leaf-cutting ants on vegetation development at Barro Colorado Island, in *Tropical Ecological Systems*, ed. FG Golley, E Medina (New York: Springer-Verlag, 1975), pp. 99–111.
65. JJ Knapp, PE Howe, A Kermarrec, Factors controlling foraging patterns in the leaf-cutting ant *Acromyrmex octospinosus*, in *Applied Myrmecology: A World Perspective*, ed. Robert K. Vander Meer, Klaus Jaffé, and Araqua Cedano (Boulder, Colo.: Westview Press, 1990), pp. 382–409.
66. RD North, CW Jackson, PE Howse 1999, Communication between the fungus garden and workers of the leaf-cutting ant, *Atta sexdens rubropilosa*, regarding choice of substrate for the fungus, *Physiol. Entomol.* 24: 127–133; and P. Ridley, PE Howse, CW Jackson 1996, Control of the behaviour of leaf-cutting ants by their "symbiotic" fungus, *Experientia* 52: 631–635.
67. R Wirth, W Beyschlag, RJ Ryel, B Hölldobler 1997, Annual foraging of the leaf-cutting ant *Atta colombica* in a semideciduous rain forest in Panama, *J. Trop. Ecol.* 13: 741–757.
68. Leafcutters particularly favor seeds with elaiosomes; see IR Leal, PS Oliveira 1998, Interactions between fungus-growing ants (Attini), fruits and seeds in cerrado vegetation in southeast Brazil, *Biotropica* 30: 170–178.
69. One desert-dwelling *Acromyrmex* drinks from nectaries, but this is not surprising, since it depends on dried vegetation for its gardens and therefore cannot feed on plant sap; see JK Wetterer, AG Himler, MM Yospin 2001, Foraging ecology of the desert leaf-cutting ant, *Acromyrmex versicolor*, in Arizona, *Sociobiology* 37: 633–650.
70. DH Feener, KAG Moss 1990, Defense against parasites by hitchhikers in leaf-cutting ants: A quantitative assessment, *Behav. Ecol. Sociobiol.* 26: 17–29.
71. DH Feener, BV Brown 1993, Oviposition behavior of an ant-parasitizing fly, *Neodohrniphora curvinervis*, and defense behavior by its leaf-cutting ant host *Atta cephalotes*, *J. Insect Behav.* 6: 675–688.
72. Hitchhiking continues at night; presumably nocturnal riders use the time to scrub clean the leaf booty; see MR Orr 1992, Parasitic flies influence foraging rhythms and caste division of labor in the leaf-cutter ant, *Atta cephalotes*, *Behav. Ecol. Sociobiol.* 30: 395–402.
73. Flavio Roces, personal communication; and F Roces, B Hölldobler 1996, Use of stridulation in foraging leaf-cutting ants: Mechanical support during cutting or short-range recruitment signal? *Behav. Ecol. Sociobiol.* 39: 293–299.
74. H Markl, Manipulation, modulation, information, cognition: Some of the riddles of communication, in *Experimental Behavioral Ecology and Sociobiology*, ed. B Hölldobler, M Lindauer (Sunderland, Mass.: Sinauer, 1985), pp. 163–194. For more on modulatory signals, see Bert Hölldobler and Edward O. Wilson, *The Superorganism* (New York: W.W. Norton, 2008), pp. 231–235.
75. F Roces, B Hölldobler 1995, Vibrational communication between hitchhikers and foragers in leaf-cutting ants *(Atta cephalotes)*, *Behav. Ecol. Sociobiol.* 37: 297–302.
76. See, e.g., CB Yackulic, OT Lewis 2007, Temporal variation in foraging activity and efficiency and the role of hitchhiking behaviour in the leaf-cutting ant, *Atta cephalotes*, *Entomol. Exp. Appl.* 125: 125–134; and EHM Vieira-Neto, FM Mundim, HL Vasconcelos 2006, Hitchhiking behaviour in leaf-cutter ants: An experimental evaluation of three hypotheses, *Insectes Soc.* 53: 326–332.
77. WOH Hughes, D Goulson 2001, Polyethism and the importance of context in the alarm reaction of the grass-cutting ant, *Atta capiguara*, *Behav. Ecol. Sociobiol.* 49: 503–508. As with the marauder ant, the vibrations can be another kind of distress signal: when a nest chamber collapses, the calls of buried ants attract the rescue squad that digs them out.
78. Relatively little aggressive behavior has been recorded between leafcutters, though Jack Longino told me of a monthlong battle between *Atta cephalotes* and *Atta colombica* in Corcovado, Costa Rica; see also MEA Whitehouse, K Jaffe 1996, Ant wars: Combat strategies, territoriality and nest defence in the leaf-cutting ant *Atta laevigata*, *Anim. Behav.* 51: 1207–1217. Such confrontations may go largely unseen in the canopy. But leaf cutting is likely to reduce the productivity of the trees on which aggressive canopy

ants of other species depend, and it is known that some of these dominant ants fight with leafcutters; see JK Wetterer 1994, Attack by *Paraponera clavata* prevents herbivory by the leaf-cutting ant, *Atta cephalotes, Biotropica* 26: 462–465; and n. 1.

79. A queen can rear workers if her garden fails, though it's unknown whether her workers can forage for a replacement fungus; see H Fernández-Marín, WT Wcislo 2005, Production of minima workers by gynes of *Atta colombica* that lack a fungal pellet, *J. Kansas Entomol. Soc.* 78: 290–292.

80. MB Dijkstra, DR Nash, JJ Boomsma 2005, Self-restraint and sterility in workers of *Acromyrmex* and *Atta* leafcutter ants, *Insectes Soc.* 52: 67–76.

81. AS Yang 2007, Thinking outside the embryo: The superorganism as a model for evodevo studies, *Biol. Theory* 2: 398–408.

82. EO Wilson 1983, Caste and division of labor in leaf-cutter ants, IV: Colony ontogeny of *A. cephalotes, Behav. Ecol. Sociobiol.* 14: 55–60.

83. JK Wetterer 1994, Ontogenetic changes in forager polymorphism and foraging ecology in the leaf-cutting ant *Atta cephalotes, Oecologia* 98: 235–238.

84. A Powell, S Shennan, MG Thomas 2009, Late Pleistocene demography and the appearance of modern human behavior, *Science* 324: 1298–1301; Elman R. Service, *Origins of the State and Civilization* (New York: W. W. Norton, 1975).

85. Overlap in generations, the reproductive division of labor, and cooperative care of the young are considered the essential attributes of the most complex, or eusocial, insect societies, which include all ant species except possibly a few that lack distinct queens.

15. The Origins of Agriculture

1. I recommend the review by TR Schultz, UG Mueller, CR Currie, SA Rehner, Reciprocal illumination: A comparison of agriculture in humans and in fungus-growing ants, in *Insect-Fungal Associations: Ecology and Evolution*, ed. FE Vega, M Blackwell (Oxford: Oxford University Press), pp. 149–190. Versions of the slovenly ant hypothesis have involved fungus growth on animal matter, vegetable matter, fecal matter, or plant roots; see UG Mueller, TR Schultz, CR Currie, RMM Adams, D Malloch 2001, The origin of the attine ant-fungus mutualism, *Q. Rev. Biol.* 76: 169–197.

2. Stephen B. Brush and Monica L. Smith, personal communications; Stephen B. Brush, *Farmers' Bounty: Locating Crop Diversity in the Contemporary World* (New Haven: Yale Univ. Press, 2004); Stephen Budiansky, *The Covenant of the Wild: Why Animals Chose Domestication* (New Haven: Yale University Press, 1999); and David Rindos, *The Origins of Agriculture* (New York: Academic Press, 1984).

3. One group of ants feeds the pellets to its larvae, such that the workers "combine the contents of the dust-bin and garbage-can and serve up the mixture as appropriate food for their young—a truly remarkable example of food-conservation"; see WM Wheeler, IW Bailey 1920, The feeding habits of pseudomyrmine and other ants, *Trans Am. Philos. Soc.* 22: 235–279.

4. With their lumpy bodies, *Proatta* also look like *Atta* leafcutters, which is how they got their name; see MW Moffett 1986, Behavior of Malayan group-predatory ant *Proatta butteli:* An Old-World relative of attine ants, *Insectes Soc.* 33: 444–457. The closest living relative of fungus-growing ants has untidy habits similar to *Proatta*; see C Rabeling, M Verhaagh, UG Mueller 2006, Behavioral ecology and natural history of *Blepharidatta brasiliensis, Insectes Soc.* 53: 300–306; and JLM Diniz, CRF Brandão, CI Yamamoto 1998, Biology of *Blepharidatta* ants, the sister group of the Attini: A possible origin of fungus-ant symbiosis, *Naturwissenschaften* 85: 270–274.

5. RP Coppinger, CK Smith 1983, The domestication of evolution, *Environ. Conserv.* 10: 283–292.

6. Feeding on fungi is not without precedent in ants: one Malayan species specializes on free-living mushrooms that it ferments in the nest before eating; see V Witte, U Maschwitz 2008, Mushroom harvesting ants in the tropical rain forest, *Naturwissenschaften* 95: 1049–1054.

7. CJ Krebs, R Boonstra, S Boutin, ARE Sinclair 2001, What drives the 10-year cycle of snowshoe hares? *Bioscience* 51: 25–35.

8. Asexual reproduction and other features of the fungus mutualism were predicted in MM Martin 1992, The evolution of insect-fungus associations: From contact to stable symbiosis, *Am. Zool.* 32: 593–605. Martin based his ideas on R Law, Evolution in a mutualistic environment, in *The Biology of Mutualism: Ecology and Evolution*, ed. DH Boucher (New York: Oxford Univ. Press, 1985), pp. 145–170.

9. The fungus's occasional success in sexual reproduction, presumably at these times, is shown by genetic evidence; see AS Mikheyev, UG Mueller, P Abbot 2006, Cryptic sex and many-to-one coevolution in the fungus-growing ant symbiosis, *Proc. Natl. Acad. Sci.* 103: 10702–10706.

10. For an exception, see, e.g., GH Perry et al. 2007, Diet and the evolution of human amylase gene copy number variation, *Nat. Genet.* 39: 1256–1260.

11. B Stadler, AFG Dixon 2005, Ecology and evolution of aphid-ant interactions, *Annu. Rev. Ecol. Evol. Syst.* 36: 345–372.

12. JS LaPolla, TR Schultz, KM Kjer, JF Bischoff 2006, Phylogenetic position of the ant genus *Acropyga* and the evolution of trophophoresy, *Insect Syst. Evol.* 37: 197–212.

13. M Dill, DJ Williams, U Maschwitz 2002, Herdsmen ants and their mealybug partners, *Abh. Senckenberg. Naturforsch. Ges.* 557: 1–373.

14. H Fernández-Martín, JK Zimmerman, SA Rehner, WT Wcislo 2006, Active use of the metapleural glands by ants in controlling fungal infection, *Proc. R. Soc. Lond. Ser. B* 273: 1689–1695. In leafcutter ants, the metapleural gland almost certainly plays a role in fungus garden health by keeping the pH low enough to selectively poison undesirable microbes; see D Ortius-Lechner, R Maile, ED Morgan, JJ Boomsma 2000, Metapleural gland secretions of the leaf-cutter ant *Acromyrmex octospinosus:* New compounds and their functional significance, *J. Chem. Ecol.* 26: 1667–1683.

15. MW Moffett 2007, Able bodies, *National Geographic,* 212: 140–151. A similar-seeming ant behavior occurs in a dissimilar situation: with a parasitic ant that climbs on other species to steal the nectar in their mandibles; see F-J Richard, A Dejean, J-P Lachaud 2004, Sugary food robbing in ants: A case of temporal cleptobiosis, *Comptes Rendus Biol.* 327: 509–517. The function of the "cleaning" behavior is unclear for ants and for most cleaning fish as well; see, e.g., R Bshary, AS Grutter 2006, Image scoring and cooperation in a cleaner fish mutualism, *Nature* 441: 975–978; and R Poulin, AS Grutter 1996, Cleaning symbioses: Proximate and adaptive explanations, *Bioscience* 46: 512–517. The *Dorymyrmex* species is probably *smithi* (Stefan Cover, personal communication).

16. F Amante 1967, Prejuízos causados pela formiga saúva em plantações de Eucalyptus e Pinus no Estado de São Paulo, *Silvicul. São Paulo* 6: 355–363; and M Autuori 1947, Contribuição para o conhecimento de saúva (*Atta* spp IV): Osauveiro depois da primeira revoada (*Atta sexdens rubripilosa*), *Arq. Inst. Biol.* 18: 39–70.

17. It's called the infrabuccal pocket; see CR Currie, AE Stuart 2001, Weeding and grooming of pathogens in agriculture by ants, *Proc. R. Soc. Lond. Ser. B* 268: 1033–1039.

18. AEF Little, T Murakami, UG Mueller, CR Currie 2006, Defending against parasites: Fungus-growing ants combine specialized behaviours and microbial symbionts to protect their fungus gardens, *Biol. Lett.* 2: 12–16.

19. The linkage between the species may be tighter for some of *Atta*'s fungus-growing relatives (which culture the bacteria on special "crypts" on their bodies) than *Atta* itself; see UG Mueller, D Dash, C Rabeling, A Rodrigues 2008, Coevolution between attine ants and actinomycete bacteria: A reevaluation, *Evolution* 62: 2894–2912; and CR Currie, M Poulsen, J Mendenhall, JJ Boomsma, J Billen 2006, Coevolved crypts and exocrine glands support mutualistic bacteria in fungus-growing ants, *Science* 311: 81–83.

20. MJF Brown, ANM Bot, AG Hart 2006, Mortality rates and division of labor in the leaf-cutting ant, *Atta colombica, J. Insect Sci.* 6: 1–8.

21. Among people, sanitation positions are made more desirable by relatively high pay and good job benefits; see Elizabeth Royte, *Garbage Land: On the Secret Trail of Trash* (London: Little, Brown, 2005).

22. ANM Bot, CR Currie, AG Hart, JJ Boomsma 2001, Waste management in leaf-cutting ants, *Ethol. Ecol. Evol.* 13: 225–237.

23. AG Hart, ANM Bot, MJF Brown 2002, A colony-level response to disease control in a leaf-cutting ant, *Naturwissenschaften* 89: 275–277.

24. HM Hull-Sanders, JJ Howard 2003, Impact of *Atta colombica* colonies on understory vegetation and light availability in a neotropical forest, *Biotropica* 35: 441–445; and AG Farji-Brener, AE Illes 2000, Do leaf-cutting ant nests make "bottom-up" gaps in neotropical rain forests? A critical review of the evidence, *Ecol. Lett.* 3: 219–227.

25. In one forest in Costa Rica all the soil is turned over by leafcutters every two or three centuries—fast work from the standpoint of a tree; see I Perfecto, J Vandermeer 1993, Distribution and turnover rate of a population of *Atta cephalotes* in a tropical rain forest in Costa Rica, *Biotropica* 25: 316–321.

26. LSL Sternberg, MC Pinzon, MZ Moreira, P Moutinho, EI Rojas, EA Herre 2007, Plants use macronutrients accumulated in leaf-cutting ant nests, *Proc. R. Soc. Lond. Ser. B* 274: 315–321; and BL Haines 1978, Element and energy flows through colonies of the leaf-cutting ant, *Atta colombica,* in Panama, *Biotropica* 10: 270–277.

27. AG Farji-Brener 2005, The effect of abandoned leaf-cutting ant nests on plant assemblage composition in a tropical rainforest of Costa Rica, *Ecoscience* 12: 554–560; and AG Farji-Brener, CA Medina 2000, The importance of where to dump the refuse: Seed banks and fine roots in nests of the leaf-cutting ants *Atta cephalotes* and *Atta colombica, Biotropica* 32: 120–126.

28. Young plants may survive for a time in the surface refuse of *Atta colombica,* where sanitation ants troll for waste but their plant-snipping sisters fear to tread; see Alejandro Farji-Brener (personal communication) and M Garrettson, JF Stetzel, BS Halpern, DJ Hearn, BT Lucey, MJ McKone 1998, Diversity and abundance of understorey plants on active and abandoned nests of leaf-cutting ants *(Atta cephalotes)* in a Costa Rican rain forest, *J. Trop. Ecol.* 14: 17–26.

29. JW Dalling, R Wirth 1998, Dispersal of *Miconia argentea* seeds by the leaf-cutting ant *Atta colombica, J. Trop. Ecol.* 14: 705–710.

30. I wouldn't be surprised if the costs to an ant plant of maintaining a symbiosis with ants can at times approach the costs in lost foliage from leafcutters and herbivores for many plants; see, e.g., M Heil, B Fiala, KE Linsenmair, G Zotz, P Menke, U Maschwitz 1997, Food body production in *Macaranga triloba:* A plant

investment in anti-herbivore defense via symbiotic ant partners, *J. Ecol.* 85: 847–861.
31. The "higher attines" include the leafcutters (*Acromyrmex* and *Atta*) and at least two other genera that use dried leaves for their gardens.
32. Even leafcutters excrete on their mulch to get the garden started, much as human farmers use compost and manure.
33. For this timeline and details, see TR Schultz, SG Brady 2008, Major evolutionary transitions in ant agriculture, *Proc. Natl. Acad. Sci.* 105: 5435–5440; UG Mueller, C Rabeling 2008, A breakthrough innovation in animal evolution, *Proc. Natl. Acad. Sci.* 105: 5287–5288.
34. The "ecological release" that results from moving of a species away from its parasites and competitors can lead to expanded resource use in what can be thought of as a "transplanted landscape"; see Edgar Anderson, *Plants, Man, and Life* (Boston: Little, Brown, 1952); and chapter 16.
35. As the earliest farming communities grew, people became smaller, weaker, and more disease ridden than their hunter-gatherer ancestors; see T. Douglas Price and Anne Birgitte Gebauer, eds., *Last Hunters, First Farmers: New Perspectives on the Prehistoric Transition to Agriculture* (Santa Fe: School of American Research, 1995). The idea of a "plant trap" is described in Robert L. O'Connell, *Ride of the Second Horseman: The Birth and Death of War* (New York: Oxford University Press, 1995).
36. ANM Bot, SA Rehner, JJ Boomsma 2001, Partial incompatibility between ants and symbiotic fungi in two sympatric species of *Acromyrmex* leaf-cutting ants, *Evolution* 55: 1980–1991.
37. These incompatibility reactions are known for the genus *Acromyrmex*; see M Poulsen, JJ Boomsma 2005, Mutualistic fungi control crop diversity in fungus-growing ants, *Science* 307: 741–744.
38. See n. 36; AS Mikheyev, UG Mueller, JJ Boomsma 2007, Population genetic signatures of diffuse co-evolution between leaf-cutting ants and their cultivar fungi, *Mol. Ecol.* 16: 209–216; and RMM Adams, UG Mueller, AK Holloway, AM Green, J Narozniak 2000, Garden sharing and garden stealing in fungus-growing ants, *Naturwissenschaften* 87: 491–493.
39. MB Dijkstra, JJ Boomsma 2003, *Gnamptogenys hartmani:* An agro-predator of *Trachymyrmex* and *Sericomyrmex* fungus-growing ants, *Naturwissenschaften* 90: 568–571; and RMM Adams, UG Mueller, TR Schultz, B Norden 2000, Agro-predation: Usurpation of attine fungus gardens by *Megalomyrmex* ants, *Naturwissenschaften* 87: 549–554.
40. T Yamaguchi 1995, Intraspecific competition through food robbing in the harvester ant, *Messor aciculatus*, and its consequences on colony survival, *Insectes Soc.* 42: 89–101; and B Hölldobler 1986, Food robbing in ants, a form of interference competition, *Oecologia* 69: 12–15.
41. J Diamond 1998, Ants, crops, and history, *Science* 281: 1974–1975. For human examples, see Jared Diamond, *Guns, Germs, and Steel: The Fates of Human Societies* (New York: Norton, 1997).
42. Among other things, the ants provide the fungus with ideal temperature, humidity, and oxygen levels. One article examines the relationship from the fungus's point of view; see UG Mueller 2002, Ant versus fungus versus mutualism: Ant cultivar conflict and the deconstruction of the attine ant-fungus symbiosis, *Am. Nat.* 160: S67–S98.

16. Armies of the Earth

1. Battle mortalities and border changes are described in ML Thomas, CM Payne-Makrisâ, AV Suarez, ND Tsutsui, DA Holway 2006, When supercolonies collide: Territorial aggression in an invasive and unicolonial social insect, *Mol. Ecol.* 15: 4303–4315.
2. Until recently the Argentine ant was known as *Iridomyrmex humilis*. In its native range, it also crosses the Argentina border into adjacent parts of Uruguay, Paraguay, and Brazil.
3. J Zee, DA Holway 2006, Nest raiding by the invasive Argentine ant on colonies of the harvester ant, *Pogonomyrmex subnitidus*, *Insectes Soc.* 53: 161–167. A more general look at competitive displacement can be found in DA Holway 1999, Competitive mechanisms underlying the displacement of native ants by the invasive Argentine ant, *Ecology* 80: 238–251; and KG Human, DM Gordon 1999, Behavioral interactions of the invasive Argentine ant with native ant species, *Insectes Soc.* 46: 159–163.
4. AV Suarez, DA Holway, TJ Case 2001, Patterns of spread in biological invasions dominated by long-distance jump dispersal: Insights from Argentine ants, *Proc. Natl. Acad. Sci.* 98: 1095–1100.
5. FR Cole, AC Medeiros, LL Loope, WW Zuehlke 1992, Effects of the Argentine ant on arthropod fauna of Hawaiian high-elevation shrubland, *Ecology* 73: 1313–1322.
6. DA Holway, E LeBrun, C Tillberg, AV Suarez 2007, Trophic ecology of invasive Argentine ants in their native and introduced ranges, *Proc. Natl. Acad. Sci.* 104: 20856–20861.
7. AV Suarez, JQ Richmond, TJ Case 2000, Prey selection in horned lizards following the invasion of Argentine ants in southern California, *Ecol. Appl.* 10: 711–725.
8. CE Christian 2001, Consequences of a biological invasion reveal the importance of mutualism for plant communities, *Nature* 413: 635–639.
9. SE Carney, MB Byerley, DA Holway 2003, Invasive ants (*Linepithema humile*) do not replace native ants as seed dispersers of *Dendromecon rigida*, *Oecologia* 135: 576–582.

10. It's not known whether any stolen pupae are ever reared as slaves, either deliberately or by accident. Both seem unlikely, given how strongly the ants abhor the scent of workers from neighboring colonies.
11. The last two rank with the Argentine ant among the hundred worst invasive species; see S Lowe, M Browne, S Boudjelas, M De Poorter 2004, 100 of the world's worst invasive alien species: A selection from the Global Invasive Species Database, Invasive Species Specialist Group, Gland, Switzerland; available as a PDF booklet at www.issg.org. Two other ants make the list: *Anoplolepis gracilipes* (originally African or Asian) and *Pheidole megacephala* (probably originally African).
12. My thanks to Alex Wild for the sports analogy. The only ants outside Argentina that have had any luck fighting the sugar ant are in Australia, aided perhaps by the severity of the climate; see ML Thomas, DA Holway 2005, Condition-specific competition between invasive Argentine ants and Australian *Iridomyrmex*, *J. Anim. Ecol.* 74: 532–542.
13. This behavior can also occur between competing species; see, e.g., TA Langen, F Tripet, P Nonacs 2000, The red and the black: Habituation and the dear-enemy phenomenon in two desert *Pheidole* ants, *Behav. Ecol. Sociobiol.* 48: 285–292.
14. David Holway, personal communication; see also n. 4.
15. The genetic diversity of Argentine ants in the southern United States suggests that their rapid expansion was effected in part through multiple invasions of different source colonies by way of several southern port cities soon after their arrival in New Orleans.
16. DA Holway, AV Suarez, TJ Case 1998, Loss of intraspecific aggression in the success of a widespread invasive social insect, *Science* 282: 949–952. This paper was written prior to Melissa Thomas's discovery of supercolonies.
17. These native habitats at best sustain patchy populations of Argentine ants; see NE Heller, KK Ingram, DM Gordon 2008, Nest connectivity and colony structure in unicolonial Argentine ants, *Insectes Soc.* 55: 397–403; and DA Holway, AV Suarez 2006, Homogenization of ant communities in Mediterranean California: The effects of urbanization and invasion, *Biol. Conserv.* 127: 319–326.
18. The relocation of nests relative to foodstuffs has been documented for another species with decentralized colonies; see chapter 9 and E van Wilgenburg, MA Elgar 2007, Colony structure and spatial distribution of food resources in the polydomous meat ant *Iridomyrmex purpureus*, *Insectes Soc.* 54: 5–10.
19. D Helbing, J Keltsch, P Molnár 1997, Modelling the evolution of human trail systems, *Nature* 388: 57–50.
20. The use of exploratory trails to integrate untouched terrain into a territory was first shown in EO Wilson 1962, Chemical communication among workers of the fire ant *Solenopsis saevissima*, 1: The organization of mass foraging, *Anim. Behav.* 10: 134–147. It's unknown if such trail cues disappear after an area has been incorporated into a territory, or whether they continue to be laid at some level, e.g., at the start of daily bouts of foraging.
21. J-L Deneubourg, S Aron, S Goss, JM Pasteels 1990, The self-organizing exploratory pattern of the Argentine ant, *J. Insect Behav.* 3: 159–168. This research was done on colony fragments of 150 to 1,200 workers and a foraging arena less than a meter square.
22. This pattern may arise if foragers travel independently but shift course each time two come into contact; see DM Gordon 1995, The expandable network of ant exploration, *Anim. Behav.* 50: 995–1007. Curiously, this study found no evidence of exploratory trails.
23. My own observations and those of George Markin suggest it might take the ants many attempts to catch large prey; see GP Markin 1970, Foraging behavior of the Argentine ant in a California citrus grove, *J. Econ. Entomol.* 63: 740–744.

17. The Immortal Society

1. Dangsheng Liang, personal communication.
2. Edward O. Wilson, *Sociobiology: The New Synthesis* (Cambridge: Harvard University Press, 1975), 7. A society can also be an amalgamation of more than one species, as discussed later in this chapter.
3. Through powerful identity labels, nationalism among people today provides for bonds and a sense of kinship even among strangers; see Karl W. Deutsch, *Nationalism and Social Communication* (Cambridge, MA: MIT Press, 1996).
4. D Liang, J Silverman 2000, "You are what you eat": Diet modifies cuticular hydrocarbons and nestmate recognition in the Argentine ant, *Linepithema humile*, *Naturwissenschaften* 87: 412–416. As this book goes to press, the identity of the compounds used by Argentine ants to identify enemies are being determined; M Brandt, E van Wilgenburg, R Sulc, KJ Shea, ND Tsutsui 2009, The scent of supercolonies: the discovery, synthesis and behavioral verification of ant colony recognition cues, *BMC Biol.* in press.
5. MA Elgar, RA Allan 2006, Chemical mimicry of the ant *Oecophylla smaragdina* by the myrmecophilous spider *Cosmophasis bitaeniata*: Is it colony-specific? *J. Ethol.* 24: 239–246. Other "guests," from silverfish to beetles, survive in ant nests by cunning use of pilfered identity signals—it's a source of amazement to me that almost none infiltrate Argentine ant colonies, suggesting their colony identity is a tough nut to crack.
6. Because of the universality of social bonding in nature, an alien coming to Earth before the evolution

of man would not have picked ants as the dominant social force, as claimed by Bert Hölldobler and Edward O. Wilson in *The Superorganism* (New York: W.W. Norton, 2008), though I like to think aliens would be as fascinated with ants as I am. Even microbes arose through the social union of smaller microbes, and those in turn through a union of complex molecules; see Lynn Margulis, *Symbiotic Planet: A New Look at Evolution* (New York: Basic Books, 1999).

7. One review suggests that treating a colony as an individual is more enlightening than treating it more narrowly as a superorganism; see A Hamilton, NR Smith, MH Haber, Social insects and the individuality thesis: Cohesion and the colony as a selectable individual, in *Organization of Insect Societies: From Genome to Sociocomplexity*, ed. Jürgen Gadau and Jennifer Fewell (Cambridge, MA: Harvard University Press, 2009), pp. 572–589. While this view has merit in some of the situations these authors discuss, parallels to organisms can both be compelling and lead to useful models (see Conclusion). For reviews of the evolution of identity, see ND Tsutsui 2004, Scents of self: The expression component of self/non-self recognition systems, *Ann. Zool. Fenn.* 41: 713–727; CM Payne, CV Tillberg, AV Suarez 2004, Recognition systems and biological invasions, *Ann. Zool. Fenn.* 41: 843–858; and John Maynard Smith and Eörs Szathmáry, *The Major Transitions in Evolution* (San Francisco: W.H. Freeman, 1995).

8. *Crematogaster levior* ants can share a garden with any of several alternative ant species. Nevertheless, they manage to recognize workers of the resident colony, distinguishing them from all other alien colonies; see J Orviel, C Errard, A Dejean 1997, Ant gardens: Interspecific recognition in parabiotic ant species, *Behav. Ecol. Sociobiol.* 40: 87–93.

9. Actually, there is no evidence for any ant species that workers distinguish between individuals in their nest by kinship or by genetic differences, and therefore no reason to expect aggression within a colony on this basis, no matter how many queens there are and how genetically diverse the nestmates might be; see DC Queller, JE Strassmann 2002, The many selves of social insects, *Science* 296: 311–313. In fact, supercolonies vary in genetic diversity, and even the more diverse of them show no sign of internal squabbling; see ND Tsutsui, AV Suarez, RK Grosberg 2003, Genetic diversity, asymmetrical aggression, and recognition in a widespread invasive species, *Proc. Natl. Acad. Sci.* 100: 1078–1083.

10. In some cases, colony odors arise primarily in queens; see A Hefetz 2007, The evolution of hydrocarbon pheromone parsimony in ants—Interplay of colony odor uniformity and odor idiosyncrasy, *Myrmecol. News* 10: 59–68.

11. ML Thomas, CM Payne-Makrisâ, AV Suarez, ND Tsutsui, DA Holway 2007, Contact between supercolonies elevates aggression in Argentine ants, *Insectes Soc.* 54: 225–233. I am reminded of human nationalism and the tendency for experienced border folk to be more outspoken and expectant of trouble than their fellow citizens. No surprise that the greatest human empires have arisen where different ethnic populations abut each other's territories, strengthening identity and solidarity—and often the hunger for conquest; see Peter Turchin, *War and Peace and War: The Rise and Fall of Empires* (New York: Plume, 2007).

12. GP Markin 1968, Nest relationships of the Argentine ant, *Iridomyrmex humilis*, *J. Kans. Entomol. Soc.* 41: 511–516.

13. Jump dispersal within a supercolony could increase genetic homogenization, however.

14. KK Ingram, DM Gordon 2003, Genetic analysis of dispersal dynamics in an invading population of Argentine ants, *Ecology* 84: 2832–2842.

15. This mental experiment is the colony equivalent of a "ring species," in which individuals of a geographically variable species breed freely everywhere, but individuals at the extremes of the range are so different that they can't interbreed.

16. It is still possible that ants might show more subtle animosity or favoritism within a colony, for example, by preferring to exchange food with genetically closer individuals (perhaps kin, but see n. 9).

17. Violence within otherwise affable supercolonies occurs in only one situation. Each spring, for reasons unclear, the workers mass-execute the queens, sparing only enough of them to maintain the high rate of colony growth. It's an exception that proves the rule: social integrity is reflected in how well the ants manage conflict when it arises. There seems to be no resistance, and the colony operates smoothly as its queens are butchered—even the queens don't protest. These mass executions were first described in GP Markin 1970, The seasonal life cycle of the Argentine ant, *Iridomyrmex humilis*, in southern California, *Ann. Entomol. Soc. Am.* 63: 1238–1242.

18. Clarence Day, *This Simian World* (New York: Alfred A. Knopf, 1920), p. 10.

19. Neil Tsutsui, quoted in M Shwartz 2004, Scientists challenge report of one Argentine ant supercolony flooding California, *Stanford Report*, 7 April.

20. ND Tsutsui, AV Suarez, DA Holway, TJ Case 2000, Reduced genetic variation and the success of an invasive species, *Proc. Natl. Acad. Sci.* 97: 5948–5953. A slightly different "genetic cleansing" hypothesis claims that, with the nests of colonists reaching extraordinary densities, any hostilities are inordinately costly, making it advantageous for the ants to evolve to lose the capacity to identify and discriminate against non-kin; see T Giraud, JS Pedersen, L Keller 2002, Evolution of supercolonies: The Argentine

ants of southern Europe, *Proc. Natl. Acad. Sci.* 99: 6075–6079. While much has been made in published reports of the high genetic diversity in Argentina compared to that in invasive populations, almost certainly this is an artifact of genes being sampled largely between colonies in Argentina versus within single colonies in California. I think it's likely that colonies of Argentine ants have the capacity to expand indefinitely, regardless of their genetic diversity.

Genetic losses could give a colony another kind of edge: workers from supercolonies with a lower genetic diversity, especially the Very Large Colony, exhibit quick attacks in combat with other supercolonies, using the "shock and awe" offensive approach we saw in the marauder ant. Perhaps their stripped-down recognition signals make the workers faster at distinguishing friend from foe; see ND Tsutsui, AV Suarez, RK Grosberg 2003, cited in n. 9.

21. The exception would be the colonies from one site in Argentina that each have a single, local nest; see NE Heller 2004, Colony structure in introduced and native populations of the invasive Argentine ant, *Linepithema humile, Insectes Soc.* 51: 378–386.

22. Argentine ants may be less inbred in Argentina, where the smaller territories mean the weak-flying males are more likely to reach neighboring colonies, potentially allowing more gene flow than we find overseas, even if most of the males are killed by workers.

23. This overlap in generations is shared by ant colonies and the cells of most organisms (see Conclusion). No wonder some scholars think our idea of an integrated, coherent self is something of an illusion. See, e.g., Daniel M. Wegner, *The Illusion of Conscious Will* (Cambridge, MA: MIT Press, 2003).

24. In most other ants that reproduce by fission or budding, a new nest develops a separate identity from its parent (see chapter 4, n. 18). The supercolonies of the Argentine ant (and likely some other invasive ant species) more closely resemble the individuals of some fungi that spread through soil for centuries, with one Oregon mat covering 10 square kilometers; see ML Smith, JN Bruhn, JB Anderson 1992, The fungus *Armillaria bulbosa* is among the largest and oldest living organisms, *Nature* 356: 428–431.

25. Linda Stone, Paul F. Lurquin, and Luigi Luca Cavalli-Sforza, *Genes, Culture, and Human Evolution* (New York: Wiley-Blackwell, 2007).

26. Species are expected to be genetically distinct, and in fact each Argentine ant supercolony has been found to have a different hydrocarbon "fingerprint"—a combination of the surface chemicals likely to be genetically determined and essential to colony (and in this case species) identity; see CW Torres, M Brandt, ND Tsutsui 2007, The role of cuticular hydrocarbons as chemical cues for nestmate recognition in the invasive Argentine ant *(Linepithema humile), Insectes Soc.* 54: 363–373.

27. I cite the biological species concept as defined in Ernst Mayr, *Principles of Systematic Zoology* (New York: McGraw Hill, 1969), p. 26. As discussed on p. 133, isolation doesn't require a mountain range: insect populations can become isolated even within a single crown of widely separated tropical trees—the space between crowns being in effect mountain ranges in miniature. Other modes of speciation can be important as well; see, e.g., Douglas J. Futuyma, *Evolution* (New York: Sinauer, 2005), pp. 353–404. In practice, admittedly, Mayr himself had a double standard in distinguishing species; see M Schilthuizen 2000, Dualism and conflicts in understanding speciation, *Bioessays* 22: 1134–1141.

28. This may generally be how distinct colonies originate in the Argentine ant, though it's possible that mating flights, thus far undetected in Argentina and abroad, occur on rare occasions. My hypothesis does not require any changes (in particular, convergent changes) in Argentine ants each time they are introduced abroad. Discussions of this ant have been muddied by confusion about what a colony is in this species; it is definitely not a single nest, as is often implied (see, e.g., PT Starks 2003, Selection for uniformity: Xenophobia and invasion success, *Trends Ecol. Evol.* 18: 159–162). Contrary to what Starks writes, Argentine ants never exhibit "indiscriminate altruism" or a "breakdown in normal nestmate discrimination behavior."

29. Early human hunter-gatherer bands were small but relatively open and fluid because of the need for exchanging mates between groups, which meant many groups shared family ties. Our identification with fixed groups developed later, after agriculture gave rise to larger, sedentary communities that were kept together by leaders demanding tributes and allegiance. Among other mammals, herds are the largest groups, but these do not have delimited memberships and are not considered societies. Their members not only do not cooperate, e.g., in rearing young, but also exercise strong self-interest, e.g., in avoiding predators.

30. The range expansion of the Argentine ant supercolonies is likely to continue: see N Roura-Pascual, AV Suarez, C Gómez, P Pons, Y Touyama, AL Wild, AT Peterson 2004, Geographical potential of Argentine ants *(Linepithema humile)* in the face of global climate change, *Proc. R. Soc. Lond. Ser. B* 271: 2527–2534.

31. In addition to the red imported fire ant, *Solenopsis invicta*, discussed here, the southern United States has another destructive but less widespread invasive fire ant from Argentina, *Solenopsis richteri*, and several native fire ant species, none of which are harmful. See Walter R. Tschinkel, *The Fire Ants* (Cambridge, MA: Harvard University Press, 2006); colonies with multiple queens described on pp. 405–411.

32. EG LeBrun, CV Tillberg, AV Suarez, PJ Folgarait, CR Smith, DA Holway 2007, An experimental study of competition between fire ants and Argentine ants in their native range, *Ecology* 88: 63–75. Fire ants and Argentine ants appear to have shifting, broadly overlapping territories in Argentina, a pattern likely made possible by their low worker densities.
33. G Buczkowski, EL Vargo, J. Silverman 2004, The diminutive supercolony: The Argentine ants of the southeastern United States, *Mol. Ecol.* 13: 2235–2242.
34. Don Mabry and Pedro Jover, personal communications.
35. D Pimentel, R Zuniga, D Morrison 2005, Update on the environmental and economic costs associated with alien-invasive species in the United States, *Ecol. Econ.* 52: 273–288.

Conclusion

1. On the other hand, carefully framed anthropomorphisms are useful in hypotheses: R Lockwood, Anthropomorphism is not a four-letter word, in *Perceptions of Animals in American Culture*, ed. RJ Hoage (Washington, DC: Smithsonian Press, 1989), pp. 41–56; GM Burghardt 1985, Animal awareness: Current perceptions and historical perspective, *Am. Psychol.* 40: 905–919; L Daston, G Mitman, *Thinking with Animals: New Perspectives on Anthropomorphism* (New York: Columbia University Press, 2005); RW Mitchell, NS Thompson, HL Miles, *Anthropomorphism, Anecdotes, and Animals* (New York: SUNY Press, 1996).
2. Ludwig Wittgenstein, *Philosophical Investigations* (Oxford: Blackwell Basil, 1958), p. 178. See also J Cole, *About Face* (Cambridge, MA: MIT Press, 1999); C Darwin, *Expression of Emotions in Man and Animals* (London: John Murray, 1872).
3. D Arendt, K Nübler-Jung 1999, Comparison of early nerve cord development in insects and vertebrates, *Development* 126: 2309–2325; R Lichtneckert, H Reichert 2005, Insights into the urbilaterian brain: Conserved genetic patterning mechanisms in insect and vertebrate brain development, *Heredity* 94: 465–477.
4. Lewis Thomas, *Lives of the Cell* (New York: Viking Press, 1974), p. 12. Douglas R. Hofstadter subsequently made the superorganism idea come alive in his *Gödel, Escher, Bach* (New York: Basic Books, 1979).
5. Compounding the problem is a lack of agreement on what a "behavior" is; see DA Levitis, WZ Lidicker, G Freund 2009, Behavioural biologists do not agree on what constitutes behaviour, *Anim. Behav.* 78: 103–110.
6. J Gautrais, G Theraulaz, J-L Deneubourg, C Anderson 2002, Emergent polyethism as a consequence of increased colony size in insect societies, *J. Theor. Biol.* 215: 363–373; R Jeanson, JH Fewell, R Gorelick, SM Bertram 2007, Emergence of increased division of labor as a function of group size, *Behav. Ecol. Sociobiol.* 62: 289–298.
7. SK Robson, JFA Traniello, Key individuals and the organization of labor in ants, in *Information Processing in Social Insects*, ed. C Detrain, J-L Deneubourg, JM Pasteels (Basel: Birkhäuser Verlag, 1999), pp. 239–259; GF Oster, EO Wilson, *Caste and Ecology in the Social Insects* (Princeton, NJ: Princeton Univ. Press, 1978). For the dishwashing example, thanks to JH Fewell 2003, Social insect networks, *Science* 301: 1867–1870.
8. JT Costa 2002, Scale models? What insect societies teach us about ourselves, *Proc. Am. Phil. Soc.* 146: 170–180.
9. Ovid, *Metamorphoses*, translated by Rolfe Humphries (Bloomington: Indiana Univ. Press, 1964), p. 173.
10. Not that one form of society is more "primitive" than another, in either ants or people; nor is change necessarily equally easy in either direction (see n. 35, chapter 15).
11. This brings to mind responses to the Hurricane Katrina disaster in New Orleans, in which decentralized local groups succeeded in providing relief where the government failed: V Bier, Hurricane Katrina as a bureaucratic nightmare, in *Risk and Disaster*, ed. RJ Daniels, DF Kettl, H Kunreuther (Philadelphia: University of Pennsylvania Press, 2006), pp. 243–254.
12. This situation of power of the majority over their leaders, referred to as a "reverse dominance hierarchy," is a modification of the typical linear hierarchy found in most apes: C Boehm, *Hierarchy in the Forest: The Evolution of Egalitarian Behavior* (Cambridge, MA: Harvard Univ. Press, 1999).
13. One World Bank official has proposed that in addressing global problems hierarchical governments be replaced by networked governance: J-F Rischard 2002, Global issues networks: Desperate times deserve innovative measures, *Wash. Q.* 26: 17–33.
14. H Reingold, Mobile media and political collective action, in *Handbook of Mobile Communication Studies*, ed. JE Katz (Cambridge, MA: MIT Press, 2008), pp. 225–239.
15. M Granovetter 1983, The strength of weak ties: A network theory revisited, *Social Theory* 1: 201–233.
16. James Carey, personal communication, and JR Carey 2001, Insect biodemography, *Annu. Rev. Entomol.* 46: 79–110.
17. PJ Wilson, *The Domestication of the Human Species* (New Haven: Yale University Press, 1988).
18. EA Langridge, NR Franks, AB Sendova-Franks 2004, Improvement in collective performance with experience in ants, *Behav. Ecol. Sociobiol.* 56: 523–529.
19. As we'll see shortly, certain ants called ponerines have workers that can serve as queens and so do not

fit this pattern, but neither do most organisms; see Leo W. Buss, *The Evolution of Individuality* (Princeton, NJ: Princeton University Press, 1987).
20. JW Pepper, MD Herron 2008, Does biology need an organism concept? *Biol. Rev.* 83: 621–627.
21. MD Herron, RE Michod 2008, Evolution of complexity in the volvocine algae: Transitions in individuality through Darwin's eye, *Evolution* 62: 436–451; K Drescher, KC Leptos, I Tuval, T Ishikawa, TJ Pedley, RE Goldstein 2009, Dancing *Volvox:* Hydrodynamic bound states of swimming algae, *Phys. Rev. Lett.* 102: 168101–168105.
22. MC McCarthy, BJ Enquist 2005, Organismal size, metabolism and the evolution of complexity in metazoans, *Evol. Ecol. Res.* 7: 681–696.
23. L Lefebvre, D Sol 2008, Brains, lifestyles and cognition: Are there general trends? *Brain Behav. Evol.* 72: 135–144.
24. M Minsky, *The Society of Mind* (New York: Simon & Schuster, 1985).
25. M Dorigo, V Trianni, E Sahin, R Gross, TH Labella, G Baldassarre, S Nolfi, J-L Deneubourg, F Mondada, D Floreano, LC Gambardella 2004, Evolving self-organizing behaviors for a swarm-bot, *Auton. Robots* 17: 223–245; MJB Krieger, J-B Billeter, L Keller 2000, Ant-like task allocation and recruitment in cooperative robots, *Nature* 406: 992–995.
26. R Wehner, T Fukushi, K Isler 2007, On being small: Brain allometry in ants, *Brain Behav. Evol.* 69: 220–228; W Gronenberg 2008, Structure and function of ant brains: Strength in numbers, *Myrmecol. News* 11: 25–36.
27. NR Franks 1989, Army ants: A collective intelligence, *Am. Sci.* 77: 138–145.
28. DS Wilson, JJ Timmel, RR Miller 2004, Cognitive cooperation: When the going gets tough, think as a group, *Human Nature* 15: 225–250.
29. See, e.g., T Monnin, C Peeters 1999, Dominance hierarchy and reproductive conflicts among subordinates in a monogynous queenless ant, *Behav. Ecol.* 10: 323–332; J Heinze 2004, Reproductive conflict in insect societies, *Adv. Study Behav.* 34: 1–58; H Helanterä, L Sundström 2007, Worker reproduction in *Formica ants, Am. Nat.* 170: E14–E25; HK Reeve, B Hölldobler 2007, The emergence of a superorganism through intergroup competition, *Proc. Natl. Acad. Sci.* 104: 9736–9740.
30. The reproductive strategies of ponerine colonies are detailed by B Hölldobler, EO Wilson, *The Superorganism* (New York: W.W. Norton, 2008).
31. S Baratte, M Cobb, C Peeters 2006, Reproductive conflicts and mutilation in queenless *Diacamma* ants, *Anim. Behav.* 72: 305–311.
32. In ant societies, as in human societies, policing can serve the self-interest of individuals, the common good, or both: Peter Nonacs, personal communication, and FLW Ratnieks, T Wenseleers 2005, Policing insect societies, *Science* 307: 54–56.
33. A Burt, R Trivers, *Genes in Conflict* (Cambridge, MA: Harvard Univ. Press, 2008).
34. A Livnat, N Pippenger 2006, An optimal brain can be composed of conflicting agents, *Proc. Natl. Acad. Sci.* 103: 3198–3202.
35. AFG Bourke 1999, Colony size, social complexity and reproductive conflict in social insects, *J. Evol. Biol.* 12: 245–257.
36. In most social species other than ants and the majority of other eusocial insects, individuals can leave a group to join another group, start a group of their own, or live alone (as do human immigrants, pilgrims, and hermits). The ants' faithfulness to their societies may serve as a measure of the strength of group selection as described in DS Wilson, EO Wilson 2007, Rethinking the theoretical foundation of sociobiology, *Quart. Rev. Biol.* 82: 327–348. Bonding is equally strong even among simple organisms such as sponges; see X Fernández-Busquets, The sponge as a model of cellular recognition, in *Sourcebook of Models for Biomedical Research,* ed. P. Michael Conn (New York: Springer, 2008), pp. 75–83.
37. NR Franks 1989, Thermoregulation in army ant bivouacs, *Physiol. Entomol.* 14: 397–404.
38. Because army ant colonies split (chapter 4), their reproduction is as if a human mother were to give birth to a child her own weight. Honeybees show a similar investment: J Tautz, *The Buzz about Bees: Biology of a Superorganism* (Heidelberg: Springer-Verlag, 2008).
39. M Maeterlinck, *The Swarm from the Life of the Bee,* trans. A Euwer (New York: Dodd, Mead, 1901), pp. 39, 45–46.
40. M Batty 2008, The size, scale, and shape of cities, *Science* 319: 769–771; LMA Bettencourt, J Lobo, D Helbing, C Kühnert, GB West 2007, Growth, innovation, scaling, and the pace of life in cities, *Proc. Natl. Acad. Sci.* 104: 7301–7306; JT Bonner, *The Evolution of Complexity by Means of Natural Selection* (Princeton, NJ: Princeton Univ. Press, 1988).
41. EO Wilson 1973, Ants of Easter Island and Juan Fernández, *Pac. Insects* 15: 285–287.
42. WJ Broad 1996, Paradise lost: Biosphere retooled as atmospheric nightmare, *New York Times,* 19 November, C1.

index

Page numbers in italic refer to illustrations.